W9-AUU-435

CONTENTS

Part One

GEMS & CRYSTALS *4*

Introduction *8*

Tracing the Story of Gems *18*

What is a Gem? *22*

Diamond *34*

Corundum *44*

Beryl *56*

Chrysoberyl & Spinel *70*

Topaz *80*

Tourmaline *88*

Zircon & Peridot *100*

Turquoise & Lapis Lazuli *108*

Opal *118*

Feldspar *126*

Jade *132*

Quartz *142*

Chalcedony & Jasper *152*

Garnet *162*

Pearls & Other Organic Gems *172*

Rare & Unusual Gemstones & Ornamental Material *188*

Glossary *202*

Reading List *204*

Part Two

MINERALS *207*

The Earth and Its Minerals *214*

Thinking like a Geologist *216*

Continents Adrift *221*

Minerals *225*

The Rock Cycle *229*

Minerals from Molten Rock *234*

Crystallization *236*

Gems from the Deep *243*

Carbonatites: Unusual Rocks with Unusual Minerals *245*

Granitic Pegmatites *249*

Agpaitic Pegmatites *262*

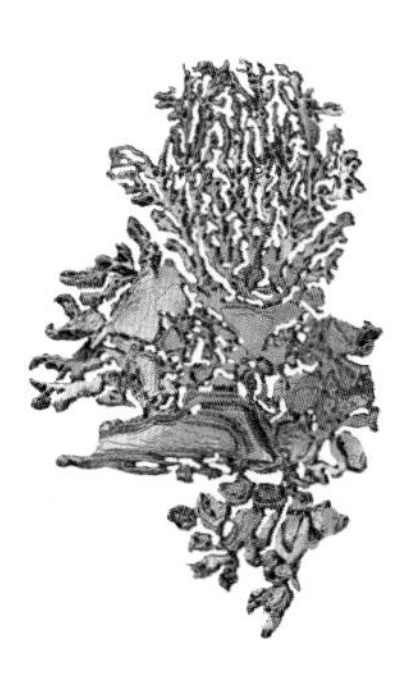

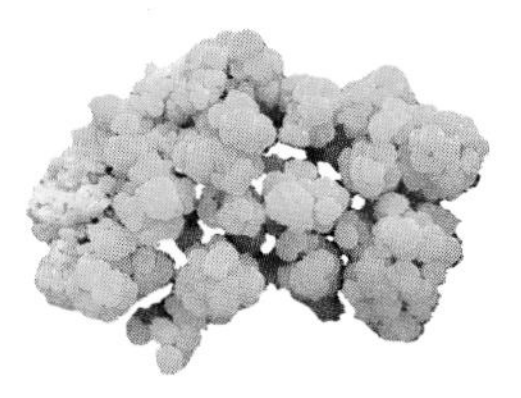

Summary of Crystallization in Magmas 267

Minerals and Water 268

Dissolution and Precipitation 271

Hydrothermal Solutions 280

Warming the Rain 283

Beyond Pegmatites 293

The Effect of Tectonism 299

Summary of Water's Role in Mineral Formation 306

Chemical Alteration 308

Equilibrium and Chemical Reactions 311

Drying Out 315

Neutralization Reactions 316

Oxidation-Reduction Reactions 319

Lights, Camera, Action! 323

Which Comes First? 341

Chemical Alteration in Igneous and Metamorphic Rocks 343

Replacement Deposits 349

Summary of the Chemical Alteration of Minerals 354

Recrystallization 356

Heat and Pressure 358

Metamorphism 360

Composition 363

Complex Recrystallization 371

Skarns: Trash or Treasure? 373

Rodingites 381

Summary of Recrystallization 386

Summary of Mineral-forming Processes 388

Interaction of Mineral-forming Processes 391

Biogenic Minerals 395

What Have We Learned? 399

Recommended Reading 402

Appendix: Some Additional Minerals and How They Form 403

Index 412

GEMS, CRYSTALS,

and

MINERALS

Part One

Anna S. Sofianides
Associate, Department of Mineral Sciences
American Museum of Natural History

and

George E. Harlow
Curator of Gems and Minerals
Department of Mineral Sciences
American Museum of Natural History

Photographs by
Erica and Harold Van Pelt

SIMON AND SCHUSTER
New York London
Toronto Sydney Tokyo Singapore

SIMON AND SCHUSTER
Simon & Schuster Building
Rockefeller Center
1230 Avenue of the Americas
New York, New York 10020

Produced by Nevramont Publishing Co., Inc.
16 East 23rd Street, New York, New York 10010

Book design by Mary Moriarty

3 4 5 6 7 8 9 10

Printed in Hong Kong by Everbest Printing Co., Ltd.
through Four Colour Imports Ltd.

Library of Congress Cataloging in Publication Data

Sofianides, Anna S.
Gems & crystals from the American Museum of Natural
History/Anna S. Sofianides and George E. Harlow:
photographs by Erica and Harold Van Pelt.
p. cm.
Bibliography: p.
Includes index.
ISBN 0-671-68704-2
1. Mineralogy—Catalogs and collections—New York (N.Y.)
2. Precious stones—Catalogs and collections—New York (N.Y.)
3. American Museum of Natural History—Catalogs. I. American
Museum of Natural History. II. Harlow, George E. III. Title.
IV. Title: Gems and crystals from the American Museum of
Natural History.
QD366.2S64 1990
549'.074747'1—dc20

89-11574
CIP

All photographs Erica and Harold Van Pelt except:

Department of Library Services, American Museum of Natural History, pages 9, 12-16 (left), 37, 54, 61 (left), 66, 82, 85, 91, 94, 97, 99, 111, 114, 129 (top), 133, 136, 138, 139, 147, 154, 192 (top), 197, 198

Dave Grimaldi, page 182

Tiffany & Co., pages 43, 173

Smithsonian Institute, page 15 (right)

Gemological Institute of America, pages 25, 27, 31, 32, 33

Acknowledgments

THERE ARE MANY PEOPLE TO WHOM THE AUTHORS ARE GREATLY INDEBTED FOR THEIR EFFORTS, ASSISTANCE, AND ADVICE.
We doubt you would be reading this book at all without the dedication, ministrations, and good humor of Nancy Creshkoff, our stylist and editor.
The moral support, valuable advice, and infinite patience of Tom Sofianides and Carole Slade, our spouses, were imperative to our completing this book.
Without the participation of Joseph J. Peters, including his painstaking review of information on the collections and assistance with many aspects of this project, it would not have come to fruition.
Peter N. Nevraumont, the producer of the book, has been very constructive with his queries, criticisms, and suggestions.
We appreciate the patience and assistance of others in the Museum's Department of Mineral Sciences, particularly Janice Yaklin and Charles Pearson, and at *Natural History*, L. Thomas Kelly and Scarlett Lovell.
The Museum's photographic studio was generous with its assistance, and we thank Jackie Beckett, Kerry Perkins, and Denis Finnin.
Many people have been generous with information: Joan Aruz, Wendy Ernst, Linda Eustis, Leonard Gorelick, Eugene Libre, George Morgan, James Pomarico, Frank Rieger, Peter Schneirla, David Seaman, James Shigley, and Nicholas Steiner.

WE ARE GRATEFUL TO THOSE WHO HAVE HELPED CREATE THE MINERAL AND GEM COLLECTIONS.
Those whose donations are featured in this book are as follows:
George Ackerman; Mrs. R.T. Armstrong; Mrs. Frank L. Babbott; Lilias A. Betts; Mrs. C.L. Bernheimer; Susan D. Bliss; Maurice Blumenthal; David A. Byers; Elizabeth Varian Cockcroft; Lawrence H. Conklin; Joseph F. Decosimo; Dr. B. Delavan; Mrs. George Bowen DeLong; Lincoln Elsworth; Alexander J. and Edith Fuller; Leila B. Grauert; Jack Greenberg; Peter Greenfield; K.B. Hamlin; Dr. George E. Harlow; Mrs. William H. Haupt; Lloyd Herman; Dr. Maurice B. Hexter; Mrs. Charles C. Kalbfleisch; Morton Kleinman; Dr. George F. Kunz; Korite Minerals, Ltd.; Vincent Kosuga; Mabel Lamb; Charles Lanier; Gerald Leach; S. Howard Leblang; Mrs. Zoe B. Larimer; Mrs. Bonnie LeClear estate; Vera Lounsbery estate; Roy Mallady; Alastair Bradley Martin; Mrs. Patrick McGinnis; Roswell Miller, Jr.; Milton E. Mohr; Dr. Arthur Montgomery; M.L. Morgenthau; John Pierpont Morgan,Sr.; John Pierpont Morgan, Jr.; Dr. Walter Mosmann; Dr. Henry Fairfield Osborn; Clara Peck estate; Phelps-Dodge Corporation; Dwight E. Potter; Arthur Rasch; Dr. Julian Reasonberg; Robinson & Sverdlik Company; Mr. and Mrs. J. Robert Rubin; Hyman Saul; Mr. and Mrs. Bernard Schiro; Elizabeth Cockcroft Schettler; Dr. Louis Schwartz; Victoria Stone estate; William Boyce Thompson estate; John Van Itallie; David Warburton; Thomas Whiteley; and several anonymous donors.

Introduction

"These gems are nice, but you know they are all fakes. The real ones are kept locked up in a vault somewhere." So I recall a visitor's remark during one of my walks through the Morgan Memorial Hall of Gems at the American Museum of Natural History. This comment is not the most disarming I have heard in that great treasure chest, but it shows some of the misconceptions that exist. If this assertion were true, the Museum's insurance agents could breathe a sigh of relief, but fortunately for the public, it is not. Everything on display is real.

More interesting to me is the level of people's appreciation. "Oh, that's the Star of India. Isn't it pretty?" This response is common but a far cry from the expert's "LOOK at that pad" (pronounced pod), in reference to our 100-carat orangy sapphire known as the variety padparadscha, "It's FANTASTIC!" Most visitors are impressed by this gem but do not know just how special it is. One of our goals here is to give you a good look at the Museum's gems and gem crystals. The more general intent is to provide information on these that is both interesting and useful.

George Frederick Kunz

Minerals and gems have been part of the Museum since it opened in 1869 in the old Arsenal Building in Central Park. There was a small mineral "cabinet" to instruct the public, but it was nothing to brag about. For their growth into international prominence, the collections awaited benefactors such as Charles L. Tiffany, Morris K. Jesup, and John Pierpont (J.P.) Morgan. George Frederick Kunz was a central figure; he was Tiffany & Co.'s gem expert from 1877 until he died in 1932 and during that time had a profound effect on the gem industry and the Museum's gem and mineral collections.

In 1889, a great Exposition Universelle was scheduled in Paris; it provided an opportunity for Tiffany & Co. to demonstrate to Europeans both American artistry in the form of jewelry and silver and North America's natural wealth through a collection of "Gems and Precious Stones" assembled by Kunz. He searched the continent and gathered a formidable array of stones, crystals, pearls, and other specimens that surprised the European audience and won the collection a gold medal. Whereas much of the Tiffany jewelry was sold then and there, the "collection" was not commercial and

was brought back to New York. This was to Kunz's liking, as he felt it should be kept intact, preferably coming to the American Museum of Natural History. The Museum's president, Morris K. Jesup, understood the value of the collection, but there was a problem of money—exactly $20,000. This was such a significant sum that the Museum's Board of Trustees felt inclined to question it or, at the very least, to haggle. After months of negotiations, the problem was ultimately resolved by J.P. Morgan, the banker, financier, and Museum Board member, who permitted $15,000 toward the purchase price to be "charged to his account," presumably at Tiffany & Co., which donated the remaining $5,000. Thus the Museum gained a significant gem collection in 1890, called the Tiffany Collection or Tiffany-Morgan Collection of Gems.

The same personalities and forces came together in 1900 for another Exposition Universelle in Paris. Morgan responded to the challenge and is thought to have supplied $1 million

An historic grouping of Kunz and Whitlock books featuring figures of crystals and objects from the Museum's collection, crystals, gems, crystal ball, and an antique crystal measuring instrument.

(Remember, this is 1900!) for George F. Kunz to search the world over for fabulous gems and specimens. The result was an even mightier exhibit, one that captured a grand prize. This collection, the Second Tiffany-Morgan (or Morgan-Tiffany) Collection of Gems, came directly to the Museum. The combined collections contained 2,176 specimens and 2,442 pearls and clearly laid claim to being the finest in North America, if not the world.

Morgan's interest in the gem and mineral collections continued. In 1901, he purchased for $100,000 one of the great private mineral collections created during the nineteenth century, that of Clarence S. Bement, a Philadelphia industrialist. This collection was not only superb in quality but so large that two railroad boxcars were required to bring the approximately 13,000 specimens to the Museum. This addition became the backbone of our mineral collection; many of its fine pieces are currently featured in the mineral and gem halls. Morgan's donations continued until his death in 1913.

(Above) Drawing of the Roosevelt Memorial by John Russell Pope (1926) from a hand-colored lantern slide.

(Left) John Pierpont Morgan

A list of noteworthy donors to the collection would be extremely long, but I would like to mention a few more. J.P. Morgan Jr. continued his father's tradition and is responsible for many of the large fine gems, particularly a group of sapphires donated in 1927. George F. Baker, a friend of the elder J.P., funded the creation of the Morgan Memorial Hall, which opened on the Museum's fourth floor on May 1, 1922. Kunz, who was not only responsible for Morgan's gifts but for those of many others, contributed numerous specimens and several collections. He was named honorary curator of precious stones in 1904—a title never bestowed before or since. William Boyce Thompson, the founder of Newmont Mining Corporation, provided a significant fund in 1940, the earnings from which permit us to purchase specimens, for example, the Harlequin Prince black opal, a fabulous 59-carat heart-shaped morganite, and the 596-pound topaz crystal. In 1951, upon the death of Gertrude Hickman Thompson, his widow, many more magnificent gems, carvings, and minerals came to the Museum.

Some have given individual stones so spectacular that each carries the donor's name: Edith Haggin DeLong (the DeLong Star Ruby), Elizabeth Cockroft Schettler (the Schettler Emerald), and Zoe B. Armstrong (the Armstrong Diamond). Harry F. Guggenheim gave both gifts and his name to the present Hall of Minerals, which—together with the Morgan Hall of Gems—opened in May of 1976.

View of the Museum from West 77th Street, from an old restored watercolor architectural rendering.

For most of the Museum's history, the mineral and gem collections have been administered by a single curator: from 1869 to 1876 by Albert S. Bickmore, the Museum's founder; from 1876 to 1917 by Louis Pope Gratacap, who in his tenure of more than forty years established the Department of Mineralogy and procedures for managing the collection; from 1918 to 1941 by Herbert P. Whitlock, who wrote numerous books on the collections; from 1936 to 1952 by Frederick H. Pough, who concentrated on developing the gem collection; from 1953 to 1965 by Brian H. Mason, an academic geologist who developed a particular interest in meteorites; and from 1965 to 1976 by D. Vincent Manson, who devoted his energies to creating the new gem and mineral exhibition halls. The collections also have many unsung heroes—the assistants to curators. In particular, Dave Seaman cared for the collections from 1950 until he retired in 1974, when Joe Peters joined us as Seaman's replacement.

In the late 1970s, the Museum recognized that one curator could not manage the collections and conduct the scientific research expected of all curators. In connection with the opening of new mineral and gem halls in 1976, the Department took a new name, Department of Mineral Sciences, and began to expand. Martin Prinz, a petrologist and now curator of the meteorite collection, became chairman; and I joined the Museum as curator of minerals and gems later that year. The Department now includes four curators who are responsible for the collections of meteorites, rocks (petrology), mineral deposits, and minerals and gems.

Louis Pope Gratacap

Herbert P. Whitlock

Frederick H. Pough

Brian H. Mason

The most notorious event in the Department's history happened on October 29, 1964, when Jack (Murph the Surf) Murphy and two accomplices made a daring robbery of the old Morgan Memorial Hall, getting away with the Star of India, the DeLong Star Ruby, the Midnight Star, the Schettler Emerald, many other stones, and virtually all of the diamonds. Having seen the movie *Topkapi,* which depicts a fantastic burglary in the Topkapi Palace Museum in Istanbul, Murphy decided that the Morgan Hall could be entered in much the same way. He and an accomplice hid on the top floor of the Museum while another circled outside in a getaway car. The two lowered themselves by rope through an open window into the old Morgan Hall, where they literally raked the stones out of the cases with a glass cleaner. The only alarm—that for the Star of India—had a dead battery. The burglars escaped easily but were so boastful about their triumph that they were quickly apprehended, and most of the stones were returned. The DeLong Star Ruby had already gone to the underworld and had to be ransomed. Thirty-five objects have never been recovered, including the uncut 15-carat Eagle Diamond from a glacial moraine near Eagle, Wisconsin, at that time the largest diamond ever found in the United States. This unique diamond crystal and others were probably cut into gems and so permanently lost—a real tragedy.

D. Vincent Manson

Martin Prinz

The minerals and gems on display number approximately 5,000, roughly 2,000 in the Morgan Hall of Gems and 3,000 in the Guggenheim Hall of Minerals. The collections actually total in excess of 90,000 minerals and 3,700 gems; that is, the majority of gems are on exhibit, but only three percent of the minerals. The reason for the vastly different percentages is the nature of the materials and the multiple goals of the Museum. The gems are by definition ornamental material, demanding to be exhibited. Mineral specimens, while sometimes manifesting spectacular crystallinity, color, or form, are frequently rather visually uninteresting—what I sometimes call "uglies." We have many of these on display to show a representative spectrum of the 3,200 mineral species, but the vast majority stay behind the scenes. The value in these "hidden" specimens is their record of Earth chemistry, of mineral-forming processes, and of the ways in which atoms can be arranged. The collections are a resource to scientists from universities and museums all over the world. The beautiful counterparts to uglies, the gems and gem crystals, can be even more valuable to science because of their unusual size and the perfection of crystallinity. My own scientific research in recent years has included studies of the jadeite variety of jade, an interest first stimulated by the beautiful jade objects. However, we try to preserve the gems and beautiful crystals for both the enjoyment and edification of Museum visitors.

Collections must continue to grow to stay alive. Much as the public relies on museums to display wealth once available to only a few, museums rely on the public to share such wealth. The American tradition of generosity provides the basis of most museum collections; certainly this is true for the American Museum of Natural History. The amount of donating fluctuates with the economy (and the tax code), but we should all hope that museum collections can grow and allow all of us to see the world's treasures.

The task of presenting a superb gem and gemstone collection is a formidable one and a rare opportunity. The superb color photographs by the esteemed photographers Harold and Erica Van Pelt speak for themselves and provide a far more vibrant portrayal of the collection than the previous one, made nearly a half-century ago in drawings and black-and-white photographs by the then curator, Herbert P. Whitlock. But the collection is much more than the sum of its images; it is a diverse resource for research and education and an archive of natural perfection. In my thirteen years as curator and in my coauthor's seventeen years of working as a gemologist with the collection, we have developed a great appreciation for the gems. The text conveys some of our knowledge of them. I have focused on the distinctive properties of the gemstones and their origins, and Anna Sofianides has contributed her wealth of information on their history and lore and on gem evaluation.

This book can serve as a concise visual guide to the gem collection at the American Museum of Natural History and a compendium of gem mineral information. However, no book or photograph can rival the real thing. Gems are visual delights, and seeing them first hand is the only way to observe their character. A sparkling faceted gem comes alive with motion—the motions of the viewer, the illumination, and the stone. The same is true for asteriated gems, cat's eyes, and opals. It is no wonder that gems have long been used for human adornment, as motion is an essence of our being. Thus I encourage you, once fortified with the images and information in our book, to visit our collection in New York City and see, while you move around the exhibit, the wonderful qualities of the gems and crystals.

George E. Harlow
February 28, 1990

Tracing the Story of Gems

Archaeology gives the earliest picture; it attempts to tell us when and where each gemstone was used, how it was fashioned, and whether it was traded. Recorded history provides insights into the early naming, classification, and everyday significance of gems but especially the stories that are so fascinating.

Early humans were decorating themselves with shells, pieces of bone, teeth, and pebbles by at least the Upper Paleolithic period (25,000-12,000 B.C.). Most of the stones used in early civilizations were opaque and soft with bright colors or beautiful patterns. Carnelian and rock crystal beads were fashioned at Jarmo in Mesopotamia (Iraq) in the seventh millennium B.C. Engraved cylinder seals appeared about two thousand years later in soft stones such as steatite and marble. Their practical function was as a means of identifying goods; when they were rolled on damp clay, a unique imprint resulted. Seals represent a significant level of technical achievement, and they were also valued as adornment and possibly a symbol of status. By the late fourth millennium in the Near East, cylinder seals were made from rock crystal quartz, a hard gemstone, in addition to the soft stones. A woman's belt from the end of the third millennium B.C. was found in Harappa, an ancient center of Indus civilization, decorated with colorful, opaque stones—red carnelian, green steatite, agate, jasper, amazonite, jade, and lapis lazuli—it represents the wealth of gems then available.

Manchu headdress pendant with Imperial jadeite, pearls, sapphires, and pink tourmalines.

ACTIVE EXPLOITATION OF THE LAPIS LAZULI MINES IN Badakshan, Afghanistan, and the turquoise mines on the Sinai Peninsula began around 5,000 years ago; and long-distance trade in gems developed. Lapis lazuli from Badakshan reached Egypt before 3000 B.C. and Sumer (Iraq) by 2500 B.C. China, India, Greece, and Rome received gemstones from the same source. By around the second millennium, Phoenician sea merchants were trading Baltic amber on the coasts of North Africa, Turkey, Cyprus, and Greece. Recent spectroscopic studies of amber beads discovered in Peloponnese graves of Mycenaean Greece (1450 B.C.) confirmed the assumption of their Baltic origin. Trade between the East and West expanded in the fourth century B.C. after the time of Alexander the Great (356-323 B.C.), resulting in an increase in the number of gemstones available.

In all civilizations, magical powers have been ascribed to gems—perhaps out of a need to explain their rarity, beauty, and strangeness in a confusing world. Color played a great role in the symbolism: gold for the sun; blue for the sky, heaven, or sea; red for blood; black for death. Colors were also associated with the planets and astrology—gems became connected here as well. Durability was also important—the unsurpassable hardness of diamond reflects in the belief that it will bring its wearer strength and invincibility. Gems have served as talismans, offering protection, preserving health, and securing wealth, love, and good luck. Of the 143 pieces of jewelry discovered on the mummy of Tuthankhamen (reigned 1361-1352 B.C.)—made of gold, carnelian, jasper, lapis lazuli, turquoise, obsidian, rock crystal, alabaster, amazonite, and jade—a few showed no sign of wear. Given the preoccupation of ancient Egyptian culture with the afterlife, these few were fashioned as amulets to avert evil and bring good luck after death. The others were worn during his lifetime and served as both adornments and talismans. The Babylonians (in contrast to the Egyptians) were not as concerned with life after death. Many of their engraved cylinder seals may have been primarily talismans worn for protection during life.

Written records provide an extensive overview of how gems have been perceived by their owners and users. George F. Kunz, in *The Curious Lore of Precious Stones*, has documented the evidence carefully. In her *Magical Jewels of the Middle Ages and the Renaissance*, jewelry historian Joan Evans offers a wealth of information through studies and translations of ancient literature, surviving inventories of jewelry (some of which note magical powers), and even court records. In one case, she writes: "One of the counts of the indictment of the Chief Justiciar Hubert de Burgh in 1232 was that he had furtively removed from the royal treasury a gem which made its wearer invincible in battle and had bestowed it upon his sovereign's enemy Llewellyn of Wales."

The supernatural powers of gems were regarded either as intrinsic virtues of the stones themselves or were attributed to figures, sigils, or magical inscriptions engraved upon them. These virtues were enumerated in the mineralogical and medical treatises of the time, known as lapidaries.

Medicinal powers of gems were first recorded in Western literature by the Greeks. These virtues, as well as astrological symbolism, were also recorded in Arabic lapidaries and, starting in the eighth century, in European lapidaries influenced by the Arabic works. In medieval Europe, gems were commonly worn as medicinal amulets or taken as potions. Before Pope Clement VII died in 1534, he had taken as medicine powdered gems valued at 40,000 ducats. Robert Boyle (1627-1691), a great advocate of experiment in natural history and author of *Some Considerations Touching the Usefulnesse of Experimental Natural Philosophy*, wrote: "I think, in Prescriptions made for the poorer sort of Patients, a Physician may well substitute cheaper ingredients in the place of these precious ones, whose Virtues are not so unquestionable as their dearnesse."

Throughout the chapters that follow, there will be mention of the chroniclers and their lapidaries. The first important references in Western literature are from the Greek Theophrastus (c. 372-287 B.C.), the successor of Aristotle. In his book *On Stones*, the oldest surviving mineralogy textbook, he described 16 minerals grouped as metals, earths, and stones (the last including gemstones). This classification remained unchallenged until the eighteenth century. He identified as physical properties color, transparency, luster, fracture, hardness, and weight and also noted the medicinal values of gems. Damigeron also wrote an early lapidary in Greek; the text was translated into Latin somewhere between the first and sixth centuries. Pliny the Elder (A.D. 23-79) compiled the knowledge of his predecessors and contemporaries to produce his 37-volume *Historia Naturalis*.

Ancient and modern carvings from China and Japan including rutilated Brazilian quartz sphere and fu dog; nephrite pi disc; lapis junk; serpentine box; yellow serpentine vase; carnelian vase; agate guppy; malachite vase; and an aquamarine-like glass vase.

Volume 37 deals with precious stones and includes "1,300 facts, romantic stories and scientific observations" about sources, mining, use, trade, and the values of gems; gem enhancements; and gem imitations. Pliny's work was influential in Europe well into the Middle Ages. Marbode, bishop of Rennes, composed his elegant lapidary in Latin hexameter in the eleventh century. Although lacking any mineralogical significance, his work is the basis of both medicinal and magical attributes that have been cited by many later writers.

Thirteenth-century works include Volmar's *Steinbüch* as well as the important *De Mineralibus* of Albertus Magnus (1206-1280). This German philosopher noted the magnetic properties of magnetite, experimented with decomposition of arsenic minerals; and described the properties, including magical virtues, of 94 minerals.

Complicating our task of finding out where initial concepts came from is the fact that "borrowing" was not uncommon. Camillus Leonardus's *Speculum Lapidum*, printed in Venice in 1502, was literally translated into Italian and republished as *Trattato delle Gemme* by Ludovico Dolce later in the same century. Another sixteenth-century work, more important than that of Leonardus, is *de Gemmis et Coloribus* by Geronimo Cardano, published in 1587. Anselmus Boetius de Boot, court physician to Holy Roman Emperor Rudolf II, wrote *Gemmarum et Lapidum Historia* (1609); in his extensive work, he provided descriptions of gems and reports on their virtues, although we find the beginnings of doubts as to the infallible powers of gems.

In addition to traditional lapidaries, travel books, such as those by Marco Polo in the thirteenth century and Jean Baptiste Tavernier in the seventeenth, provide information about the use of gems and their sources. Garcia de Orta (1565), Portuguese physician to the viceroy of Goa in India, described diamond mines there, observed mining practices, and reported gems' virtues—he flatly denied a then-commonplace belief that diamond was poison, having seen workers swallow the gem in order to smuggle it.

With the development of empiricism and scientific inquiry in the seventeenth and eighteenth centuries, the study of gemstones as manifestations of a strictly physical universe began. Concepts of chemistry, optics, and crystallography developed along with a desire to categorize so that definitions and tests could begin to differentiate among all objects.

Today we view gems from a very different perspective from that of a few hundred years ago, but we still have much to learn. Color, the great deceiver in the transparent stones, is still a subject filled with questions about specific causes in each gemstone. New gemstones and new treatments of the old are discovered and add to the diversity. The beauty of the challenge is the gems themselves; they offer a wonderfully exciting stimulus for exploring nature.

What Is a Gem?

The purpose of this chapter is to answer this question and to discuss the attributes that distinguish gems. The rest of the book examines the gemstones; it starts with the traditional "precious" stones, then moves through the colored stones, ending with organic gems, rare and unusual gemstones, and ornamental material. However—aside from the "precious" four: diamond, ruby, sapphire, and emerald—there is neither a system nor agreement on how to order gems, because beauty, their hallmark, is a matter of taste and culture. Pearls, jade, and opal are highly regarded, but ranking them—who knows? Moreover, tastes and availability change with time; today's ranking could be noticeably different in a decade. You undoubtedly have your own favorites, but browse through all the chapters for some beautiful surprises.

Getting back to the title question: to us, a gem is a gemstone that has been fashioned—cut, shaped, and/or polished—to enhance its natural beauty. The gemstone is the raw material or "rough"; the gem is the finished product. Most gemstones are minerals, but some are rocks, and a few are the organic products of once-living animals or plants. A gem ruby is created from a piece of the mineral corundum; lapis and jade are rocks; and pearls, amber, and jet are organic products.

Myriad colored stones, including topaz, amethyst, aquamarine, morganite, chrysoberyl, peridot, smoky quartz, citrine, calcite, rhodochrosite, kunzite, and fluorite; varying from 14.91 to 454 cts.

Important Gem Properties

THE MOST SIGNIFICANT CHARACTERISTIC OF A GEMstone is its visual beauty; after this come durability and rarity, but without beauty, the others mean little. The beauty of ruby, emerald, and turquoise lies in magnificent, intense colors, while that of diamond is the complete absence of color combined with high brilliance. Flawless transparency is critical for the beauty of diamond, aquamarine, and topaz, while inclusions account for the presence of a star in ruby and sapphire and the cat's eye. The lively play of color in opal and the pleasing iridescence in labradorite are unique for these gems, as are the numerous patterns in agate. To understand what gives a gem its most important characteristic, visual beauty, we need to examine the ways in which a mineral interacts with light.

Light, Vision, and Color

The source of color is light, its interaction with an object, and our ability to perceive the result. The color we see is the light that is reflected or transmitted and not absorbed. The causes of color in gemstones are many and varied.

If a mineral's color is inherent, it is called *idiochromatic*, "self-colored." Malachite, copper carbonate, is always green because copper causes the color and is intrinsic to the mineral.

Minerals that owe their color to physical effects, such as internal boundaries and contaminants, are called *pseudochromatic*, "false-colored." Jasper, a form of quartz with extremely fine grain size, can contain small particles of iron oxide particles (hematite) that make it brick red. Physical scattering of light, described a little later, produces the play of colors in precious opal.

Allochromatic, "other-colored," minerals are generally colorless and transparent in their pure state but develop color with minor changes in crystal composition or from structural imperfections. Such gemstones are the most numerous, intriguing, and difficult to identify by color alone. Substitution of certain transition elements for aluminum in corundum yields a variety of colors: some iron and titanium causes blue sapphire; a little iron alone results in yellow sapphire; a little chromium produces ruby. (Transition elements are chemical elements in the middle of the periodic table whose electron energy transitions can be stimulated by visible light, thus yielding color.) The same element can result in different colors; a minor substitution of chromium for aluminum in colorless beryl produces the spectacular emerald. Other transition elements important in causing color are manganese, copper, and vanadium.

Damage and/or mistakes in the crystal can cause color. Smoky quartz is the result of radiation damage to the crystal. It can be produced by naturally-occurring radioactive minerals adjacent to a quartz crystal or by bombardment with subatomic particles from a nuclear reactor.

Color in some crystals changes with their orientation; the phenomenon is called *pleochroism*, "more coloring." Sapphires and rubies and pink

(Opposite) Gemstone crystals including tourmaline, aquamarine, morganite, heliodor, topaz, kunzite, spodumene and citrine.

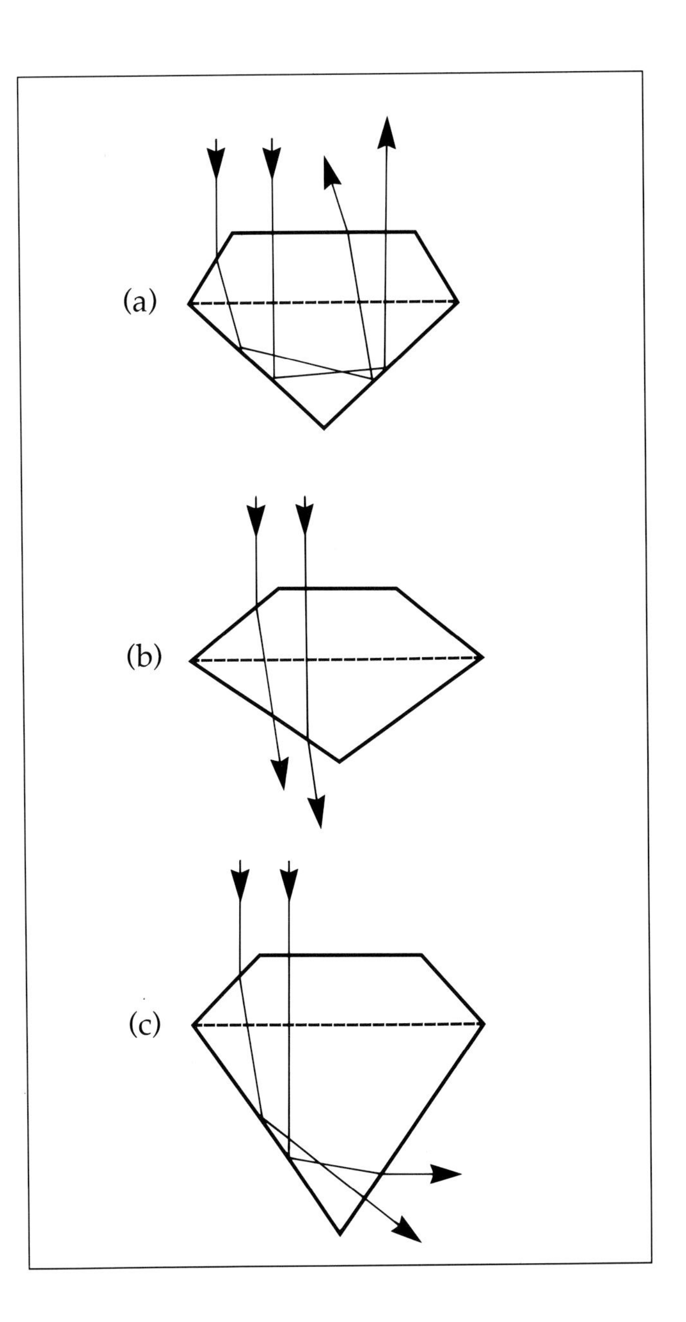

(Right) A gem properly proportioned with respect to its refractive index will reflect back the light that enters it (a), yielding maximum brilliance. Light "leaks" out of a gem cut too shallow (b) or too deep (c), diminishing brilliance.

spodumene are more deeply colored when viewed down the prism axis. Tourmaline gems can have two different colors, depending upon the direction you look through the gem or crystal. Orientation is very important to the appropriate fashioning of pleochroic gemstones.

A few gems, notably the ruby, have the property of fluorescence; they can absorb blue and ultraviolet light and reradiate some of the energy in a redder portion of the spectrum. The result is a more intense color with an extra glow that dazzles the eye.

Gems' Sparkle

Another important pair of related optical properties of a gem is the way it reflects and refracts light. Luster is the reflection or scattering of light from a gem's surface; it can range from metallic to vitreous to resinous to earthy. High luster requires both a smooth surface and a high reflectivity. Polishing of all gems is important in part to improve luster.

Brilliance, on the other hand, is the reflection of light from inside a faceted gem. ("Life" and "liveliness" are used synonymously for brilliance.) This quality is a function of both the cut and the refractive index (R.I.). The R.I. is actually a measure of the velocity of light in the gemstone but is manifested by the degree to which light is bent when entering a substance at an angle and the critical angle at which light is reflected instead. The angles of cut, and thus a gem's proportions, are specifically gauged to the R.I. of each gemstone so that the faceted gem will reflect back from inside the light that enters. All minerals except those of the highest symmetry—cubic—actually have two or three R.I.s and are called "birefringent." Reflectivity is positively correlated with refractive index, and both increase with a substance's density. Thus it is not an accident that the fine transparent gemstones like diamond and sapphire are denser than most minerals. (Density is measured in terms of specific gravity, S.G., the weight of a substance relative to that of an equal volume of water.)

Fire develops in a gem from the phenomenon known as *dispersion*. The component colors in white light are bent to varying degrees during refraction, with the consequent separation of colors into the rainbow. For two different gems of the same size and cut, the one with greater dispersion will display a better spectrum of colors, or fire; the diamond will have better fire than the quartz gem. In colored gemstones, dispersion is usually masked by the predominant color, so that this property becomes unimportant.

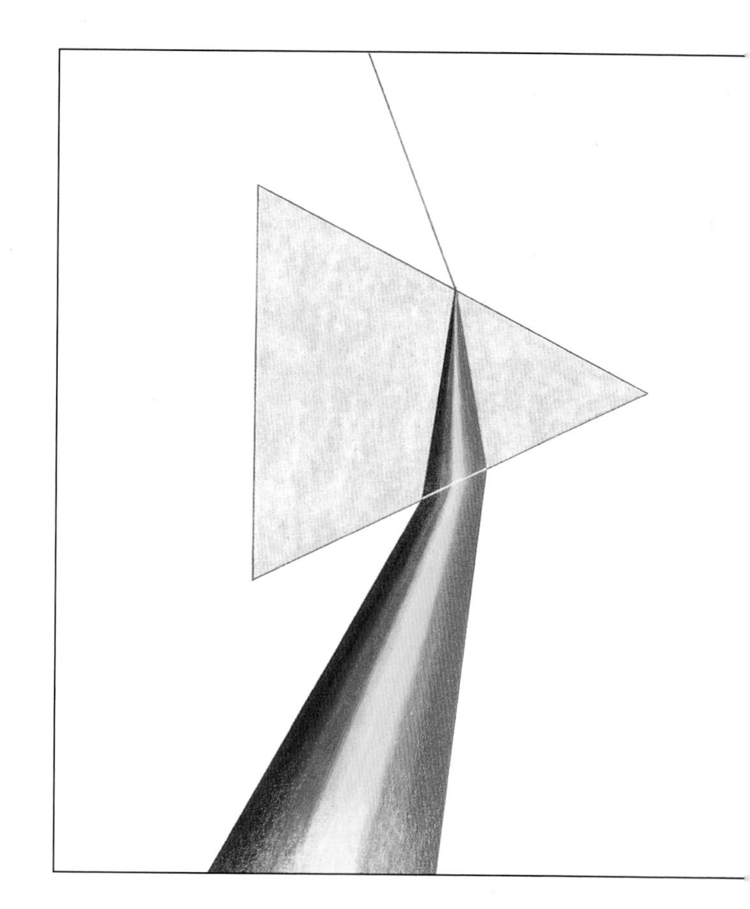

Scattering of Light

Small to submicroscopic features can produce some surprising visual effects in gemstones. Reflections from parallel layers of transparent materials cause pearly luster. The pearl is built up of concentric layers and gives this luster its name. The cat's eye effect, chatoyancy (a literal translation from French), is a band of reflected light that appears in certain gemstones. The cat's eye is produced by many straight parallel fibrous inclusions that scatter light perpendicular to their long direction. In corundum, three directions of needles can occur, yielding multiple chatoyancy in the form of a six-rayed star. This is called *asterism*. The star is visible only when the stone is viewed down the axis of intersection of the inclusions.

Scattering from very small features often imparts colors. In moonstone, thin layers and small elliptical bodies scatter blue light most effectively and yield the characteristic pale blue sheen. Small particles arranged in a periodic pattern will scatter individual colors by optical diffraction, which is observed commonly in bird feathers and butterfly wings. The best example in gemstones is precious opal. A number of other similar scattering phenomena produce colors in stones like fire agate.

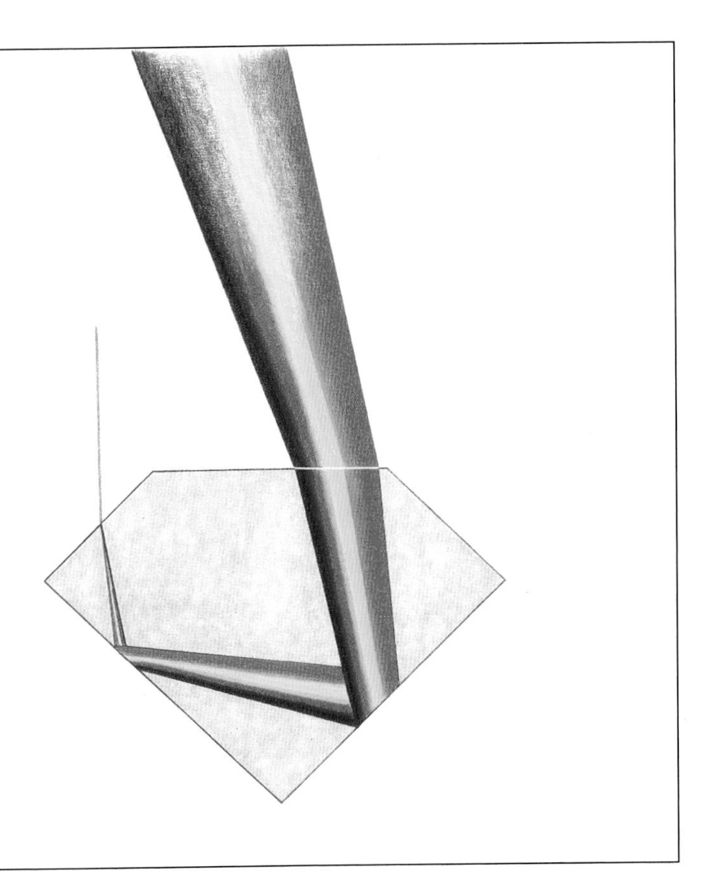

White light separates into component colors on passing through (opposite) a glass prism and (above) a colorless gem due to dispersion—the result for the gem is "fire."

Durability

Durability, the most important physical attribute in gauging a gemstone's merit, has three aspects: hardness, toughness, and stability. Hardness is the resistance to being scratched and is literally a measure of the strength of the chemical bonds in a substance. In 1822, German mineralogist Friedrich Mohs proposed a scale of hardness consisting of 10 minerals ranked in order of their ability to scratch one another; 1 is the softest, and 10 is the hardest. The scale is relatively linear; that is, each mineral is nearly one value of hardness greater than the previous one. Diamond, 10, is anomalous; it should have a value more like 100 to show its hardness relative to the others. Diamond is held together with extremely strong chemical bonding.

Quartz, with Mohs hardness of 7, is a common component of dust, so that gemstones softer than 7 are subject to scratching, particularly in rings, where abrasion is commonplace.

Toughness is a gem's resistance to cracking, chipping, and actually breaking. A chief threat to crystals is planes of weakness, representing directions in the crystal structure with relatively fewer or weaker chemical bonds. The result is cleavage—splitting along a plane. Diamond, the hardest mineral, lacks toughness because of its octahedral cleavage planes. Most diamonds in engagement settings will show small chipped corners after years of exposure to everyday wear and tear. Topaz, with a hardness of 8, also lacks toughness. It has one perfect cleavage and therefore is difficult to facet. Some gemstones fracture easily as a result of internal stress, which lowers their strength. Both opals and obsidian can chip easily due to physical or thermal shock. Nephrite jade, with a hardness between 6 and 6.5, is the toughest of gemstones. With a strong interlocking network of fibrous crystals, it can be fashioned into the most intricate shapes. Another tough gem is the pearl; one will not break if dropped on a hard floor, although the gem's hardness is only about 3.

Stability, the resistance to chemical or structural change from deteriorating forces, is an important factor in a gemstone's durability. Opals contain water, and some lose it in dry air; the result is cracks, or crazing, from loss of volume. Pearls are damaged by acids, alcohol, and perfume. Porous gemstones like turquoise can pick up oils and coloring from the skin. The color of some kunzite and amethyst fades on exposure to sunlight. However, the majority of gems are stable in most conditions the wearer is likely to place them.

Mohs Hardness Scale

1	*Talc*
2	*Gypsum*
3	*Calcite*
4	*Fluorite*
5	*Apatite*
6	*Orthoclase*
7	*Quartz*
8	*Topaz*
9	*Corundum*
10	*Diamond*

Where Gems Come From

GEMSTONES ARE UNCOMMON IN THE MINERAL KINGdom and require unusual geological conditions for their creation. They can form at various depths within the Earth's crust or even below in the mantle. All three classes of rock-forming environments—igneous, metamorphic, and sedimentary—produce gemstones, although the first two are predominant. Important gemstone sources, or occurrences, are gem pegmatites. Crystals that are measured in inches and feet occur in granitic pegmatites. They crystallize from the molten rock, magma, as the final step after a large quantity of granitic rock has already solidified. The residual magma becomes rich in volatiles such as fluorine, boron, lithium, beryllium, and water. The volatiles promote the growth of large crystals and are also components of aquamarines, tourmalines, and topaz, for example. The pegmatites of Minas Gerais in Brazil, the Ural Mountains of the Soviet Union, the Malagasy Republic, and San Diego County, California, are remarkable for their gem-quality crystals.

Another environment in which gemstones are found that has nothing to do with their original formation is placers or alluvial (river) deposits. Minerals released by weathering of rocks exposed at the Earth's surface wash into rivers (and onto beaches), where they concentrate as gravels; less durable minerals break up and wash away. Dense minerals are most effectively concentrated in this way. Alluvial gemstones are often superior to those found in solid rock because only the strongest, most perfect specimens survive the abrasive transport.

Gems and the Market

Practical questions also determine which minerals and rocks can be used as gems. Does a gemstone occur in pieces of sufficient size to fashion a reasonable gem? Many minerals could be used as gems if they occurred in clear crystals weighing several carats. Olivine is a common mineral, but geological occurrences of the green peridot crystals of adequate size are rare.

To be commercial, the gemstone abundance must be sufficient to underwrite the cost of developing a demand. Inadequate supply leads to unviable economics. Today the issue is often the depletion of sources. Alexandrites were never plentiful, but now the supply is so scarce that few examples ever reach the mass market. Consequently, even with a spectacular, though short-lived, find at Lavra de Hematita in Brazil in 1987, alexandrite is almost unknown except by collectors and gem experts.

Is a gemstone sufficiently rare to have status? Because gems are often the hallmark of social status and wealth, a degree of rarity is important. A gemstone may not lose its appeal with overabundance, but its value will certainly be affected.

CARAT, THE STANDARD UNIT OF MASS FOR GEMS

The abbreviation for carat is ct.

1 ct.	=	*0.2 grams*		
5 cts.	=	*1.0 grams*	=	*0.035 ounce (Avoirdupois)*
141.75 cts.	=	*28.35 grams*	=	*1 ounce*

Do not confuse carat with karat, the unit of measure of gold purity. Both terms probably originate from the Middle Eastern word for the seed of the carob tree (Arabic* quirat*). The seeds have remarkably uniform weight and were used to balance the scales in the ancient markets.

Evaluating Gems

EVALUATION OF GEMS IS A SEARCH FOR AND COMPARISON with perfection. There is a great range in qualities of some properties in each gemstone and no absolute code by which to compare different gems.

Color. A fine colored gem must have a good depth of color, not so pale as to be uncertain and not so deep as to appear black. Different color saturation can mean a remarkable difference in the price of two gems. The color should be uniform, not blotchy or strongly zoned. For some gemstones like amethyst, finding uniformity is the most serious problem in locating a fine sample. In multicolored gemstones, sharp, straight boundaries to color change are important rather than uniformity.

Clarity. This characteristic is important in most gems, less so in others. Two varieties of beryl have very different limits to what is acceptable. A fine aquamarine should be virtually without flaw, but a fine emerald will almost certainly have a few small inclusions. In fact, an emerald free of inclusions is suspected of being synthetic. Flawless transparency—freedom from inclusions and cracks—is critical to the beauty of gems like diamond and topaz.

Weight. The weight of a gem is always a factor in determining price. The value of ruby, diamond, and emerald increases dramatically with weight because large crystals are rare. This increase is an actual rise in the price per carat. Topaz, aquamarine, and rock crystal increase far less in value with increasing size because large crystals of these gemstones are relatively abundant. When a stone exceeds the size where it can be readily set in a piece of jewelry, its unit price plateaus or even begins to fall.

Cutting and Polishing. For an opaque gemstone, only the surface properties are important. Such stones are rarely faceted and are more frequently polished to obtain a smooth, rounded surface. The cabochon ("bald head"), a rounded top usually with a flat base, is used for translucent and opaque gemstones and gems displaying

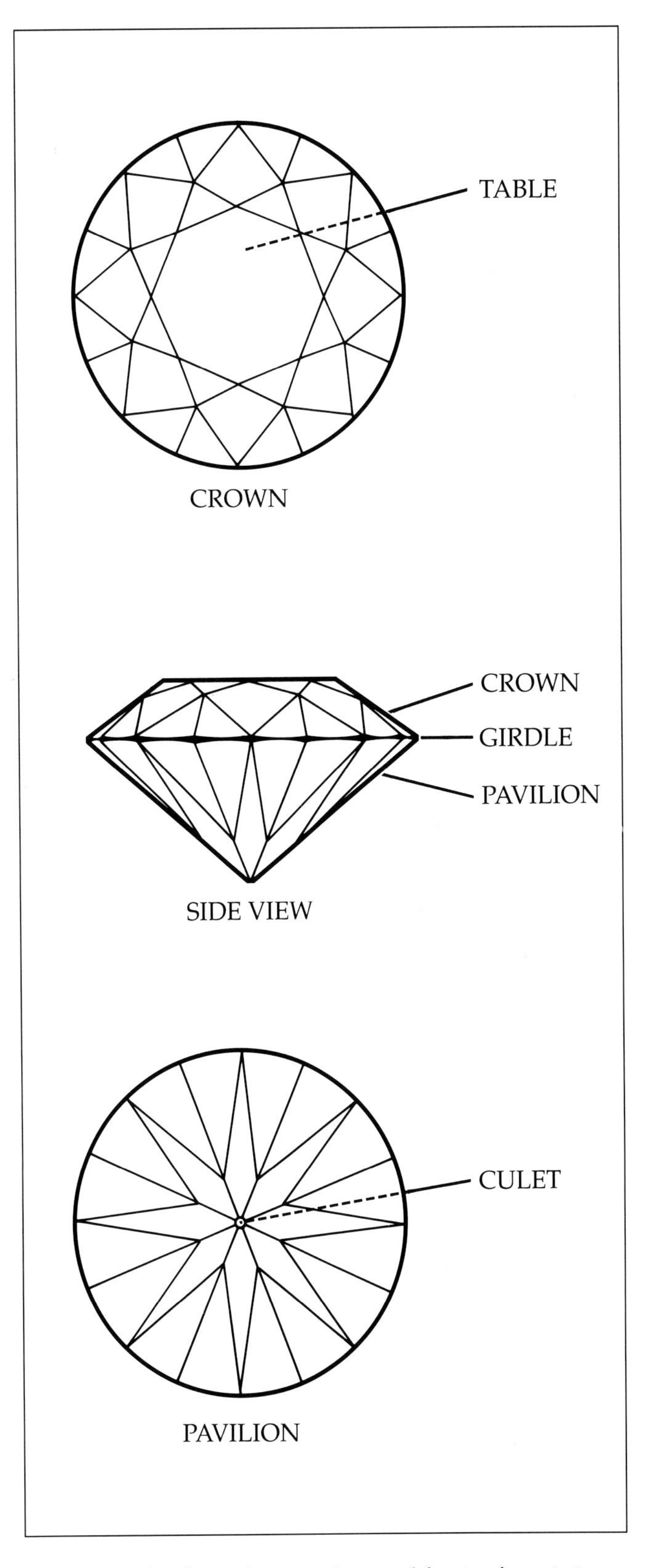

The names for the various portions and facets of a cut stone diagrammed for a round brilliant.

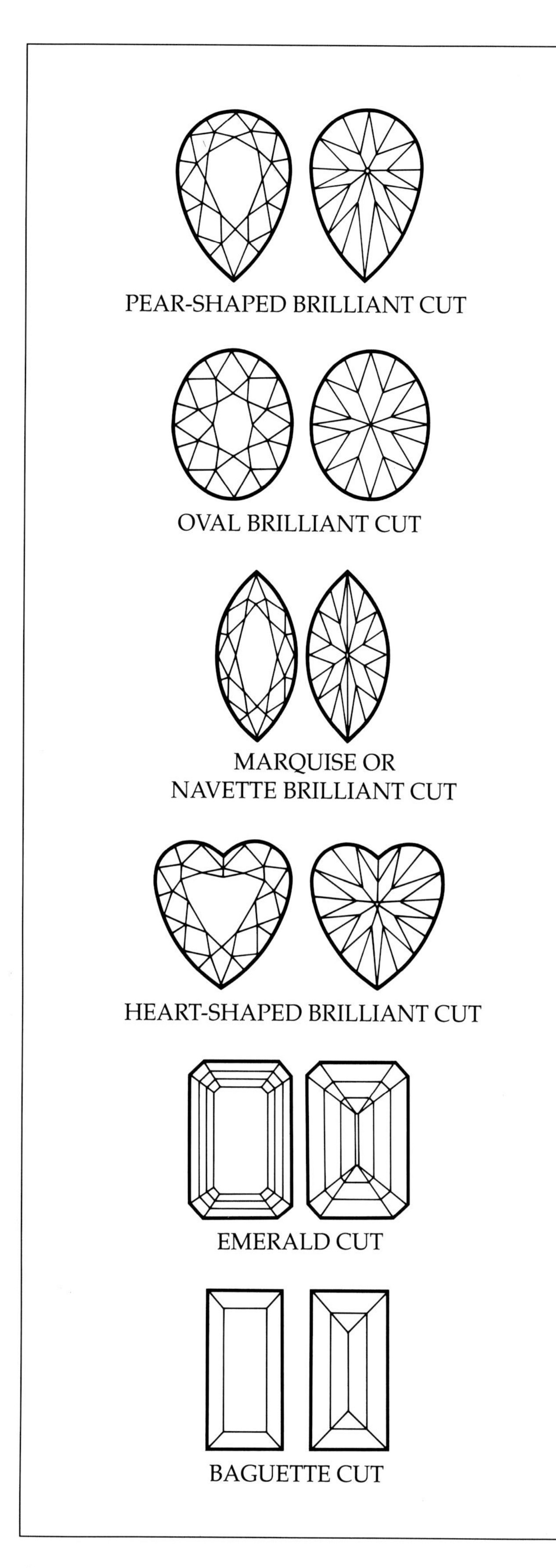

Some popular gem cuts, viewed from the table (top) and pavilion (bottom)

optical phenomena such as chatoyancy, asterism, and play of color.

Transparent gemstones are faceted. The process consists of cutting with an abrasive (usually diamond) saw, grinding with abrasives, and polishing facets. Cut also means the shape or style in which a gem is fashioned. Faceting and proper proportioning are essential for revealing the full beauty of transparent minerals, particularly a diamond's fire. Diamond faceting first appeared in the fourteenth century, but intense study of methods was stimulated by the great nineteenth-century discoveries of diamonds in South Africa. We now know the exact angles which must be present between facets to cause all light incident on the gem to be completely reflected for maximum brilliance. The round brilliant cut with its modifications (oval, pear, marquise, and heart) and the step (emerald) cut are the most popular.

The quality of cutting and polishing is another factor in the evaluation of all gems, although particularly significant for diamond. Many dealers will buy poorly faceted or proportioned stones and have them recut with reduction in weight but dramatic increase in value.

Gem Enhancement. During the last decade, enhancement of gemstones by chemical and physical means has become very common. Irradiation is being used to enhance or change the color of many stones. For some gems, chemical treatment or impregnation is used; procedures include bleaching, oiling, waxing, plastic impregnations, and dyeing. Heat treatment improves the color and clarity of some gems. Not all types of treatment can be detected at present. In 1989, a resolution was adopted by the members of the International Colored Gemstone Association for disclosure of gemstone enhancement upon request by customers.

Gem Substitutes. Substitutes are substances that so resemble a gemstone's properties that mistaken identity can occur. A relatively inexpensive gemstone may be substituted for a more valuable one, such as citrine quartz substituted for precious topaz. This practice is fraudulent.

Man-made substitutes fall into two categories: synthetics and simulants. A synthetic is the exact same substance as the natural mineral but grown in the laboratory. Synthetic gemstones have been produced commercially since the end of the nineteenth century; the first was synthetic ruby. Initially, there was fear that less expensive synthetics would dilute the market for natural stones and, thus, reduce the latter's value; this has never occurred. Distinguishing the natural from the synthetic gem is not always easy. Natural stones usually contain some inclusions that aid identification, whereas synthetics sometimes manifest evidence, such as lines and bubbles, of their synthesis. Very sophisticated techniques may be required to differentiate the nearly perfect natural gem and the synthetic.

Simulants usually have no natural counterpart but have optical properties that closely resemble those of a natural gem. Cubic zirconia—zirconinum oxide—is so inexpensive, so readily available, and so resemblant of diamond that it has replaced virtually all previously used diamond simulants.

Imitation gems have been in use since antiquity. They may be glass, plastic, or composite stones consisting of two (doublets) or three (triplets) parts. These parts may be genuine, imitation, or both. A clever imitation of Imperial jade is a triplet made of a hollow, colorless jadeite cabochon filled with a green jellylike substance cemented to a flat jadeite back. The color is magnificent but eventually disappears when the green substance dries. Opal is commonly made in doublets and triplets that utilize thin seams of the gem material and at the same time protect this fragile gem. Except in the instance of opal, composite stones generally have been replaced by synthetics.

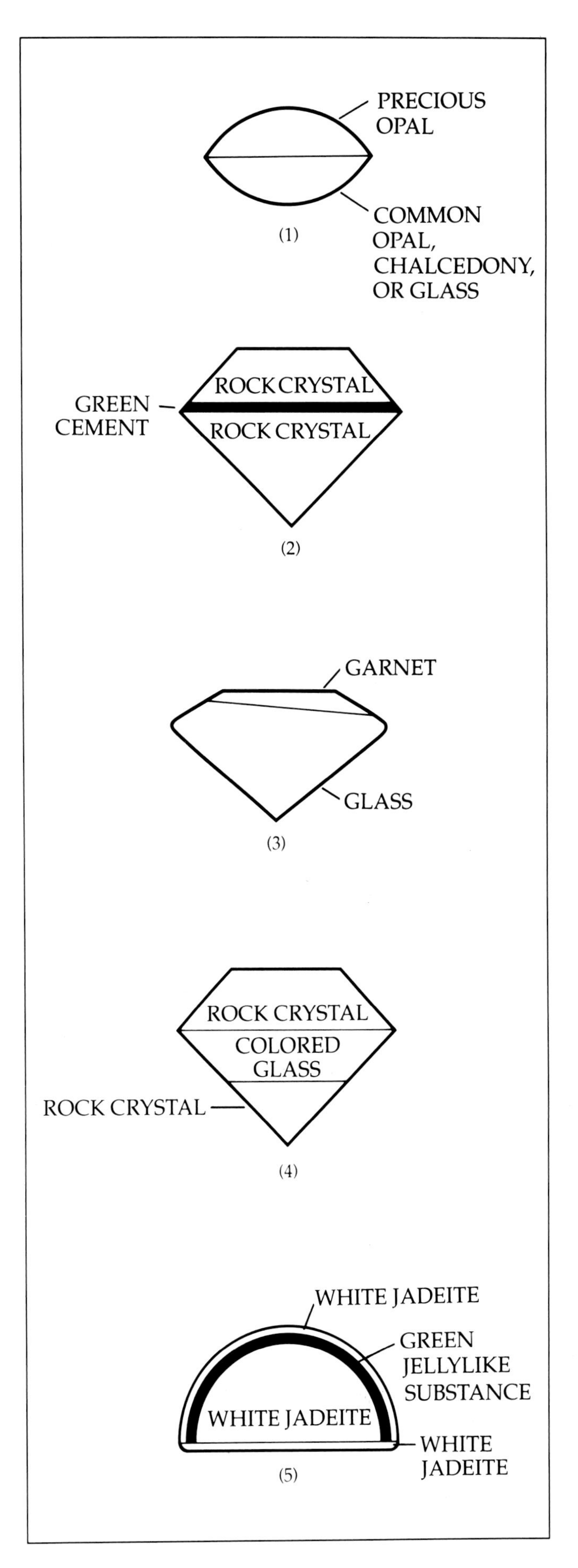

Several examples of composite gems: (1) Opal doublet; (2) Soudé emerald doublet; (3) Garnet doublet; (4) A triplet; and (5) A jadeite triplet.

Diamond

Diamond is the ultimate example of a gemstone with dual values. On the one hand, it is the peerless model for the colorless gem with its superior brilliance and fire and, on the other, it is the hardest of all substances; only diamond abrasive can fashion a diamond. Diamond as an ancient gem with magical properties is tied to India, its earliest source. There its hardness was appreciated, but the wellspring for diamond's value was the magical virtues that were believed to derive from the crystal form and optical properties—the ideal stone was a clear, transparent, fiery octahedron. The octahedron is the only crystal habit of diamond that manifests the "dazzling division of colors" that epitomizes the mineral's ideal high dispersion. Although diamond octahedra are not uncommon, ones that are sufficiently perfect to show fire are a rarity. This problem of scarcity in obtaining "perfection" might have been solved by reshaping (cleaving and polishing) poorly formed diamonds to bring out their sparkle and improve octahedral form. However, any alteration of a diamond was taboo, destined to destroy its magical properties. It is a wonder that the connection of fiery optics to the popularly perceived magical virtues was ever made.

Naturally-colored diamonds from the Aurora collection from sources around the world. (On loan from Aurora Gems, Inc.)

Properties

DIAMOND'S HARDNESS IS PRODUCED BY THE EXTREMEly strong chemical bonds each carbon extends to four adjacent neighbors. This configuration yields a crystal structure of high symmetry as expressed by octahedron- and cube-shaped crystals. However, one direction in the structure is crossed by relatively few bonds, which results in a perfect cleavage plane that repeats via symmetry to define an octahedron. Diamonds are routinely split along this cleavage to form smaller pieces for cutting gems. Cleavage is also diamond's main failing as a gemstone—a faceted stone can break or chip. Other results of diamond's strong and compact structure are a relatively high density, high refractive index, and high dispersion. These two optical properties give diamonds their ideal brilliance and fire.

We usually think of diamonds as colorless gems, but most are slightly to noticeably yellowish. Intense, attractive colors like canary yellow, pink, green, blue, purplish, and the rare red are called "fancy" colors. Color is due to the presence and distribution of minor amounts of nitrogen (most yellows) or boron (blues) in diamond crystals. Defects and radiation damage also contribute to color; several colors can be produced by artificial irradiation, sometimes followed by heat treatment. A bluish daylight fluorescence is relatively common in diamonds, and fancy colored stones are renowned for their spectacular range of fluorescence in ultraviolet light.

Chemically diamond is pure carbon, the same as graphite, a soft mineral used in pencils. The dramatically different properties of the two minerals are caused by the chemical bonding of carbon atoms—all strong bonds in diamond, some weak ones in graphite. Diamonds are extremely stable under ordinary conditions but—unlike other crystalline gemstones—when heated red hot in air, will literally vanish to form carbon dioxide.

DIAMOND DATA

Carbon:	*C*
Crystal symmetry:	*Cubic*
Cleavage:	*Perfect in four directions, defining an octahedron*
Hardness:	*10*
Specific gravity:	*3.514*
R.I.:	*2.417 (high), superb brilliance*
Dispersion:	*High, superb fire*

(See pages 22–29—"What Is a Gem?"—for explanation of data and definition of terms.)

A large 25-ct. octahedral diamond crystal in kimberlite from the Premier Mine, Kimberley, South Africa. (On loan from Mrs. Charles W. Engelhard.)

Historic Notes

DIAMONDS WERE REVERED AND HIGHLY VALUED AS talismans as far back as 800 B.C. in India, and for more than twenty-five centuries, that country was the only supplier of the gemstone. Diamonds and their reputation for metaphysical powers arrived in Rome around the first century B.C. Pliny describes this substance that will scratch all others, confirming the name for the hardest of substances—*adamas*. The word is Greek, meaning "I tame" or "I subdue," and well suits diamond's hardness.

During the first century A.D., prominent Romans wore uncut diamonds set in gold rings as talismans, and leading figures were awarded diamonds by the emperor; specimens of the period are yellowish or brownish, possibly indicating that India's best stones were not exported and almost certainly that the stones available were worn for their powers rather than their beauty.

According to Pliny, diamonds were known only to kings; indeed, very few kings possessed them even as late as the thirteenth century. Louis IX of France (1214-1270) issued an order forbidding women, including queens and princesses, to wear them. Agnes Sorel, the mistress of King Charles VII (1422-1450) dared to wear diamonds, and she is credited with popularizing them in the French court. During the second half of the century, the gem was more frequently worn but only by royalty. The first diamond engagement ring was given to Mary of Burgundy by Hapsburg Emperor Maximilian I in 1477.

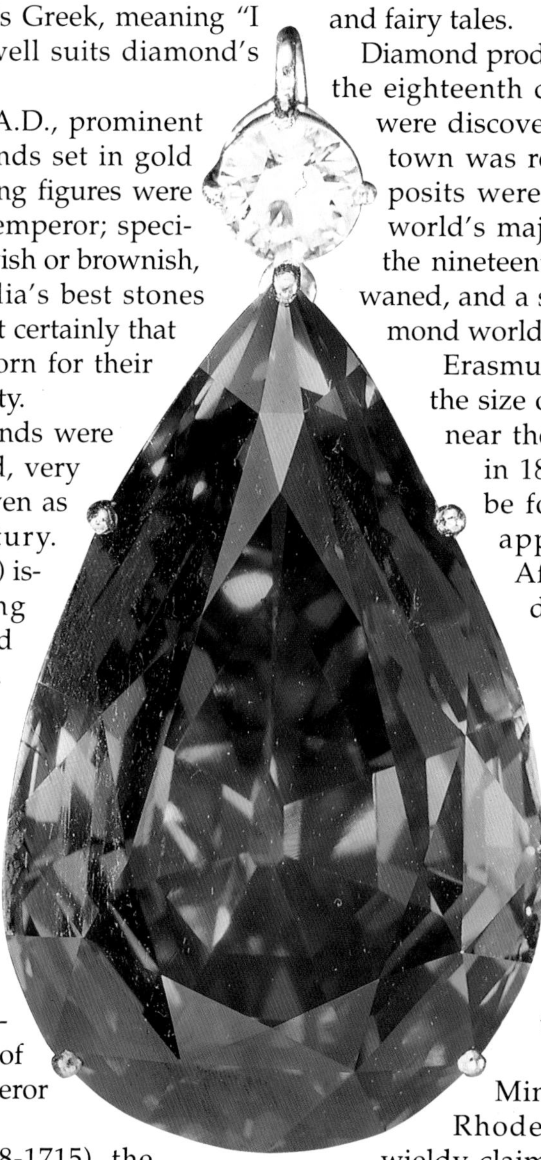

Louis XIV of France (1638-1715), the Sun King, collected many fine-quality diamonds. Jean Baptiste Tavernier, jeweler, merchant, and traveler, was influential in bringing diamonds to the court's attention. He made six trips to the Orient, visited Indian diamond mines, and brought back fabulous stones.

Cutting began in the fourteenth century in India and Europe, and diamond became a gem in the modern sense, treasured for its sparkling beauty. During the eighteenth century, diamonds became gems "par excellence," although exclusively for the super rich. Monarchs maneuvered for possession of them; the histories of famous and notable diamonds read like adventure stories and fairy tales.

Diamond production in India began to wane in the eighteenth century, but in 1725, diamonds were discovered at Tejuco in Brazil, and the town was renamed Diamantina. Other deposits were found, and Brazil became the world's major supplier. Toward the end of the nineteenth century, Brazilian production waned, and a series of events changed the diamond world dramatically.

Erasmus Jacobs' son found a diamond the size of a marble on the De Kalk farm near the Orange River in South Africa in 1866. It was the first diamond to be found in South Africa and was appropriately named "Eureka." After several other discoveries, a diamond rush began. In 1871, the De Beer brothers discovered diamonds on their farm, and the Kimberley mine began production in what proved to be the richest deposit ever found. Unmechanized mining produced a huge crater known as "The Big Hole." Bucket load by bucket load, 25 million tons of earth were excavated to recover about 3 tons of diamonds.

In 1888, De Beers Consolidated Mines, Ltd., was formed by Cecil Rhodes, who consolidated the unwieldy claims at Kimberley, thus establishing the De Beers monopoly. (More than 100 years later, De Beers still has the monopoly and, through its Central Selling Organization, controls 80 percent of the wholesale market.) A host of diamond discoveries have taken place in this century in Africa, the Soviet Union, China, and Australia.

With these nineteenth- and twentieth-century discoveries, diamonds have reached a popularity never before enjoyed. Ownership of the gem that until the fifteenth century had been reserved for royalty has become a realistic goal for the average person.

THE KOH-I-NOOR

This magnificent gem has the longest history of all the famous diamonds. In 1304, it was in the possession of the rajahs of Malwa. In 1526, it fell into the hands of the founder of the Mogul dynasty and was passed down the line to all the great Moguls until 1739, when Nadir Shah of Persia invaded India. All of the treasures of the Moguls fell into his hands, except the great diamond. He was told that the emperor had the stone hidden in his turban. So, in accordance with an Oriental custom, he invited his vanquished opponent to a feast where turbans would be exchanged. Later, in private, Nadir Shah unrolled the turban, the gem tumbled out, and Nadir is supposed to have exclaimed, "Koh-i-Noor!" (Mountain of Light) when the stone tumbled to the floor.

Later, a Persian king fled with it to the Sikh court. After the Sikh wars, the gem was taken by the East India Company as part of the indemnity levied in 1849 and was subsequently presented to Queen Victoria. She had the 186-carat gem recut to a 108.93-carat oval brilliant. The Koh-i-Noor, set in the Queen Mother's crown, is on view in the Tower of London.

Legends and Lore

HINDUS DIVIDED DIAMONDS INTO FOUR CATEGORIES according to four major castes; each category brought special good to its possessor: the Brahmin, power, friendship, and wealth; Kshatriya, everlasting youth; Vaisya, success; Sudra, good fortune.

Diamond is protective of its owner, according to a mid-fifth-century Sanskrit manuscript. It wards off serpents, fire, poison, sickness, thieves, flood, and evil spirits. Another Hindu belief was that a flawed stone has quite opposite effects; it could deprive even the god Indra of his highest heaven and could cause lameness, jaundice, pleurisy, and even leprosy.

The virtues of diamond are legion. The stone provided fortitude, courage, and victory in battle, and it stood for constancy and purity and enhanced love between husband and wife. Writing in the fourteenth century, Sir John Mandeville noted that diamond loses its magical power because of the sin of its wearer. Two centuries later, Geronimo Cardano was cautious; whereas diamond might make its owner fearless, fear and prudence might contribute more to well-being and survival, he noted. In addition, he stated that diamond's brilliance irritates the mind just as the sun irritates the eye.

For hundreds of years, the belief that diamonds had gender persisted. Theophrastus (c. 372-287 B.C.) divided each species into male (the dark colored stones) and female (the light colored stones). As late as 1566, François Ruet described two diamonds that produced offspring.

During the Middle Ages, physicians debated whether diamond was a potent poison or an antidote to poison. The poison theory was refuted by Portuguese Garcia de Orta, physician to the viceroy of Goa; in describing the Indian mines in 1565, he noted that slaves working in the mines swallowed diamonds in order to steal them and showed no ill effects.

(Opposite) The Golden Maharaja, a 65.60-ct. diamond, was owned by one of the richest maharajahs when shown at the 1937 World's Fair in Paris and at the 1939 New York World's Fair. The gem was featured in the Morgan Hall from 1976 to 1990 as an anonymous loan.

The Lounsbery diamond necklace, designed by Richard Lounsbery for his wife Vera and executed by Cartier of Paris. The necklace contains over 100 diamonds, fashioned in rose, brilliant, pendaloque, and modified single cuts.

The 14.11-ct. emerald-cut Armstrong Diamond.

Single and twinned diamond crystals; maximum dimension is 1 cm (3/8 in.).

Occurrences

DIAMONDS REQUIRE PRESSURES IN EXCESS OF 50,000 atmospheres to grow. The pressure corresponds to a depth of more than 90 miles (about 150 kilometers), which is within the Earth's upper mantle. Kimberlite, an unusual volcanic rock, is the mantle-sourced host for most diamonds. No kimberlite "volcanos" have erupted in over 60 million years, but if one did, the event would be devastating. A velocity exceeding the speed of sound, 740 miles per hour, is required to bring diamond up from the mantle intact. Kimberlites form somewhat carrot-shaped vertical volcanic vents, called "pipes," near the surface.

Initially, kimberlite weathers into a surface layer of "yellow ground" and a lower layer of "blue ground" that will contain intact diamonds. Further erosion carries diamonds into streams, rivers, and eventually beaches where, due to high density, diamonds become concentrated in placers. Many such deposits were discovered in the search for placer gold.

Australia is presently the world's largest producer of diamonds. It is the major producer of the rare and highly valued pink diamonds as well as much industrial-grade diamond. Its Argyle Mines contain 50 percent of the world's proven reserves.

Africa is the diamond continent. Zaire is the second largest diamond producer and was the first for thirty years, until the mid-1980s. Botswana ranks high in diamond production and is particularly important, since more than 50 percent of its output is of gem quality. Namibia is known for its productive beach placer deposits of diamonds, which are 90 to 95 percent of gem quality. South Africa has dropped from first to fifth place in rough diamond output. The giant Premier Mine has yielded many of the finest and most famous of them. Of the 20 largest diamonds found so far, 10 were mined in South Africa. Other important African sources are Ghana, Sierra Leone, Tanzania, Angola, and Lesotho.

The Soviet Union is also a major diamond producer. The sources are primary, with the richest pipe being the well-known Mir in central Siberia.

India's alluvial deposits were the first ever exploited, and many great historic diamonds were found there. The name Golconda, the ancient source, is synonymous with mine wealth. Indian production declined in the eighteenth century, although there is still some small-scale mining. Brazil and Venezuela are old but presently minor diamond suppliers, and China is a new one.

Evaluation

A DIAMOND'S VALUE DEPENDS ON "THE FOUR C'S": carat weight, color, clarity, and cut.

Carat Weight. A 2-carat diamond costs more than twice as much as two 1-carat diamonds of the same quality and substantially more than four times four 1/2-carat diamonds. This fact is a manifestation of the greater rarity, and value, of larger diamonds.

Color. Absolutely colorless "white" stones are graded on the Gemological Institute of America's color grading scale as D. The letter grades run down from D, E, and F on to S through X, at the bottom, for noticeably yellowish stones. To judge a diamond's color, look at it through its side against a white background. The gem should not be examined in direct sunlight, because fluorescence can mask diamond's true color. The most valued natural fancy diamonds are bright red, but all vivid colors command high prices. Diamonds that are colored by irradiation and heating—green, yellow, golden brown, blue, purple and red—require a disclosure.

Clarity. Most diamonds contain natural inclusions; by international agreement, a diamond is regarded as flawless (FL) if no inclusions are visible to the trained eye through a lens or loupe with tenfold magnification. Increasingly visible inclusions diminish the quality and grade of a stone. Recently, vaporizing inclusions with a laser and filling of cleavages and fractures to reduce their visibility have been reported. Such practices must be disclosed.

Cut. Cutting brings out the full brilliance and fire of a diamond. The diamond should be faceted so that the maximum amount of light entering the stone reflects from the back facets and emerges back through the top. Poorly proportioned stones lose a lot of light through the back facets. The round brilliant cut was developed for diamond and is most popular, because it displays the most brilliance. The oval, pear, and marquise cuts, which are modified brilliant cuts, appear larger than a round brilliant of the same weight but do not attain the same level of brilliance. The emerald cut, also called the step or trap cut, also yields reduced brilliance and, so, is often used for large flawless diamonds that would be blindingly brilliant if cut round.

Synthetic Diamonds and Diamond Simulant

Gem-quality synthetic diamonds greater than a carat in size were first made in 1970 but only recently have been grown in commercial quantities. These synthetics have been more expensive than natural diamonds. Cubic zirconia, synthetic zirconium oxide, is now the principal diamond simulant and is produced in both colored and colorless varieties.

(Above) The Tiffany Diamond is a 128.54-ct. canary yellow diamond from Kimberley, South Africa. This superb gem was selected for the firm by George F. Kunz in 1879 and cut in Paris. The stone was featured at the Museum in the exhibit "Tiffany: 150 Years of Gems and Jewelry" in 1988.

Corundum

Sapphire and ruby are the gem varieties of the mineral corundum, but few people anticipate the myriad of colors displayed in these gemstones. Ruby is red, as everyone knows, and sapphire comes in all colors except red: pink, orange, yellow, brown, green, blue, purple, violet, black, and colorless. (Sapphire colors other than blue are termed "fancy" colors.) These are all represented in the Museum as in no other public display—our suite of large sapphires, some exceeding 100 carats, is famous. The huge 563-carat Star of India sapphire is one of the Morgan gifts. Its name suggests a story—one might speculate that, after being mined in Sri Lanka in the sixteenth century, it circulated among the treasuries of Indian potentates. If we only knew the sights this stone may have seen! However, George F. Kunz recorded only this enigmatic statement: "[It] has a more or less indefinite historic record of some three centuries and many wanderings." How the gem came into Kunz's hands is unrecorded, but rumor has it that a royal owner needed cash without publicity. An alternate, but doubtful, story is that Kunz had the stone fashioned in New York City in 1900—so much for romance! No matter, the Star of India is magnificent.

The Star of India, the most famous gem in the Museum—is the largest gem-quality blue star sapphire in the world. The 563.35-ct. stone is nearly flawless and exhibits a perfect star. The almost spherical stone also displays a good star on its back side.

Properties

SPECTACULAR COLORS, COMBINED WITH GREAT DURAbility and reasonable abundance, have made sapphires and rubies important gemstones for centuries. Corundum has a tightly bonded structure that results in high density and great hardness, second only to diamond. The mineral is very stable chemically and essentially has no cleavage.

When aluminum is replaced in corundum by transition elements, the great range of colors in sapphire and ruby are produced. Corundum gems are pleochroic, with colors being more intense when the crystal is viewed down the trigonal axis. Some stones manifest color-change pleochroism from purple to blue or, like alexandrite, from blue green to red.

Titanium is incorporated into the crystal structure of corundum at high temperature and, during slow geologic cooling, it crystallizes separately as fine, silky fibrous inclusions of rutile (TiO_2). The result is chatoyancy that is multiplied by corundum's trigonal symmetry into six- and occasionally twelve-rayed star sapphires or rubies. These inclusions also make the gemstone translucent. Heat treatment "dissolves" the silk back into the corundum structure and generally results in a transparent, more intensely colored stone. We could heat the Star of India and create a fabulous blue cabochon—perish the thought!

Corundum crystals are typically hexagonal prisms that are somewhat tapered, often tabular, and frequently marked with triangular striations on the ends.

The Midnight Star, 116.75 cts., is notable for its deep purple violet color. The stone was found in Sri Lanka.

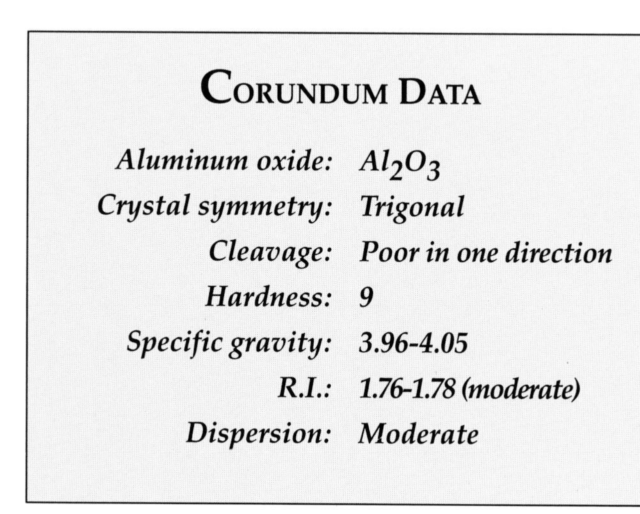

CORUNDUM DATA

Aluminum oxide:	Al_2O_3
Crystal symmetry:	*Trigonal*
Cleavage:	*Poor in one direction*
Hardness:	*9*
Specific gravity:	*3.96-4.05*
R.I.:	*1.76-1.78 (moderate)*
Dispersion:	*Moderate*

GEMSTONE CORUNDUM VARIETIES, COLORS, AND SOURCES OF COLOR

Ruby: ***Intense red—chromium***

Sapphire: ***Blue—iron + titanium***

Other than blue (called fancy sapphires)

Padparadscha: *Orange—chromium + ferric iron*

Alexandrite-like: *Vanadium*

Yellow: *Ferric iron (or defects)*

Green: *Ferrous + ferric iron (and defects)*

Colorless: *Pure, no substitutions*

Small Sri Lankan sapphires, ranging from 2.00 to 16.90 cts.

Historic Notes

PRECIOUS STONES WERE GROUPED BY COLOR IN ANTIQUITY. *Carbunculus* was the Latin term that Pliny the Elder used for transparent red stones. Before 1800, red spinel, red garnet, and ruby were all termed ruby, deriving from the Latin *rubeus,* meaning "red." A number of famous "rubies" have turned out to be spinels; the Black Prince Ruby, the Timur Ruby, and the Côte de Bretagne Ruby are examples. The Catherine the Great Ruby, considered for years to be the largest ruby in Europe, was identified as tourmaline. *Sapphirus* means "blue," and until the Middle Ages, this name was used for lapis lazuli.

The history of sapphire dates back to the seventh century B.C., when it was used by ancient Etruscans. Following centuries saw the gemstone used in Greece, Egypt, and Rome. Rome obtained rubies from what is now Sri Lanka and from India, where ruby was the most valued of gems and called "king" and "leader" of precious stones.

Marco Polo's travels took him to the "Island of Serendib" (Sri Lanka), and his thirteenth-century *Book of Marvels* gives high praise to both stones. He tells the story of a Sinhalese king, a ruby, and the Chinese Emperor Kublai Khan. The ruby was huge—4 inches—and Kublai Khan offered an entire city in exchange for it. The king refused, saying that he would not give up his prize for all the treasures in the world. Nothing more is known of this stone, and its reported size leads one to speculate whether it really was a ruby.

Sapphire was a favorite stone for rings and brooches of the medieval kings in Europe, and, beginning in the eleventh century, it also became the preferred stone for ecclesiastic rings as well. By the time of the Renaissance, both ruby and sapphire had found favor with the wealthy; indeed, only the wealthy could afford them. Benvenuto Cellini, writing in 1560, stated that the price of ruby was eight times that of diamond. And ruby is still generally the most valuable gemstone.

(Above) Typical Montana "Yogo" rough and cut sapphires in blue and violet. The thin tabular gems range from .75 to 2.25 cts. in weight.

(Opposite) The superb 100-ct. Sri Lankan stone is the largest and finest padparadscha sapphire on public display. The varietal name originates from a Sanskrit word for the orangy pink color of the revered lotus.

A large Burma ruby cabochon, 47 cts., a round 1.38-ct. ruby from North Carolina, and a 1.87-ct. ruby from Tanzania.

Legends and Lore

THE POWERS THAT HAVE BEEN ASCRIBED TO RUBY OVER the centuries are innumerable. Early Burmese thought the stone would bestow invulnerability when it was actually inserted into the owner's flesh. During the Middle Ages, ruby was believed to have an inner fire that could not be concealed. Camillus Leonardus scorned those who denied the magical powers of precious stones and in the sixteenth century wrote that ruby would preserve its owner's health, remove evil thoughts, control amorous desires, dissipate pestilential vapors, and reconcile disputes. Another belief was that ruby could warn its owner of impending misfortune or calamity by becoming dull and dark. Catherine of Aragon (1485-1536), the first wife of Henry VIII, is said to have foretold her own downfall in perceiving the darkening of her ruby.

Sapphire's powers were equally sweeping. In a treatise ascribed to Damigeron, sapphire protected kings from harm and envy. Its powers included the capacities of banishing fraud and preventing terror, according to Marbode in the eleventh century. Two hundred years later, a French manuscript reported that the stone had the power of preventing poverty, while another lapidary of the same period stated that sapphire makes a stupid man wise and an irritable man good-tempered.

The star sapphire has been called "the stone of destiny"; its three crossed lines represented faith, hope, and destiny. Still another legend refers to these gems as sparks from the Star of Bethlehem. To the Germans, it was "Siegstein," meaning "victory stone," according to De Boot, writing in 1609.

The Hindus, Burmese, and Ceylonese (Sinhalese) recognized a relationship between sapphire and ruby long before the Europeans did. To them, the colorless sapphire was an unripe ruby; if buried in the ground, it would mature and turn red—a belief documented in a sixteenth-century manuscript by Garcia de Orta, the Portuguese physician to the viceroy of Goa. Flawed stones were considered overripe.

A ruby crystal, 4 cm (1 1/2 in.) long, in white marble, from Jagdalak, Afghanistan.

Occurrences

THE PRIMARY SOURCES ARE OF TWO MAJOR TYPES. First, high-temperature metamorphism of claystones and dirty limestones can form all corundum gemstones. Second, sapphires are found in some quartz-free igneous rocks. After weathering from primary sources, corundum gemstones concentrate in placer gravels—the most important commercial deposits.

The earliest sources for both ruby and sapphire are placers in Sri Lanka. Mining began before Buddha's time (624-544 B.C.) near Ratnapura (Sinhalese for "City of Gems"). The rubies found here are paler than Burmese stones. This is the only source of the rare lotus-colored padparadscha sapphire and of the finest star sapphires. The blue sapphires are usually light in color.

Today the world's major sources of rubies are alluvial deposits in Thailand near the Cambodian border. The rubies are mostly dark and brownish red. Dark blue, some green, and only occasionally very fine blue sapphires are also found here.

The world's finest rubies come from the Mogok valley of Upper Burma—the finest deep red (pigeon blood) rubies, occasionally called "Burma rubies" regardless of source, and also very dark or pale rubies. The deposits also yield sapphires and many other gems. The first known record of the Mogok mines is dated A.D. 1496, but indirect evidence suggests much earlier mine operation. The mines' present output is limited.

Australia is the world's largest source of sapphire. In 1987, about 75 percent of the world's output came from New South Wales and Queensland. The sapphires are alluvial, weathered from basalt. The blue stones are generally dark and have an inky appearance.

The major American source of sapphire is Yogo Gulch in Montana, which was discovered in 1895 and has been worked intermittently. The stones are small and waferlike and clear blue to violet. Sapphire gravels occur along the Missouri River near Helena, Montana.

Other important sources of ruby and sapphire are Pakistan, Afghanistan, India, and East Africa.

(Above) The DeLong Star Ruby, 100.3 cts., one of the great star rubies, was discovered in Burma during the 1930s and donated in 1938 by Mrs. George Bowen DeLong. In 1964, the ruby was stolen in "The Great Jewel Robbery." Ten months of intricate negotiations involving underworld figures and a ransom of $25,000 followed before the famous ruby was returned.

(Opposite) A fabulous assortment of Sri Lankan sapphires, ranging in weight from 3.50 cts. to 188 cts. The large 112-ct. yellow stone (upper left) is on loan from the Precious Stone Company, New York City.

Two intergrown water-worn sapphire crystals, 5 cm (2 in.) long, from Sri Lanka.

Evaluation

COLOR QUALITY IS MOST IMPORTANT TO RUBIES AND sapphires. Rubies with intense and uniform red to slightly purplish red—the so-called pigeon-blood color—are the most valuable. Medium-deep cornflower-blue sapphires are the most highly prized, and evenness of color is extremely significant. Orange and alexandrite-like sapphire of good quality command very high prices.

Flaws diminish the value of both gems; however, a fine-colored ruby is of high quality even if a minor flaw is present. To improve color and transparency, about 90 percent of new-mined sapphires are heat-treated, with permanent results.

Star rubies and sapphires must be at least translucent to be of gem quality (with the exception of black star sapphires). The star must have well-defined, sharp, straight rays that intersect at the center of the stone.

Large rubies are rarer than large diamonds, emeralds, and sapphires. Thus the value of ruby, even more than of other gems, increases with weight.

Many other gems look like ruby and sapphire and are readily confused with and substituted for them. Synthetic corundums have been on the market since 1902 and are widely used in less expensive jewelry.

GEMSTONES CONFUSED WITH GEM CORUNDUM AND TRADE NAMES

Ruby: *Spinel (Balas ruby), pyrope garnet (Cape and Arizona ruby), red tourmaline (Siberial ruby), and pink topaz (Brazilian ruby)*

Blue Sapphire: *Benitoite, cordierite, kyanite, spinel and synthetic spinel, and tanzanite*

Green Sapphire: *Zircon*

Yellow Sapphire: *Chrysoberyl*

Beryl

Beryl is a mineral with several colorful gemstone varieties—aquamarine, morganite, heliodor—but the preeminent one is emerald. A popular misconception is that the Mogul, Ottoman, and Persian emerald treasures come from Oriental deposits. In fact, there are no such sources; these fabulous gems are from Colombian mines, Spanish loot from the New World. Spain needed money and found buyers among the Mogul nobility of India. The Museum's Schettler Emerald is an example; it was cut in India during the period of Mogul domination and probably worn as headgear or a sleeve ornament by a Hindu prince. Its uncut counterpart is the Patricia Emerald. The largest gem-quality crystal on record from the famous Colombian Chivor Mine was discovered in 1920 and sold the following year for $60,000. Only a few large fine crystals have been preserved in museums and bank vaults—emeralds are so valuable as gems that the crystals rarely escape being cut.

The Patricia Emerald is a twelve-sided crystal, 6.6 cm (2 3/5 in.) long, from the Chivor Mine, Colombia, and named for the mine owner's daughter. It weighs 126 grams and is famous for its crystal perfection and superb color as well as its size.

Aquamarine, morganite, and heliodor from various localities, ranging in weight from the 11.38-ct. emerald-cut golden heliodor to the 390.25-ct. aquamarine.

BERYL DATA

Beryllium aluminum silicate:	$Be_3Al_2Si_6O_{18}$
Crystal symmetry:	***Hexagonal***
Cleavage:	***None***
Hardness:	***7.5-8***
Specific gravity:	***2.63-2.91***
R.I.:	***1.566-1.602 (low)***
Dispersion:	***Low***

Properties

BEAUTIFUL COLOR DISTINGUISHES THE BERYL VARIETIES as gemstones. Beryl is a hard mineral but has only moderate brilliance and little fire. Elemental substitutions for aluminum in the crystal structure are the most common sources of color. However, there is a cavity along the sixfold axis in the structure that frequently accepts a chromophoric metal like iron as in the case of aquamarine and some green beryl. Color-zoned crystals are possible, the most interesting being bicolor moganite-aquamarine.

Crystals form distinctive hexagonal prisms that, when from pegmatites, can be among the largest gemstone crystals. Inclusions vary with occurrence and variety. Emeralds are typically heavily flawed with cracks and inclusions of fluids and minerals from rocks in which they grow; these are called "jardin" (garden), the inclusion patterns resembling leaves and branches. The other beryl varieties usually have greater clarity, most commonly containing small parallel fluid-filled tubes that have the appearance of rain when a crystal is strongly illuminated. If these inclusions are fibrous, they can yield chatoyancy in a polished gem.

This Chinese carving of a sitting goddess 10.5 cm (4 1/4 in.) high is the finest and largest morganite carving known.

(Opposite) Well-formed emerald crystal 6.5 cm (2 1/2 in.) long, from Takowaja, Ural Mountains, the Soviet Union.

(Above) Heliodor crystal 10.2 cm (4 in.) in height, from Minas Gerais.

(Right) A platinum pendant set with diamonds and containing an emerald, 1.6 cm (5/8 in.) long.

Gemstone Beryl Varieties, Colors, and Sources of Color

Emerald:	*Intense green or bluish green—chromium and/or vanadium*
Aquamarine:	*Greenish blue or light blue—iron*
Morganite:	*Pink, peach, or purple pink—manganese*
Heliodor or Golden beryl:	*Golden yellow to golden green—iron*
Green beryl:	*Light green or too pale to qualify as emerald—iron, chromium, and/or vanadium*
Red beryl or Bixbite:	*Raspberry red—manganese*
Colorless beryl or Goshenite:	*Pure beryl, sometimes with cesium*

Historic Notes

OF THE GEM BERYLS, EMERALD HAS THE LONGEST HIStory. The term *emerald* derives from the Greek *smaragdos*, for which there are conflicting meanings and antecedents. The earliest emeralds, extracted by Egyptians, date from the Ptolemaic era (323-30 B.C.); however, tools dating to Rameses II (c. 1300 B.C.) or even earlier have been found in tunnels at Sikait and Zabara, the location of the emerald mines. These mines were the West's principal source for emeralds until the sixteenth century, although another source was known to the Celts and Romans—Habachtal, south of present-day Salzburg in Austria. (The Archbishop of Salzburg had the deposits worked during the Middle Ages.)

Prior to the Spanish invasions in South America, emeralds from Colombia were traded and prized from Mexico to Chile. Almost from the time of their arrival, the conquistadors observed native rulers wearing them: in 1533 alone, Pizarro sent back four chests of emeralds to Spain. Finding no source in Peru, the Spaniards looked farther north and in 1537 discovered Chivor in what is now Colombia. Following a skirmish with the Muzo Indians, from which the Spanish were forced to retreat, a soldier found an emerald and delivered it to his commander. The Spanish returned in force, defeated the Indians, and took over the Muzo mine. The Colombian gems were larger and of finer quality than any seen in Europe and Asia before the Conquest. They totally supplanted the Egyptian emeralds because of their superior quality.

The histories of aquamarine, heliodor, and morganite are more recent. The first documented use of aquamarine is by the Greeks between 480 and 300 B.C. The gem has been very popular since the seventeenth century. *Heliodor* (golden beryl) derives from two Greek words meaning "sun" and "gift"; this gemstone has also been known since antiquity. It has rarely been used in jewelry, for its color is not outstanding among other yellow gems. Morganite is the latest gem to be recognized in the beryl group. First mined in Madagascar (the Malagasy Republic) in 1902, the gem was named after J.P. Morgan by George F. Kunz.

(Left) Heliodor crystal 7.5 cm (3 in.) long from Siberia in the Soviet Union and a 59.01-ct. cut stone from Sri Lanka.

(Below) A square-cut 278.25-ct. morganite from Minas Gerais in Brazil next to a matrix specimen consisting of a perfect morganite crystal (6.5 cm across) with a gem-quality bicolor elbaite from San Diego County, California.

(Opposite) This is a 5.28 kg (11.6 lb.) fragment from the largest aquamarine crystal ever found. The hexagonal prism, weighing 110.5 kg (243 lb.) and measuring 48.3 by 40.6 cm (19 by 16 in.), was discovered near Marambaia, Brazil, in 1910 and was so clear that, looking down the long axis, one could read a newspaper through it. The gemstone was cut in Idar-Oberstein, Germany, and yielded about 200,000 cts. of gems. A 47.39 ct. stone from Siberia, USSR, is shown for scale.

(Right) The official trapping of a vizier of Morocco (c. 1750). Aquamarines set in gold and surrounded by small diamonds, rubies, sapphires, and red garnets.

(Below) A delicate aquamarine necklace with pearls and diamonds dates from the early twentieth century.

(Opposite) Aquamarine brooch, earrings, and ring in platinum with diamonds. The simple, clean lines and the bold design of the swirl brooch are typical of the Art Deco style that developed during the 1930s. The large emerald-cut stone weighs 41.25 cts.

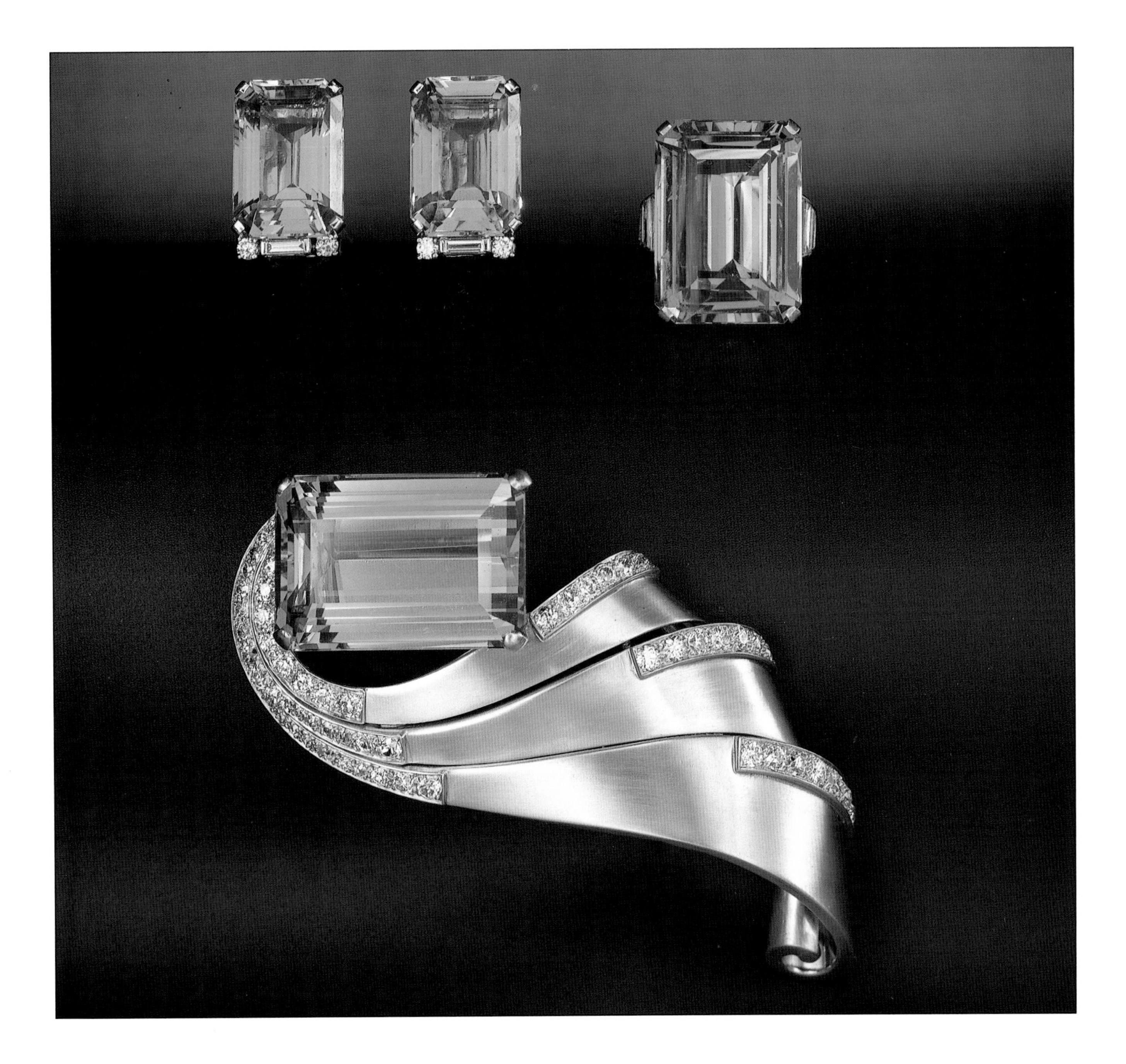

Legends and Lore

TO THE ROMANS, EMERALD SYMBOLIZED THE REPRODUCTIVE forces of nature and was dedicated to Venus; to the early Christians, it represented resurrection. In the fourth century B.C., Theophrastus noted its power to rest and relieve the eyes. Far later, Anselmus de Boot (1609) recommended emerald as the most powerful amulet to prevent epilepsy, stop bleeding, cure dysentery and fever, and avert panic.

In addition, an emerald was thought to give its owner the ability to foretell the future. According to Marbode, writing in the eleventh century, emerald improves memory, makes its owner eloquent and persuasive, and brings him joy. On the other hand, emerald was considered an enemy of sexual passion, and, in the thirteenth century, Albertus Magnus wrote that when King Bela of Hungary embraced his wife, his magnificent emerald broke into three pieces.

Aquamarine derives from two Latin words meaning "water" and "sea." Aquamarine amulets were thought to render sailors fearless and protect them from adversities at sea, especially if the stone were engraved with Poseidon on a chariot. The stone was a symbol of happiness and eternal youth, and according to Christian symbolism, it signified moderation and control of the passions to its owners. In medieval Europe, heliodor was believed to cure laziness.

An aquamarine cat's eye weighing 145 cts., from Minas Gerais, Brazil.

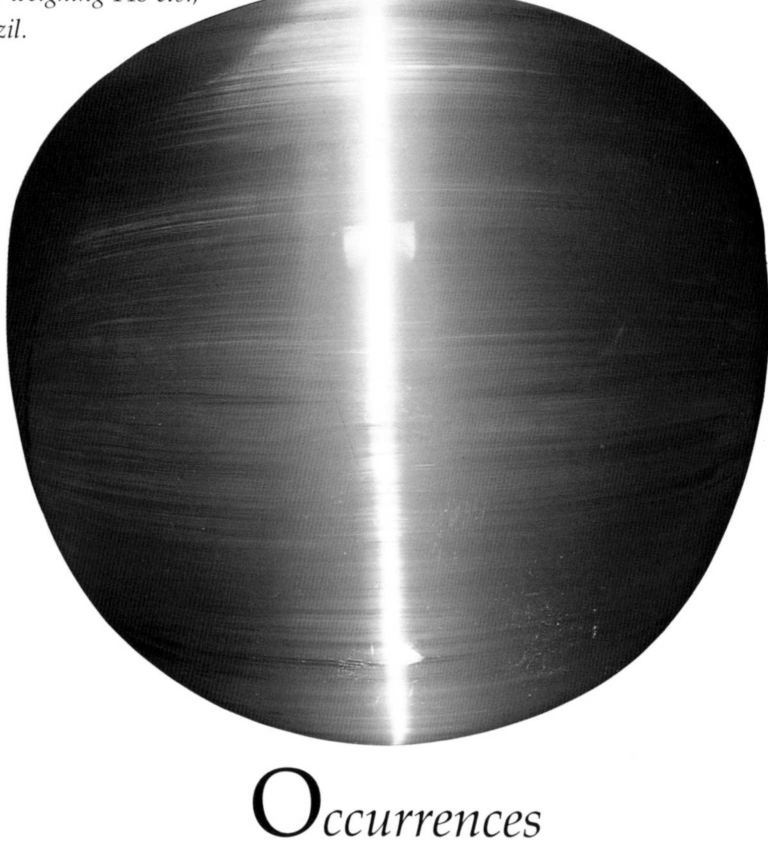

Occurrences

EMERALDS ARE MOST FREQUENTLY FOUND IN METAmorphosed shales, particularly mica schists—the reason for some emeralds containing mica inclusions. In Colombia, emerald occurs in calcite veins in black shale (Muzo) and in quartz veins in limestone (Chivor). Aquamarine, morganite, and heliodor are found as well-formed crystals in pegmatites. Beryls are not sufficiently dense to concentrate in placers and are normally mined from the primary source or its weathered equivalent.

Colombia is the world's largest emerald producer with about 100 mines in operation. Muzo and Chivor are the two principal mines. Muzo yields the world's finest emeralds. Mining operations have continued there since the Spanish Conquest almost without interruption. In general, Chivor emeralds are less flawed but do not have as velvety an appearance as those from Muzo.

Mining of emeralds in Russia began shortly after a gem crystal was discovered in 1830 by a peasant in the Ural Mountains northwest of Ekaterinburg (Sverdlovsk). He took his find to the lapidary factory in Ekaterinburg, where geologist Jakov I. Kakovin recognized it as an emerald. Some years later, Kakovin's office was searched, and a large emerald crystal was found. Kakovin was sent to prison, where he committed suicide. The 2226-gram crystal now resides in the Mineralogical Museum of the Academy of Sciences in Moscow. The Russian emeralds vary from fine deep green (but heavily included and flawed) to yellowish green (and less included).

Zambia emerged as a source of emerald in 1977. Emerald is also found in North Carolina. The best-known locality is around Hiddenite, where emeralds were discovered in 1880. Other occurrences are in Brazil and Pakistan.

Brazil is the principal source of aquamarine. More than 80 percent of the country's aquamarine comes from an area around Teofilo Otoni in the western part of Minas Gerais. The Malagasy Republic is known for the rich blue color of its aquamarine, which resembles sapphire. Aquamarine is also found in the Soviet Union in the Ural Mountains, in Transbaikalia, and in Siberia. Other occurrences of aquamarine include China, Pakistan, Afghanistan, and Maine, Idaho, and California.

The major sources of morganite are San Diego County in California, Minas Gerais in Brazil, and the Malagasy Republic. Heliodor is found in Minas Gerais and Goias, Brazil; the Ukraine in the Soviet Union; and also in Connecticut and Maine.

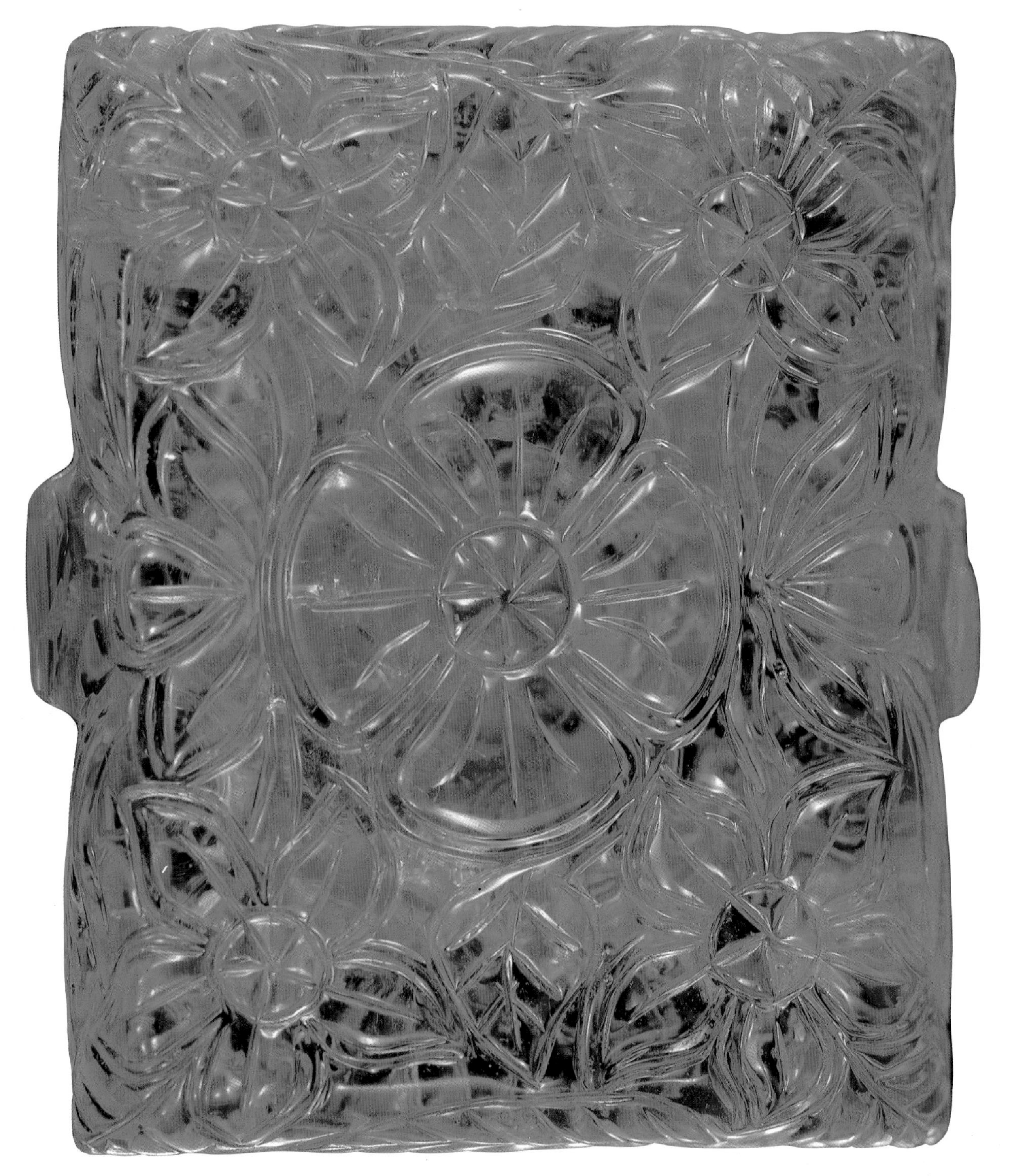

The Schettler Emerald weighs 87.62 cts., and its longest dimension is 3.5 cm (1 3/8 in.). It is engraved on both sides in a flower-and-leaf pattern and is probably from Muzo, Colombia.

An etched aquamarine crystal 8 cm (3 1/8 in.) tall, associated with white albite, from the Dusso area in Pakistan.

Evaluation

EMERALDS WITH A RICH, VELVETY GREEN UNIFORM color and a minimum of flaws are considered the finest quality. Skillful cutting can both minimize the visibility of inclusions and bring out the gem's best color. Emerald is often oiled to conceal cracks, and sometimes dye is added to improve the color—a fraudulent practice. Synthetic emeralds have been produced since 1934. Their prices are higher than those of other synthetics and even some natural emeralds.

Intensity of color and clarity are the most essential considerations in evaluating aquamarine, morganite, and heliodor. Aquamarine should be bright sky blue or sapphire blue. Aquamarines with intense color are becoming very scarce, and their price has increased substantially. The bright sky blue shade is now produced by heat-treating greenish yellow, greenish, and even brownish beryls. The color change is permanent. Gem-quality aquamarine is generally free from inclusions. Morganite should be deep purple pink, but peach colored gems are next best. Heliodor with a deep yellow to yellow green color is desirable.

A superb 58.79-ct. morganite from the Malagasy Republic.

GEMSTONES CONFUSED WITH GEM BERYLS

Emerald: ***Demantoid and tsavorite garnets, Imperial jade, tourmaline, peridot, green zircon, and hiddenite***

Aquamarine: ***Blue topaz, euclase, kyanite, apatite, sapphire, tourmaline, and zircon***

Morganite: ***Kunzite, tourmaline, topaz, sapphire, spinel, and rhodolite garnet***

Chrysoberyl & Spinel

CHRYSOBERYL

Alexandrite and cat's eye are renowned for their eye-catching properties, but few people know that they are varieties of a mineral that is more common in its transparent yellow green gemstone form. All three are chrysoberyl. Cat's eye chrysoberyl is *the* cat's eye; when spotlighted, the gem exhibits a band of light that opens and closes as the stone is turned. Alexandrite changes from red in incandescent light to green in daylight. It is *the* color-change gemstone. By comparison, ordinary chrysoberyl, though fine in its own right, seems a "poor relation."

An 85.0-ct. cat's eye from an unknown locality.

A chrysoberyl trilling, consisting of three twinned crystals, from Espirito Santo, Brazil; it measures 8 cm (3 1/8 in.) across—front and side view.

Properties

CHRYSOBERYL IS AMONG THE MOST BRILLIANT GEMstones and only surpassed in hardness by diamond and corundum. The common variety is transparent yellowish green to greenish yellow and pale brown, the result of small amounts of iron replacing aluminum. Chromium is the coloring agent in green chrysoberyl and alexandrite. Alexandrite's pleochroism, which is the same as its color-change, is evident when the gemstone is viewed from perpendicular directions. The cat's eye is caused by fine needle inclusions of rutile (TiO_2) in one direction. Yellow, brownish, and green cat's eye are most common; alexandrite cat's eye is very rare.

Individual chrysoberyl crystals as rectangular prisms are rare. Instead, intergrowths (twins) called "trillings" or "sixlings" with near-hexagonal symmetry are more abundant.

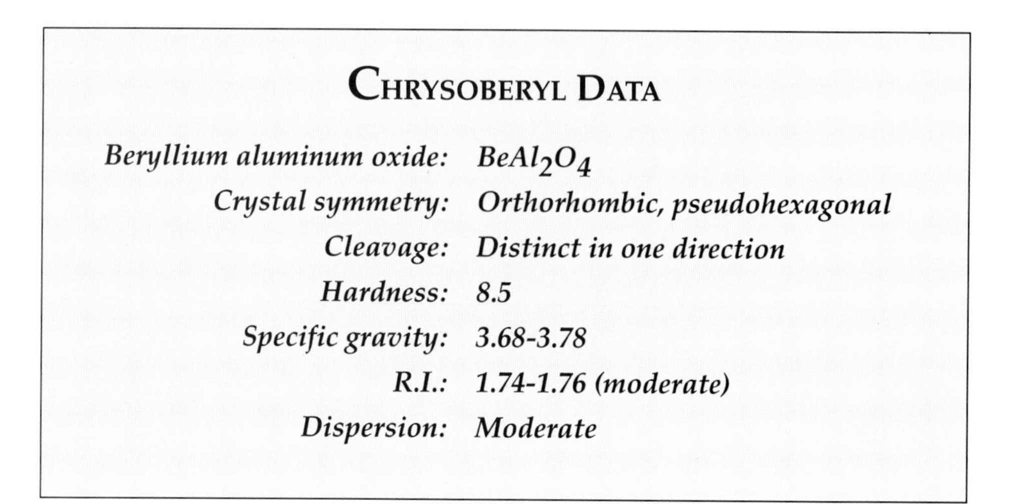

CHRYSOBERYL DATA

Beryllium aluminum oxide:	***$BeAl_2O_4$***
Crystal symmetry:	***Orthorhombic, pseudohexagonal***
Cleavage:	***Distinct in one direction***
Hardness:	***8.5***
Specific gravity:	***3.68-3.78***
R.I.:	***1.74-1.76 (moderate)***
Dispersion:	***Moderate***

Historic Notes

THE TERM *CHRYSOBERYL* DERIVES FROM THE GREEK *chrysos*, referring to the stone's golden color, and the mineral beryl. Until 1789, when A.G. Werner, a famous German geologist, identified chrysoberyl as a mineral species, the stone was erroneously assumed to be a variety of beryl.

Cat's eye has the longest history of the chrysoberyl varieties. Although known in Rome by the end of the first century, it had been treasured even earlier in the Orient, where chatoyant stones have always had admirers. The gem was forgotten in the West until the late nineteenth century, when the Duke of Connaught gave a cat's eye betrothal ring to Princess Louise Margaret of Prussia. The gem's popularity—and price—rose immediately; Ceylon (Sri Lanka) had difficulty in keeping up with the demand. Cat's eye is currently a fashionable ring stone, particularly in Japan and Hong Kong.

Alexandrite was discovered in 1830 in an emerald mine near Ekaterinburg (Sverdlovsk) in the Russisan Ural Mountains on the birthday of the then heir apparent Czar Alexander II, for whom the gem was named. The naming was doubly appropriate not only because of the discovery date but because the chameleonlike green and red colors were the same as those of the Russian Imperial Guard.

The third and common variety of chrysoberyl, the transparent greenish yellow form, was found in Sri Lanka and Brazil. The Brazilians called it "crisolita" and named a city in its honor. Exported to Europe, the gem became popular and was used in eighteenth- and nineteenth-century Spanish and Portuguese jewelry. This variety of chrysoberyl was in great demand during the Victorian and Edwardian eras but is now overshadowed by alexandrite and cat's eye.

Chrysoberyls from Sri Lanka and Brazil, ranging in weight from 8.9 to 74.44 cts.

Legends and Lore

THE NATIVES OF SRI LANKA BELIEVED THAT CAT'S EYE protected its wearer from evil spirits. According to Hindu lore, it preserved an owner's health and guarded him or her against poverty. An Oriental belief was that, if pressed against the forehead at a point between the eyes, the gem would endow foresight.

Alexandrite has been regarded as a stone of good omen in Russia and is the only gem accorded the role of a talisman as recently as the nineteenth century.

An 8.9-ct. alexandrite from Sri Lanka, showing its color change—in incandescent light (red) and in daylight (green).

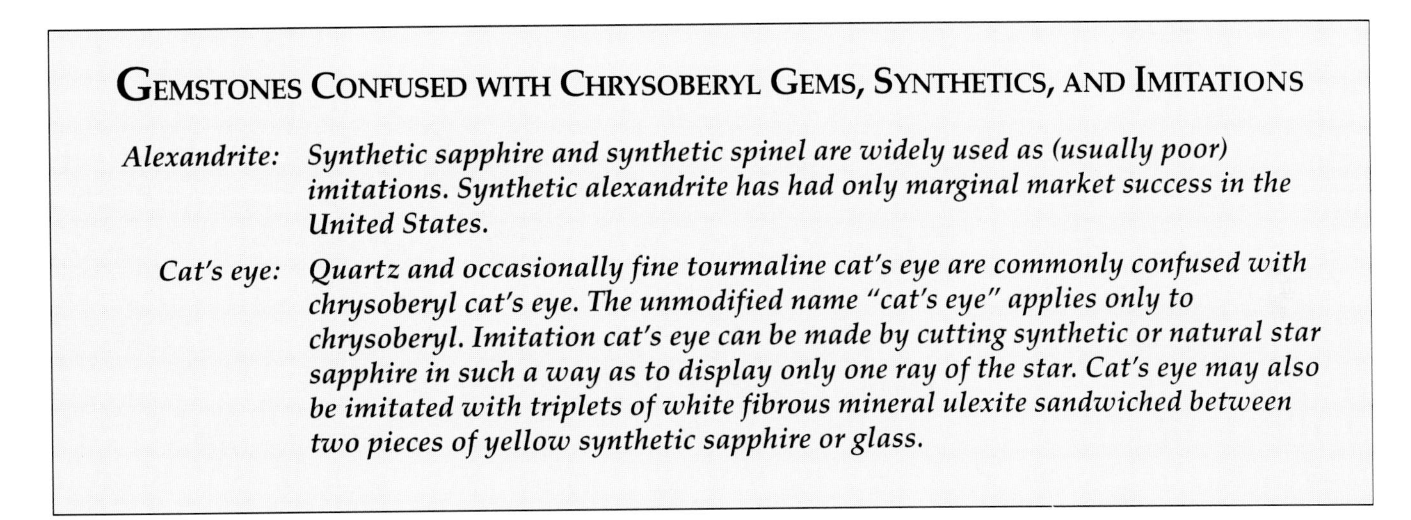

GEMSTONES CONFUSED WITH CHRYSOBERYL GEMS, SYNTHETICS, AND IMITATIONS

Alexandrite: *Synthetic sapphire and synthetic spinel are widely used as (usually poor) imitations. Synthetic alexandrite has had only marginal market success in the United States.*

Cat's eye: *Quartz and occasionally fine tourmaline cat's eye are commonly confused with chrysoberyl cat's eye. The unmodified name "cat's eye" applies only to chrysoberyl. Imitation cat's eye can be made by cutting synthetic or natural star sapphire in such a way as to display only one ray of the star. Cat's eye may also be imitated with triplets of white fibrous mineral ulexite sandwiched between two pieces of yellow synthetic sapphire or glass.*

Occurrences

CHRYSOBERYL CRYSTALLIZES IN AND AROUND pegmatites rich in beryllium, but deposits commonly form as alluvial concentrations from weathered pegmatites.

The major source of all chrysoberyl varieties is Minas Gerais in Brazil. Here, at Lavra de Hematita in 1987, the world's largest find of alexandrite produced, in less than three months, 50 kilograms of fine gems—some up to 30 carats. The Soviet Union alexandrite deposits near Sverdlovsk and one discovered later near the Sanarka River in the southern Urals are now apparently exhausted. Sri Lanka produces cat's eye and alexandrite from its gem gravels. Generally, the alexandrites are larger than the Russian stones and have more attractive daytime green color. However, the Russian alexandrite has a better color change and a finer red color under artifical light. Other occurrences of alexandrite are in Zimbabwe, Tanzania, and Burma.

Evaluation

ONLY CHRYSOBERYL WITH A DISTINCT COLOR CHANGE is alexandrite; either the red or green must be a good hue. Flaws diminish the value. Fine-quality stones exceeding 5 carats are both rare and expensive. Alexandrite cat's eye is one of the rarest and most costly of gems.

Cat's eye is the most valuable chatoyant gemstone. A rich honey yellow brings the highest price; the green stones follow in value. The chatoyant band must be sharp, narrow, and positioned in the center of the cabochon. In the finest stones, when the band is at right angles to the light, the half of the stone facing the light should appear milky; the other half should have a rich honey color. The eye should also open widely in oblique illumination and close sharply in direct illumination. Stones that approach transparency and exhibit a sharp eye command the highest price. Fine stones over 20 carats are rare and expensive. Of the common transparent variety of chrysoberyl, intensely colored yellowish green stones are the most popular and highly valued.

A 70.99-ct. spinel from Sri Lanka.

SPINEL

For centuries, most gem spinels were thought to be rubies or sapphires—a reasonable assumption, because both spinel and corundum are found in the same deposits and have similar properties. A famous example is the Black Prince's Ruby. The stone's known history began in 1367, when it was taken from the treasury of the King of Grenada by victorious Dom Pedro of Castile. He presented the "ruby" to the Black Prince, son of Edward III, in repayment for services at the Battle of Nájera in northern Spain. As part of the break-up of the crown jewels by the Commonwealth in the 1650s, the gem was sold (inventoried at four pounds Sterling) and somehow returned to the monarchy during the Restoration. The two-inch irregular spinel resides at the center of the British Imperial Crown.

Properties

LIKE THE CORUNDUM GEMSTONES WITH WHICH IT IS confused, spinel is known for its many colors and durability. Spinel is slightly softer than corundum because the magnesium to oxygen bonds in spinel are not quite as strong as the aluminum to oxygen bonds found in both minerals. This hardness, combined with no weak plane in its crystal structure, makes the gemstone very durable.

In another comparison to corundum, the presence of magnesium in addition to aluminum in spinel permits a wider range of chemical substitutions, transforming the pure colorless spinel into many possible colors but, surprisingly, not as many colors as corundum. Names have been applied to many of the different varieties, but now they are simply referred to by their color—red spinel, green spinel, etc. Synthetic spinels can be produced in an even greater range of colors by adding elements such as cobalt, manganese, and vanadium in amounts or combinations not found in nature.

Infrequently, spinels contain fine needles in three perpendicular directions that manifest four- and six-rayed stars in cabochon stones.

With cubic symmetry, spinel crystals are predominantly in the form of an octahedron. The term *spinel* may refer to this shape; the Latin *spina* means "thorn." Although spinels have probably been recognized by their octahedral crystals for many centuries, the distinction between ruby and spinel as different mineral species was not made until 1783 by Romé de Lisle.

SPINEL DATA

Magnesium aluminum oxide:	*$MgAl_2O_4$*
Crystal symmetry:	*Cubic*
Cleavage:	*None*
Hardness:	*8*
Specific gravity:	*3.58-4.06*
R.I.:	*1.714-1.75 (moderate)*
Dispersion:	*Moderate*

Spinel is also the name of a mineral group of multiple oxides with the same crystal structure. **Gahnite,** *with zinc replacing magnesium, is a blue spinel gemstone.*

Historic Notes

THE EARLIEST RED SPINEL USED AS AN ORNAMENT WAS found in a Buddhist tomb near Kabul, Afghanistan, and dates from about 100 B.C. Red spinel was also used by the Romans of the first century B.C. Blue spinels have been found in England dating from the Roman period (51 B.C.-A.D. 400), and a ring set with a pale green octahedral spinel from the Eastern Roman Empire has been described.

Mining of spinel began in Badakshan, Afghanistan, sometime between A.D. 750 and 950. The locality was first described by the Arab geographer Istakhri in 951 and later by Marco Polo. Many of the historic spinels (Balas rubies) were probably mined here.

Of the spinels, the red stones have had the longest and most dramatic histories. Like the Black Prince's Ruby, the Timur Ruby's known history goes back to the fourteenth century. The stone bears six Persian inscriptions, the oldest identifying it as being in the possession of the Tartar conqueror Timur (Tamerlane) in 1398. Over the years, the stone was traded or plundered in India until the East India Company took possession in 1849 and presented it to Queen Victoria two years later. The stone is now in the private collection of Queen Elizabeth II.

Red spinels were used in Renaissance jewelry in Europe and became popular during the eighteenth century. Elegant pendants, earrings, and brooches set with spinels and diamonds survive from the former Russian and French crown jewels. In Moscow, a deep red 412.25-carat spinel surmounts the Great Imperial Crown commissioned for the coronation of Catherine the Great. The world's largest collection of spinels, including a record-holding 500-carat stone, is part of the former crown jewels of Iran.

Legends and Lore

THE HINDUS CONSIDERED SPINELS TO BE RUBIES AND divided them according to caste. The members of each of the four major castes should wear the appropriate stone in order to benefit from its virtues: The Brahmin priestly caste—true ruby; Kshatriya (knights and warriors)—rubicelle; Vaisya (landowners and merchants)—ruby spinel; Sudra (laborers and artisans)—Balas ruby.

In ancient and medieval times, when color had strong symbolism, red spinel and other red stones were considered cures for hemorrhages and all inflammatory diseases—as well as prescriptions to soothe inflamed emotions, eliminating anger and conflict. An Indian belief, reported by an Armenian writer of the seventeenth century, is that powdered spinel taken in a potion eliminates dark forebodings and brings happiness.

Occurrences

GEM SPINELS, LIKE CORUNDUM, FORM IN HIGHLY METAmorphosed claystones and especially dirty limestones transformed into marbles. Most spinels are produced from alluvial concentrations from weathered primary sources.

The area around Mogok in Upper Burma is the source of the finest-quality spinels—rose, pink, orange, blue, and violet colors—which occur as water-worn pebbles and some as perfect octahedra. Sri Lanka's gem gravels are located in the southwest part of the island around Ratnapura ("City of Gems"). There the spinels, always water-worn, are generally blue, violet, or black (named "ceylonite" after the country's former name). Red, orange, and pink spinels are rare. Other occurrences are in Thailand, Pakistan, Afghanistan, and the Pamir Mountains of the Soviet Union.

GEMSTONE SPINEL COLORS, SOURCES OF COLORS, AND FORMER NAMES

Red: *Chromium (deep red—ruby spinel; rose red—Balas ruby)*
Purple red: *Chromium + iron (almandine spinel)*
Orange: *Chromium + vanadium (rubicelle)*
Blue: *Iron and/or cobalt (sapphire spinel and gahnospinel)*
Green: *Iron (chlorspinel)*

Evaluation

COLOR, CLARITY, AND WEIGHT ARE IMPORTANT considerations in appraising spinel. Red spinels are the most valuable, the most highly prized ones being orange red and intense red to purplish red. Because spinel is often flawless, its clarity is of great importance—much more important for spinel than for ruby, which is rarely found without inclusions. Stones weighing more than 5 carats are uncommon—the large ancient stones are no longer found.

Spinel possesses both beauty and durability, but confusion with the cheap synthetic version and insufficient supply prevent it from enjoying the popularity that it merits.

(Opposite) A 4.03-ct. cut spinel and an octahedral crystal 1 cm (3/8 in.) across in marble. Both are from Mogok, Burma.

(Above) Ring with 9.5-ct. spinel from Mogok, Burma. Other spinels, from Sri Lanka, range from 1.89 to 46.48 cts.

GEMSTONES CONFUSED WITH SPINEL AND SYNTHETIC SPINEL

Spinel can be mistaken for ruby, sapphire, pyrope garnet, amethyst, and zircon. Synthetic spinel is manufactured in large quantities, is inexpensive, and is used to imitate ruby, sapphire, emerald, aquamarine, peridot, alexandrite, and diamond.

Topaz

Topaz is renowned for its capacity to form large gem-quality crystals. Pliny in the first century remarked: "Topazos of all precious stones is the largest. In this, it excels all others." While in Minas Gerais, Brazil, in 1938, Allan Caplan, a New York mineral dealer, noticed with some excitement exceptionally large topaz crystals for sale. Upon learning the news, many United States museums began competing to obtain the "giants." The Smithsonian picked a topaz weighing 156 pounds; then the Cranbrook Institute got a slightly smaller one. Harvard, followed by the American Museum, wanted one, but no more were available.

During his fourth trip to Brazil in 1940, Caplan learned of three prodigious specimens in transit to Rio de Janeiro. On the basis of photographs alone, he bought them and returned home to await the arrival of his prizes, which were revealed to weigh 596, 300, and 225 pounds!

Finally, after many months, the crates arrived. At the U.S. Customs Office, eagerly and anxiously, Caplan approached the largest crate and removed the top boards. To his horror, he saw a great broken surface staring up at him from the packing material! Bitterly disappointed, he had the crate resealed and all three specimens sent to the American Museum for evaluation. There a group assembled for the grand opening. The moment arrived. The crate, now turned top side down, was pried open, and the superbly terminated "top" of the great crystal came into view. All in attendance breathed a collective sigh of relief as they gazed at the largest fine topaz crystal in "captivity"—and ever since a proud possession of the Museum.

Imperial topaz crystal, 5.5 cm (2 1/8 in.) long from Ouro Preto, Minas Gerais, Brazil, and a 16.95-ct. cut stone from Sri Lanka.

Properties

TOPAZ IS REVERED FOR ITS COLOR, CLARITY, AND HARDNESS. Strong chemical bonding makes it relatively dense and the hardest silicate mineral. However, a weak plane in the crystal structure occupied by fluorine and hydroxyl is the source of one excellent cleavage. This cleavage is topaz's major failing as a gemstone and demands great care in cutting and handling.

A common misconception is that all topaz is yellow. In fact, pure topaz is colorless, and colors include blue, pale green, and the spectrum from yellow through the familiar sherry orange to pink and even the most rare red. Chromium substitutes for aluminum, producing red and some pink topaz, but most other colors are a result of minor atomic substitutions and defects in the crystal followed by radiation damage. Some of these colors are unstable and can fade; some browns fade totally in sunlight, and some sherry orange stones become pink upon heating. High-energy irradiation of colorless topaz followed by heat treatment at moderate temperature yields blue stones with stable color. Natural blue topaz appears to be created by an identical process, so there is no way to distinguish between natural and "created" blues. However, intense natural blue stones are not known, so it is a good bet that deep blue color in gem topaz is artificially produced.

Topaz crystals typically form prisms having a diamond-shaped cross-section and a pyramidal top; the cleavage cuts straight through the prism.

The world's largest topaz crystal, 271 kg (596 lb.), is from Minas Gerais, Brazil.

Both the 47.55-ct. pink topaz and the 47.75-ct. Imperial topaz are from Minas Gerais in Brazil.

Topaz Data

Aluminum silicate fluoride hydroxide:	$Al_2SiO_4(F,OH)_2$
Crystal symmetry:	*Orthorhombic*
Cleavage:	*Perfect in one direction*
Hardness:	*8*
Specific gravity:	*3.5-3.6*
R.I.:	*1.606-1.644 (moderate)*
Dispersion:	*Low*

Historic Notes

HOW TOPAZ CAME BY ITS NAME IS UNCERTAIN. *Topazos*, which means "to seek," was the name of a fog-bound, hard-to-find island in the Red Sea (Zabargad). Pliny the Elder considered topazos to be a green stone from that island. However, since it is the green gemstone peridot—not topaz—that is found there, Pliny must have been describing peridot in this case. The Sanskrit *tapaz* means "fire" and seems a more appropriate possibility for the derivation of *topaz*.

Topaz was used in ancient Egypt and Rome; the Romans obtained their topaz from Ceylon (Sri Lanka), an early and continuing source for the gemstone. In the seventeenth century, Jean Baptiste Tavernier mentions both the stone and the location in accounts of his buying trips in the Orient.

During the Middle Ages in Europe, topaz was not particularly popular, although it was occasionally used in ecclesiastical or royal jewelry. But by the eighteenth century in Spain and France, the gem enjoyed increased popularity and, together with diamond, was set in many magnificent pieces of jewelry. Early in the next century, topaz and amethyst were the most stylish gems for earrings and necklaces in both France and England. Topaz continued to be one of the most popular gems during the Victorian era and later became a favorite stone of the Art Deco jewelers. Commonly regarded as the finest yellow stone, its popularity persists.

(Left) The largest faceted red topaz in the world is an oval brilliant-cut gem of 70.40 cts. with an unusual natural deep red color. It is either from Brazil or from the Soviet Union.

(Opposite) The Brazilian Princess was fashioned from a 34 kg (75 lb.) crystal discovered in Minas Gerais, Brazil. Cut in a square cushion form with 221 facets, it measures 14.5 cm (5 3/4 in.) across and weighs 21,005 cts. This gem is astounding both because of its size (the largest faceted blue topaz) and its nearly complete freedom from flaws.

Legends and Lore

DURING THE MIDDLE AGES, TOPAZ WAS THOUGHT TO strengthen the mind and prevent mental disorders as well as sudden death. Marbode's eleventh-century poetic treatise recommends it as a cure for weak vision. The prescription called for immersing the gem in wine for three days and three nights, followed by application of the topaz to the afflicted eye. A topaz engraved with the figure of a falcon could help its bearer cultivate the goodwill of kings, princes, and magnates, according to Ragiel's thirteenth-century *The Book of Wings*. Still later, topaz was recommended by Geronimo Cardano as a cure for madness, a means of increasing one's wisdom and prudence, and a coolant for both boiling water and excessive anger.

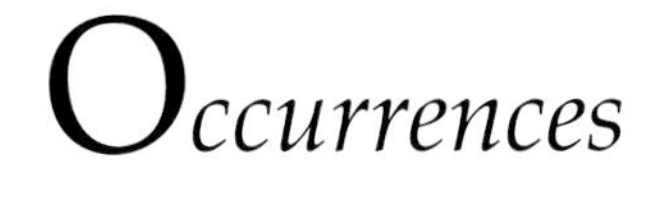

Occurrences

TOPAZ IS FOUND PRINCIPALLY IN GEM PEGMATITES, where fluorine is often abundant. This volatile-rich environment stimulates the growth of large crystals. Weathering of these pegmatites releases topazes into streams and rivers; the gemstones concentrate in alluvial gravels.

Minas Gerais in Brazil is the world's largest producer of topaz—blue, colorless, and sherry-colored. Topaz was discovered there first near Ouro Prêto in 1735, the primary source of sherry-colored topaz. In the Soviet Union's Ural Mountains, topaz is found northeast of Sverdlovsk in Mursinka and Alabashka and at Sanarka in the southern Urals. Another important Soviet site is Volini in the Ukraine. Topaz is mined north of Katlang in Pakistan in veins of coarse-grained calcite and quartz in marble that yield topaz of many colors, including a rare pink and reddish brown.

Topazes from various localities, ranging in weight from 17.16 cts to 375 cts.

Evaluation

TWO FACTORS SHOULD BE CONSIDERED IN EVALUATING topaz: color and clarity. The most valued color is a rare, almost unavailable, red. Imperial topaz, a sherry-colored stone, has always been the most popular. Both sherry-colored (brownish yellow, orangy yellow, and reddish brown) and pink topaz command high prices. Precious topaz is yellow topaz, a term commonly used to distinguish it from other topaz colors and from citrine. Light blue and pale yellow stones are of less value. The value diminishes significantly when the stone is flawed.

Gemstones Confused with Topaz and Trade Names

Tourmaline, sapphire, chrysoberyl, and the rare danburite, andalusite, and apatite can be confused with topaz. Yellow quartz, or citrine, which lacks topaz's velvety appearance, brilliance, and rich color, is occasionally sold as topaz—an unethical practice. Equally unethical is the substitution of "treated" blue topaz for the rarer and more expensive aquamarine.

Misleading trade names are frequently used in substituting other less valuable stones for precious topaz: they include Spanish, Saxon, and Bohemian topaz for citrine quartz; Smoky, Burnt, and Scotch topaz for smoky quartz; and Oriental topaz for yellow sapphire.

Tourmaline

One spring morning in 1876, a young man walked briskly and confidently into Tiffany & Co. In the director's office, he unfolded a gem paper and placed what he later called a "drop of green light" on the desk. The "light" was a sparkling faceted green tourmaline from Maine, which spoke for itself. Both men admired its quality and beauty. Charles Tiffany bought it immediately, much to the delight of George F. Kunz. Within a year, the twenty-year-old gemologist had embarked on his illustrious career at the company. As the preeminent gem expert of his day, he championed lesser-known colored stones. Tourmalines were his favorites among American stones, and the timing of his visit was opportune. Several United States sites yielding commercially viable quantities of gem tourmaline had just been or were about to be discovered. Kunz collected tourmaline for many institutions, including the Museum—his Maine, Connecticut, and California tourmalines abound in the collections.

Bicolored tourmalines from Mesa Grande in California: three superb "pencil" crystals. The longest is 9.5 cm (3 3/4 in.). The cabochon is 22.40 cts., and the cut stone is 30.50 cts.

Properties

TOURMALINE IS A MINERAL GROUP WHOSE MEMBERS display the broadest spectrum of gemstone colors. There are ten mineral species in the group, but only three have found use as gems. These are sufficiently durable (hard and free from cleavage) to be fine gemstones. Named for the Isle of Elba, where it was first found, elbaite is the tourmaline most often used in jewelry; dravite and uvite are less common and rarely of the appropriate quality but are also gemstones. Gem elbaite can be pink to red, blue, green, violet to red purple, yellow, orange, brown, black, and colorless. Bicolored crystals with one end green and the other pink are common. Crystals with a pink core and green rind are called "watermelon."

Color in gem tourmaline is primarily a result of substitutions of transition elements for other metals in the crystal structure. There are few useful generalities relating color to a specific chemical element, but pink is usually due to manganese, and green is attributed to ferrous iron, chromium, or vanadium. Color may be improved by heating and/or irradiation, but the changes are not always permanent.

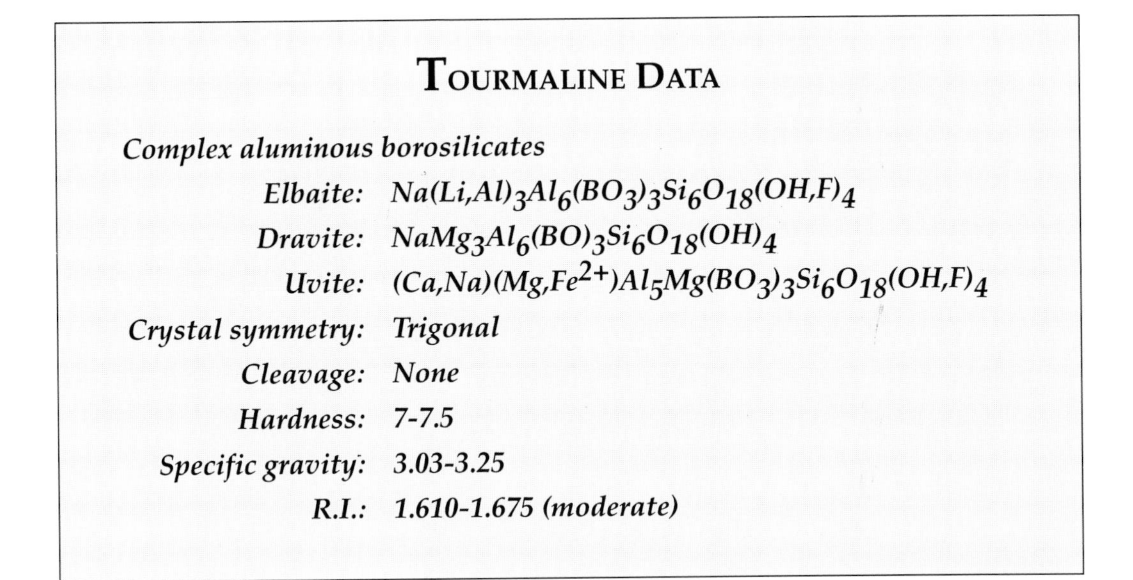

Tourmaline Data

Complex aluminous borosilicates

Elbaite:	*$Na(Li,Al)_3Al_6(BO_3)_3Si_6O_{18}(OH,F)_4$*
Dravite:	*$NaMg_3Al_6(BO)_3Si_6O_{18}(OH)_4$*
Uvite:	*$(Ca,Na)(Mg,Fe^{2+})Al_5Mg(BO_3)_3Si_6O_{18}(OH,F)_4$*
Crystal symmetry:	*Trigonal*
Cleavage:	*None*
Hardness:	*7-7.5*
Specific gravity:	*3.03-3.25*
R.I.:	*1.610-1.675 (moderate)*

(Opposite) Faceted elbaites in different colors ranging in weight from 127.7 cts. to 1.270 cts., from various localities.

(Right) A superb bicolored elbaite crystal group 10.5 cm (4 1/8 in.) high from Tourmaline Queen Mine in Pala, California.

Elbaite crystals are often recognizable by their prismatic form, typically elongated like pencils, with cross-sections that range between hexagonal and trigonal; crystals are often sufficiently perfect and clear to be natural gems. Color zoning is created by changes in chemical composition in response to changing conditions during crystal growth. Elbaite crystals grow fastest at their ends, changing color as they grow and forming pencil-like bicolor crystals; layers of growth from core to rim produce the watermelons. Elbaite is the principal gem mineral that can be and is cut into bicolored and multicolored stones.

Most tourmalines are strongly pleochroic, an important factor in cutting gems; when viewed along the prism axis, the color is deeper or even different from that seen through the side of the crystal. Rare alexandrite-like tourmalines appear yellowish or brownish green in daylight and orange red in incandescent light. Crystals occasionally grow with fluid-filled tubes parallel to the long prismatic direction. If these are sufficiently numerous and narrow, like fibers, chatoyancy will be manifested from a properly fashioned gem.

The crystal structure of tourmaline lacks a center of symmetry; structural elements are much like arrows that point in one direction along the crystal's trigonal axis. As a result, when some crystals are heated, a positive charge develops at one end and a negative charge at the other; the reverse occurs upon cooling. This property, the pyroelectric effect, was first observed in gem tourmaline. A comparable electric charging is developed if pressure is applied to the ends of a crystal. This property, piezoelectricity, has important industrial and electronic applications. Among common gems, only tourmaline and quartz possess the properties of piezoelectricity and pyroelectricity.

Some Tourmaline Varieties and Colors

Rubellite: ***Pink to red***

Siberite: ***Violet to red purple***

Indicolite: ***Blue***

Verdelite: ***Green***

Achroite: ***Colorless***

The new convention is simply to label tourmalines by their color, but old habits die hard.

Historic Notes

A LONG-HELD BELIEF THAT TOURMALINE FROM THE Orient was imported by Greece and Rome was recently confirmed when a fine convex intaglio depicting the head of Alexander the Great (now in the Ashmolean Museum in England) was identified as a zoned purple yellow tourmaline. A minute inscription indicates India as the place of origin and the date of its carving as third or second century B.C. Another much later documented piece is a gold ring of Nordic origin from A.D. 1000 set with a pink tourmaline cabochon. Examination of surviving early jewelry will probably reveal more tourmalines, confirming that they have been used as a gem material for over 2,000 years.

The term *carbunculus* was applied to red transparent gems, including ruby, spinel, garnet, and probably red tourmaline from Pliny's time in the first century A.D. through the Middle Ages. Green tourmaline was exported from Brazil to Europe in

(Above) Bicolored elbaite from Brazil—carved rhinoceros, 85 cm (3 1/3 in.) long.

(Opposite) A polished slice of elbaite with liddicoatite rim, 13.2 cm (5 1/4 in.) wide, from the Malagasy Republic. Liddicoatite is a rare tourmaline species named in 1977 after gemologist R.T. Liddicoat.

Elbaite crystal measuring 20.5 cm (8 in.) with quartz crystal from Pala, California.

Gemstones Frequently Confused with Tourmaline

These include topaz, beryl (emerald, aquamarine, morganite, and golden beryl), spodumene (kunzite and hiddenite), peridot, andalusite, and apatite. Dark green synthetic spinel is sold as synthetic tourmaline.

the early sixteenth century; it was known as Brazilian emerald.

The Chinese valued red and pink tourmaline and made small carved tourmaline ornaments for headdresses and girdles, and certain mandarins wore badges or buttons on their caps to signify their rank. Gustavus III of Sweden chose an incredibly large "ruby" as a gift to Catherine the Great when he was on a state visit to Russia in 1777; it is, in fact, a stunning red Burmese tourmaline carved in China.

In 1703, a packet of stones from Sri Lanka labeled "turmali" or "toramalli" (a Sinhalese word applied to any unidentified yellow, green, or brown stone and meaning "something little out of the earth") arrived at a Dutch lapidary. According to one story, children playing with some of the "pebbles" outside a gem worker's shop noted that when warmed by the sun, the little stones attracted ashes and straws much as a magnet attracts iron filings. So the stones were called "Aschenstrekkers," or "ash-drawers." (As they are natural dust collectors, the Museum's tourmalines require frequent cleaning as a result of their daily heating by lights in the exhibit cases.) This discovery of tourmaline's pyroelectric property set off a spate of investigations resulting in observations that only certain gemstones, but with various colors, possessed this property. Finally in 1801, all the information came together with the recognition of the tourmaline "family."

During the eighteenth century, the principal sources of the mineral were Burma, Russia, Sri Lanka, and Brazil, but late one afternoon in 1820, two Maine schoolboys, Elijah L. Hamlin and Ezekiel Holmes, happened on a brilliant green crystal sparkling in the roots of an overturned tree as they returned from a hike on Mount Mica. The stone was identified as tourmaline, and, starting in 1822 with Mount Mica and later at other Maine locations, mines were opened that could provide sufficient red and green tourmaline to create a market—and value. With the Maine discoveries and several in California, the United States became for a time the world's major supplier of tourmaline. Among the purchasers was Tz'hsi (1835-1908), the dowager empress of China, who sent personal emissaries to California to purchase the favored red variety of the gem.

Legends and Lore

WITHOUT MANY CENTURIES OF ASSOCIATED HISTORY, the newcomer tourmaline gems do not appear in lore. Even George F. Kunz, popularizer of tourmaline and the man who introduced the stone to Tiffany & Co., opposed acceptance of the gem as the alternate birthstone for the month of October on that account. With the recent revival of mysticism, tourmaline has become a favorite of New Age adherents, who believe the mineral's pyroelectric and piezoelectric properties produce powerful amplification of psychic energy and neutralization of negative energies.

Occurrences

TOURMALINE IS A RELATIVELY COMMON MINERAL, THE most common boron-bearing silicate. Gem tourmaline is virtually restricted to pegmatites—rich in volatile elements like boron, beryllium, and lithium. Pegmatites yield not only crystals of elbaite but other gem minerals that contain these elements—gemstones such as spodumene and beryl.

Brazil is the world's major source of tourmaline. The pegmatitic region in the eastern part of Minas Gerais yields green, pink, red, and watermelon stones. Tourmaline is also produced in the states of Bahia, Ceara, Goias, Paraiba (blue), and Rio Grande de Norte. The United States also ranks high as a tourmaline supplier. The 400 pegmatitic dikes of Mesa Grande in San Diego County, California, produced 120 tons of gem tourmaline beween 1902 and 1911. Production reached its peak in 1910; however, with increased Brazilian supply and the fall of the last Chinese dynasty in 1912, California elbaite lost its markets, and many mines closed in 1914. Mining has been reactivated within the last 20 years. Most California elbaite is pink and pure in color, but it lacks the exceptional clarity of Maine elbaite. Pegmatites in eastern Maine, such as those at Newry, have been sporadically productive of the state's gem—tourmaline. Connecticut was a source at the beginning of the century. Other important sources include the Malagasy Republic, Sri Lanka (uvite), the Soviet Union, and Mozambique.

Tourmaline crystals from Mount Mica in Maine. The green tourmaline (upper left) is the original Maine find. Elijah Hamlin had it set in a watch charm bearing the inscription "Primus." It was given to the Museum by his great-granddaughter, K.B. Hamlin. The largest crystal is 5.3 cm (2 in.) long.

Multicolored elbaite crystal 7.3 cm (2 7/8 in.) long, from Alto Ligohna, Mozambique.

(Right) A cluster of bicolored elbaite crystals 23.0 cm (9 in.) long from Nuristan, Afghanistan.

(Opposite) Pink elbaites from Pala, California—a crystal 10.0 cm (4 in.) long and a 419.5-ct. cut stone.

Evaluation

PURITY AND INTENSITY OF COLOR AND CLARITY ARE the most important qualities to consider. The most valued tourmalines are raspberry red, followed by medium-dark emerald green, and intense blue. Bicolored and multicolored gems are next in value. Cat's eye gems are valuable if the eye is well defined and the fibrous cavities causing it are not coarse.

During the past ten to fifteen years, tourmaline jewelry has been in great demand. Tourmaline crystals are also avidly sought by mineral collectors for the splendor of their colors and forms. George Kunz's faith in his "drop of green light" has been amply justified.

Zircon & Peridot

ZIRCON

In the 1920s, a new blue gemstone suddenly appeared on the market. Endowed with spectacular brilliance, it was an immediate hit. The gems were zircons, normally brown to green—but not blue. George F. Kunz, the legendary Tiffany gemologist, immediately suspected trickery; not only were these extraordinary stones available in abundance but available all over the world! Upon Kunz's behest, a colleague made inquiries during a trip to Siam (Thailand) and learned that a large deposit of unattractive brown zircon had stimulated color-improvement experimentation by local entrepreneurs. Heating in an oxygen-free environment had turned the drab material into "new" blue stones, which were sent to outlets worldwide. When the deception was revealed, the market simply accepted the information, and the demand for the new gems continued unabated.

Zircons from Sri Lanka and Thailand, ranging in weight from 7.76 to 40.19 cts.

Properties

SUPERIOR BRILLIANCE; GOOD DISPERSION, OR FIRE; clarity; and a breadth of colors stand out as zircon's fine gemstone qualities. Natural zircons range from colorless to pale yellow or green when they initially crystallize. The colors are a result of minor amounts of thorium and uranium that replace zirconium in the crystal structure. But with geologic time, uranium and thorium emissions cause radiation damage, which can be so severe that the original structure is obliterated. A glasslike substance develops with colors ranging from red to brown, orange, and yellow. Heat treatment can restore the structure and color or create new colors, including yellow, blue and colorless. Colorless zircons imitate diamond's optics better than any other gemstone mineral; their refractive indices approach diamond's, and the dispersive fire is nearly as good.

Many zircons are very brittle; a slight knock will remove a corner or even split the stone. This fragility is the result of internal stress either from radiation damage or from heat treatment. Zircon crystals are distinctive because of their square cross-section and pyramid terminations owing to the tetragonal symmetry.

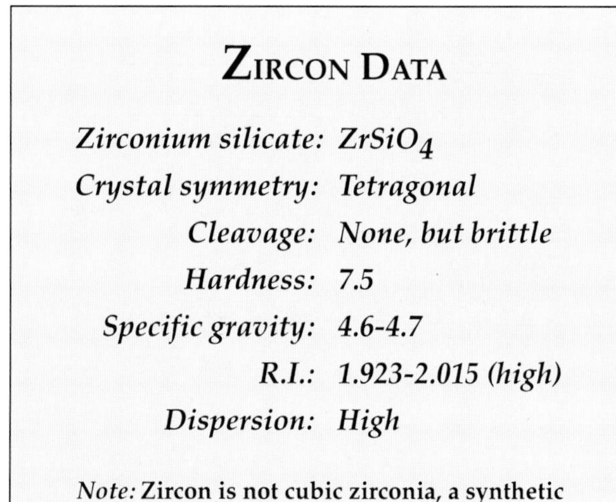

ZIRCON DATA

Zirconium silicate:	*$ZrSiO_4$*
Crystal symmetry:	*Tetragonal*
Cleavage:	*None, but brittle*
Hardness:	*7.5*
Specific gravity:	*4.6-4.7*
R.I.:	*1.923-2.015 (high)*
Dispersion:	*High*

Note: **Zircon is not cubic zirconia, a synthetic diamond simulant—zirconium oxide, ZrO_2.**

Historic Notes

THE ARABIC *ZAR* AND *GUN* MEAN "GOLD" AND "color" and may be the source of the word we use. The terms *hyacinth* and *jacinth* were used in Europe for reddish brown and orange red stones and applied to zircons and other minerals with similar color.

The gem was in use in Greece and Italy as far back as the sixth century A.D. Gem zircons were marketed as diamonds some time after faceting began in the fourteenth century. Colorless zircon was mined at Puy en Velay in France in 1590 and sold as Diamond of France. Later, Ceylonese (Sri Lankan) colorless zircon was sold as Matura diamond (named for the locality where it was found).

Reddish brown zircon became moderately popular in Europe during the last century, but currently the most commonly used zircons are light blue, golden brown, and colorless.

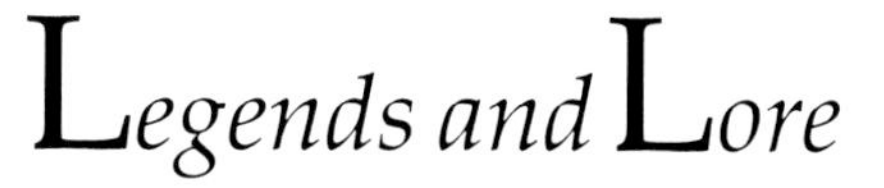

Legends and Lore

ZIRCON IS AMONG THE GEMS UTILIZED IN THE KALPA Tree of Hindu religion. The tree was a symbolical offering to the gods and described by nineteenth-century Hindu poets as a glowing mass of precious stones, including sapphire, diamond, and topaz. Green zircon represented the tree's foliage.

As an amulet for travelers, zircon protected its wearer from disease and injury, ensured good sleep and a cordial welcome everywhere, according to the eleventh-century writings of Marbode. Five centuries later, the stone rendered its owner prudent in practical matters (and thus promised financial success) and guaranteed that he or she would never be struck by lightning, according to Geronimo Cardano. By the seventeenth century, the belief in magical properties of gems in general appeared to be waning. Anselmus de Boot declared that gems cannot of themselves produce supernatural effects. Nevertheless, he believed in zircon's power to prevent plague.

Occurrences

ZIRCON IS A COMMON MINOR CONSTITUENT OF igneous rocks, particularly granites, and, to a lesser extent, of metamorphic rocks. Gemstone crystals are rare and found mainly in coarse-grained rocks (pegmatites) or in fissures. Zircons concentrate in alluvial and beach deposits.

The Chanthaburi area in Thailand, the Palin area in Cambodia, and the southern part of Vietnam near the Cambodian border are the major sources of zircon. It occurs as water-worn pebbles in gem gravels which are seldom more than 10 feet deep. Bangkok is the world's cutting and marketing center for zircon. Blue, colorless, golden yellow, orange, and red stones, almost all of them heat-treated, are exported from there.

Sri Lanka, where zircon is also found in gem gravels, is the next most important source. Other occurrences are in Burma, France, Norway, Australia, and Canada.

Round brilliant-cut zircon from Thailand; weighing 208.65 cts., it is the largest known blue zircon on public display.

Evaluation

COLOR AND CLARITY ARE THE MOST IMPORTANT CONsiderations in evaluating zircon. The most prized and rare color is red; next is pure, intense blue and sky blue. Colorless, orange, brown, and yellow are less valued. Any visible flaw diminishes the value substantially.

Most zircons have been heat-treated. The color of some of the heat-treated stones may change, and the brittleness of some treated gems is also a negative factor. The beauty of this gem—expressed in the variety of its colors and in its clarity, brilliance, and fire—makes it popular today. It is also reasonably priced in comparison with most other gems.

PERIDOT

For three millennia, a small, desolate, and forbidding island in the Red Sea has been exploited for the gemstone. Nothing grows on this spot of land, there is no fresh water, and the brutal heat relents only in the middle of winter. From the port of Râs Banâs in Egypt, small boats are still used to cross the more than thirty miles of shark-infested water to reach the island. The beaches near the deposits are green with tiny gem crystals. March up Peridotite Hill through the ancient diggings, and you find the fissures lined with complete and fractured gem crystals measuring from millimeters to several centimeters. The island is Zabargad, Arabic for *olivine*—the gemstone peridot. For peridot, Zabargad has been the most important and illustrious source.

Properties

Olive to lime green color is the most important quality for peridots. This characteristic color is caused by iron; the color saturation increases with iron content, but too much iron yields a brownish tinge. Most peridot is about 90 percent forsterite and the rest fayalite. The transparent gemstone has reasonably good properties: moderate durability and brilliance with a slightly greasy-looking luster.

Peridot Data

A variety of forsterite, Mg_2SiO_4, which with fayalite, Fe_2SiO_4, constitutes a complete series (solid solution) in the olivine group of minerals.

Magnesium iron silicate:	*$(Mg,Fe)_2SiO_4$*
Crystal symmetry:	*Orthorhombic*
Cleavage:	*Imperfect in two directions*
Hardness:	*7*
Specific gravity:	*3.22-3.45*
R.I.:	*1.635-1.690 (moderate)*

Peridot crystal 4.1 cm (1 5/8 in.) long and a 10.92-ct. cut stone, both from Zabargad Island, Egypt.

A 164.16-ct. peridot from Burma (top); the other gems are 61.55 and 95.19 cts., both from Zabargad Island.

Historic Notes

THE EGYPTIANS FASHIONED PERIDOT BEADS AS EARLY as 1580-1350 B.C. Second- and first-century B.C. writings of the Greek geographers Agatharchides and Strabo described Zabargad and its mining operations. In the third and fourth centuries in Greece and Rome, the gemstone was used for intaglios, rings, inlays, and pendants.

During the Middle Ages, the Crusaders brought peridot back to Europe; some of these gems are preserved in European cathedrals. Peridot was highly prized late during the Ottoman Empire (1300-1918). Turkish sultans amassed the world's largest collection of them. In Istanbul's Topkapi Museum, there is a gold throne decorated with 955 peridot cabochons ranging up to an inch across, peridots in turban ornaments and on jeweled boxes, and literally thousands of loose peridots.

When the term *peridot* was first used is

uncertain; French jewelers used it long before French mineralogist R.J. Haüy (1743-1822) applied it to the mineral. Yellow green peridot is sometimes called *chrysolite,* a term deriving from the Greek words meaning "gold" and "stone."

During the nineteenth century, the peridot became popular in both Europe and the United States, and production on Zabargad was active during the first half of this century.

Legends and Lore

FROM EARLY TIMES THROUGH THE MIDDLE AGES, peridot was considered a symbol of the sun. An early Greek manuscript on precious stones tells us that peridot bestows royal dignity on its wearer. Another belief was that the stone would protect its owner from evil spirits; in order to do so, the gem must be pierced, strung on the hair of an ass, then tied around the wearer's left arm, a procedure outlined by Marbode. A thirteenth-century English manuscript states that if a torchbearer, sign of the sun, is engraved on the gem, it will bring wealth to its owners.

Occurrences

FORSTERITE IS COMMON IN BASALTS AND PREDOMINANT in peridotite rock, but large unfractured peridot crystals are rare. At Zabargad, Egypt, the peridotite verges on being a pegmatite. Zabargad is presently inactive, awaiting better times in the Middle East.

The major source of peridot for the last fifteen years has been peridotite on the San Carlos Indian Reservation in Arizona. The stones are small, rarely exceeding 5 carats. The only source of large masses of fine-quality peridot is Upper Burma, near Mogok. Other occurrences are in Minas Gerais in Brazil, Sunnmore (formerly Söndmore) in Norway, China, and Kenya.

Evaluation

THE GREENER THE PERIDOT, THE HIGHER ITS VALUE. A tinge of brown diminishes its price, and any flaws make the stone undesirable. Usually, the price per carat does not increase with size. Peridot may be confused with tourmaline, green zircon, green garnets, chrysoberyl, diopside, moldavite (a tektite—natural glass), and sinhalite.

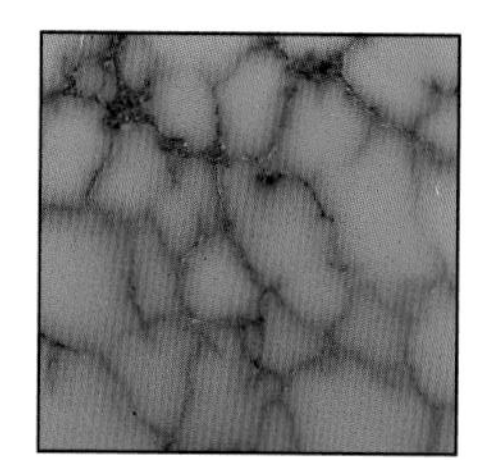

Turquoise & Lapis Lazuli

TURQUOISE

Turquoise is a gemstone with two probable firsts—first to be mined and first to be imitated. Indirect evidence suggests that the Wadi Maghara and Serabit El Khadem mines on the Sinai Peninsula were in production before 3100 B.C. Egyptian turquoise beads dating to 4000 B.C. have been found at al Badari. Surviving records from the time of King Semerkhet (c. 2923-2915 B.C., during the First Dynasty) document extensive mining operations that employed thousands of laborers and continued until about 1000 B.C. By 3100 B.C., either supplies were not meeting demand or a cheaper substitute was desired, because imitations (soapstone glazed blue and green—a form of faïence) are found as artifacts of this period.

A 90.20-ct. cabochon of turquoise with spiderweb matrix from Santa Rita in New Mexico and a 93.98-ct. high cabochon from Iran.

Properties

COLOR IS TURQUOISE'S SUPERLATIVE GEM PROPERTY; the mineral's other properties are less than ideal. The gemstone usually forms in aggregates of submicroscopic crystals that make it opaque. Turquoise is relatively soft and subject to scratching. Its porosity makes it discolor by absorption of oils and pigments, and friability can lead to easy breakage; only the most compact varieties resist these tendencies. The sky blue color is intrinsic, a result of copper. Iron in turquoise leads to greener tones. Ochre or brown black veining is common, the result of oxide staining or inclusion of adjacent rock fragments during turquoise's formation.

The turquoise Chou Dog was carved in China from Tibetan material and is 6.1 cm (2 3/8 in.) long.

TURQUOISE DATA

Copper aluminum phosphate:	$CuAl_6(PO_4)_4(OH)_8 \cdot 5H_2O$
Crystal symmetry:	*Triclinic (normally cryptocrystalline)*
Cleavage:	*Not observed in massive gemstone form*
Hardness:	*5-6*
Specific gravity:	*2.6-2.8*
R.I.:	*1.62 average*

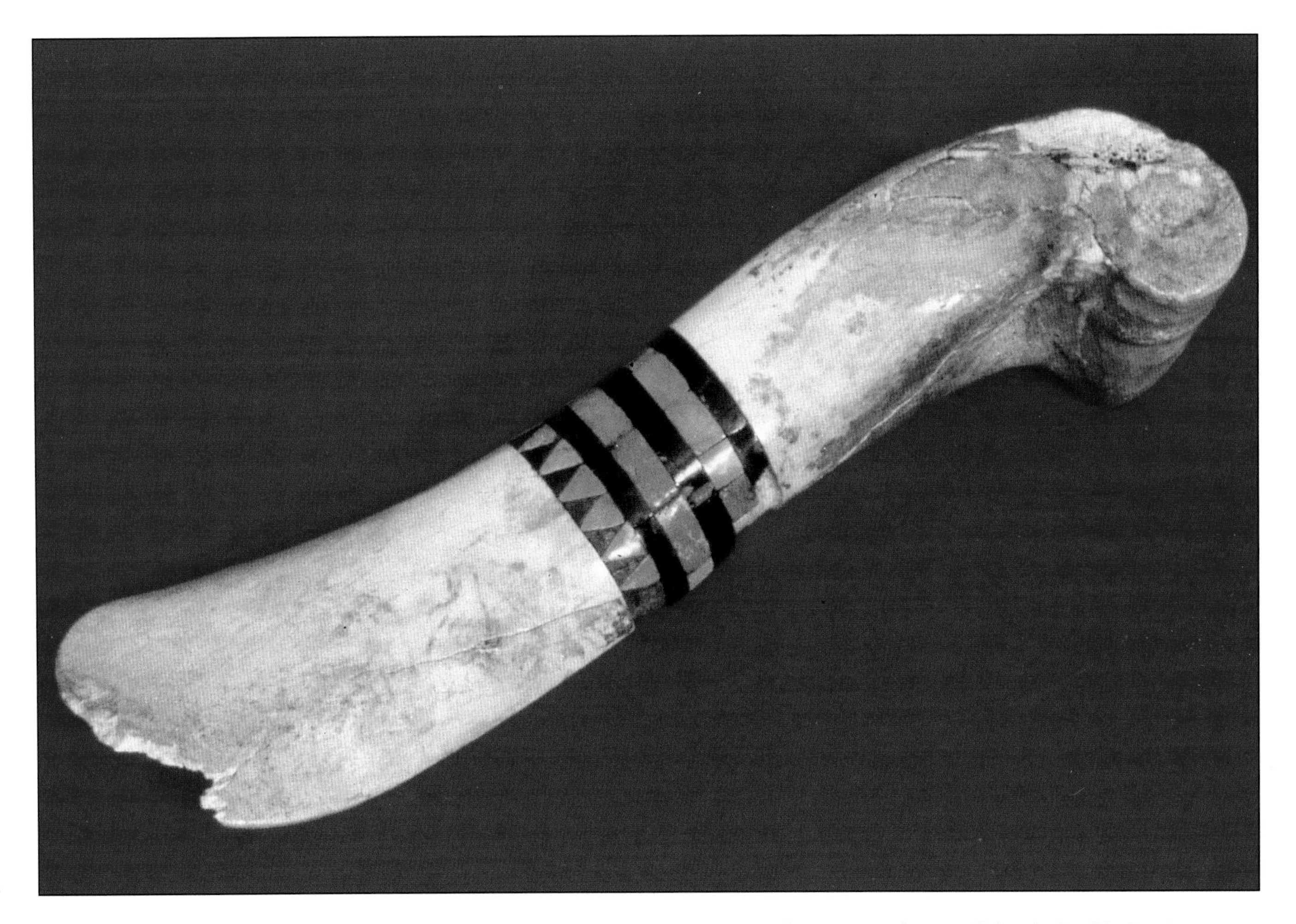

Bone scraper inlaid with turquoise and jet, 15 cm (6 in.) long, found during the 1896 Hyde Expedition in Pueblo Bonito, Chaco Cañon, New Mexico.

Historic Notes

THE NAME *TURQUOISE* DID NOT COME INTO USE UNTIL the thirteenth century. Pliny used the term *callais,* derived from the Greek *kalos lithos,* meaning "beautiful stone." Purchased by Venetian merchants in Turkish bazaars for European trade, the blue stone was called "pierre turquoise" by the French recipients, meaning "Turkish stone."

The first uses of turquoise were in Mesopotamia (Iraq), where beads dating from about 5000 B.C. have been found. Turquoise is Iran's national gemstone. It has decorated thrones, daggers, sword hilts, horse trappings, bowls, cups, and ornamental objects. High officials once wore turquoise seals decorated with pearls and rubies. Large stones were embellished with gold scrollwork to hide the imperfections. After the seventh century A.D., turquoise pieces decorated with passages from the Koran or Persian proverbs in incised gilt characters were treasured as amulets. Turquoise has been the most cherished gem in Tibet as well, with a role comparable to that of jade in China.

Turquoise was frequently set in Siberian jewelry of the fifth and sixth centuries B.C. and in pieces from southern Russia of a slightly later date. The ancient Greeks and Romans engraved turquoise for ring stones and pendants and also used it as beads, but certainly it was not one of their favorite stones. In Europe, it became more popular during the Middle Ages for decoration of vessels and the covers of manuscripts. Popularity of the stone for personal adornment grew during the Renaissance; by the seventeenth century, "no man considered his hand well-adorned" unless

A polished variscite sphere 7.5 cm (2 15/16 in.) in diameter, two cabochons, and a gold pin with three cabochons from Fairfield, Utah. Variscite is often mistaken for turquoise.

he wore a turquoise, according to Anselmus de Boot. In the following centuries, turquoise remained popular, gracing royal crowns as well as modest jewels. In Europe, it has been the most popular opaque gemstone.

The gem's history in the Americas began about 1,000 years ago with the initiation of turquoise mining at Mount Chalchihuitl in Cerrillos, New Mexico. Native Americans, using hand tools, quarried the entire mountain; all that remains on the north side is a pit about 200 feet across and up to 130 feet deep.

Turquoise has been found in burial sites in Argentina, Bolivia, Chile, Peru, Mexico, Central America, and the southwestern United States. The Incas carved beads and figurines and crafted beautiful turquoise inlays. The Aztecs used turquoise in mosaic pendants and ritual masks. The Zuni, Hopi, Pueblo, and Navajo all made magnificent necklaces, ear pendants, and rings. At Pueblo Bonito in northwestern New Mexico, nearly 9,000 beads and pendants were found near a single skeleton. All told, 24,932 beads were found in these burial sites.

Legends and Lore

IN PERSIA (IRAN), ONE WHO COULD SEE THE REFLECtion of a new moon on a turquoise was certain to have good luck and be protected from evil. Hindus had a comparable belief: if an individual could look at a new moon and immediately after at a piece of turquoise, great wealth would surely follow. To the Navajos, a piece of turquoise thrown into a river (while a prayer to the rain god was being spoken) would ensure the blessing of rain. A turquoise attached to a gun or bow would guarantee accurate aim, according to Apache lore.

The belief that turquoise would protect its owner from falling—especially from a horse—is first recorded in the thirteenth century. The virtue traces to turquoise's use in Persia and Samarkand as a horse amulet. Legend also has it that, by changing color, turquoise reveals a wife's infidelity.

Occurrences

TURQUOISE CRYSTALLIZES AS VEINS AND NODULES near the water table in semiarid to arid environments. Its chemical stockpiles are the adjacent rocks, which are leached by rain and ground water; thus turquoise is often associated with weathered igneous rocks containing primary copper minerals.

Before World War I, turquoise production from nearly 100 mines was Iran's most important industry. Following World War II, output declined and ceased altogether with the revolution. Turquoise is found in Nevada, Arizona, Colorado, New Mexico, and California, the primary producers of turquoise today. Much of the turquoise is a byproduct of copper mining. Most American turquoise is light in color, porous, and chalky, usually with matrix, and only 10 percent of the turquoise mined is of gem quality. Other occurrences are in Armenia and Kazakhstan in the Soviet Union, China, Australia, Tibet, Chile, Mexico, and Brazil.

Evaluation

THE INTENSITY AND EVENNESS OF COLOR AND QUALIty of polish affect the value of turquoise. The very rare, intense sky blue (robin's egg blue) is most desired. Turquoise with matrix is generally less valuable than stones without it. Of matrix turquoise, the spiderweb variety is the most valuable.

A fine polish is possible only with stones that are hard, relatively nonporous, and compact. Very pale and chalky turquoise is sometimes impregnated with oil, paraffin, liquid plastic, glycerin, or sodium silicate to enhance its color and ability to take a good polish. Occasionally, this turquoise is sold as stabilized turquoise or turquolite. Some turquoise is even painted on the surface with blue dye and then coated with clear plastic. Much of the turquoise on the market has been treated in these ways, and some may change color.

GEMSTONES CONFUSED WITH TURQUOISE, IMITATIONS, AND SYNTHETICS

Chrysocolla, chrysocolla quartz, odontolite (a naturally stained fossilized bone), variscite, and malachite are easily confused with turquoise. Turquoise may be imitated with glass, porcelain, plastic, enamel, stained chalcedony, dyed howlite, blue-dyed and plastic-treated marble, and doublets. Artificial products are sold with names such as Viennese turquoise, Hamburger turquoise, and Neolith. Synthetic turquoise has been produced and marketed in France since 1970.

LAPIS LAZULI

Lapis lazuli is probably the original blue gemstone and has an ancient history of exploitation in Afghanistan. The gemstone has been so important to Afghanistan that lapis figured in United States foreign policy. In 1985, during Senate Armed Services Committee hearings on the war in Afghanistan, some testimony indicated that lapis lazuli was an important source of cash for the mujahadeen to buy weapons to battle the Soviet-backed regime. In fact, the Kabul government was also trying to raise cash by selling quantities of the blue stone. The result for the market has been an unusual abundance of lapis available in recent years. Unfortunately, although there may be tons to pick from, only a minute proportion has the fine and uniform blue color with only occasional golden flecks that distinguish the most desirable lapis lazuli.

(Opposite) Polished lapis lazuli boulder from Afghanistan, 13.5 cm (5 1/4 in.) high.

(Below) Lapis lazuli Chinese junk, 15.2 cm (6 in.) high.

Lapis Lazuli Data

A rock composed principally of the mineral lazurite [a sodium aluminosilicate containing sulfur, chlorine, and hydroxyl— $(Na,Ca)_{7-8}(Al,Si)_{12}O_{24}[(SO_4),Cl_2,(OH)_2]$ with variable amounts of pyrite (the brassy flecks) and white calcite

Cleavage:	*Not relevant for a rock*
Hardness:	5-5.5
Specific gravity:	2.7-2.9
R.I.:	*About 1.5 (opaque)*

Properties

AS LAPIS LAZULI IS OPAQUE, THE MOST IMPORTANT qualities are color and a moderate durability. Lazurite is not very hard, but fine-grained lapis is reasonably tough. The intrinsic blue color of lazurite is caused by sulfur, an interesting and unusual case where a nonmetallic element yields a strong color. Lapis was ground into the pigment ultramarine until an artificial replacement was developed in 1828.

IN EGYPT, CARVED LAPIS BEADS, SCARABS, PENDANTS, and lapis-inlaid jewelry date prior to 3100 B.C. The stone was esteemed as a gem and amulet. Ground into powder, it was used as a medicine and a cosmetic—the first eye shadow.

The tomb of Queen Pu-abi (2500 B.C.) in the city of Ur in Sumer contained adornments rich with lapis—three gold headdresses, two bead necklaces, a gold choker, a silver pin, and a gold inlaid ring.

During the time of Confucius (c. 551-479 B.C.), the Chinese carved lapis hair and belt ornaments. As early as the fourth century B.C., the Greeks used lapis for carving scarabs and scaraboids, and it was described by Theophrastus. In Rome, lapis was fashioned into intaglios, plain ring stones, beads, and inlays. The ancient Greeks and Romans used the term *sapphirus*; *lapis lazuli* did not come into use until the Middle Ages. *Lazulus* means "blue stone" in Latin and derives from the ancient Persian *lazhuward*, meaning "blue," and the Arabic *lazaward*, meaning "heaven," "sky," or simply "blue in general."

The stone was a favorite material for carving *objets d'art* during the Renaissance in Europe. When Catherine the Great was told that lapis had been discovered near Lake Baikal, she ordered that mining be commenced immediately. In the following year, 1787, the empress decorated a room in her palace in Tzarskoye Selo (now Pushkin) with the stone. Sections of walls, doors, fireplaces, and even mirror frames were made of lapis.

Today, lapis is favored for beads, ring stones, and pendants and is a preferred stone for men's jewelry.

Legends and Lore

TO THE BUDDHISTS, LAPIS BROUGHT ITS OWNER peace of mind and equanimity and dispelled evil thoughts. In *De Materia Medica* (c. A.D. 55), Greek physician and pharmacologist Dioscorides recommended lapis as an antidote for the bite of a poisonous snake. By the thirteenth century, broadened curative powers were attributed to lapis; Albertus Magnus advised using lapis for intermittent fever and melancholy in his mineralogical treatise.

GEMSTONES CONFUSED WITH LAPIS LAZULI, SUBSTITUTES, AND IMITATIONS

Sodalite, azurite, lazulite, and dumortierite may be confused with lapis lazuli. The most common substitute for lapis is blue-dyed chalcedony, sold as German lapis and Swiss lapis. Lapis has been imitated with synthetic spinel with gold forced into surface holes to simulate inclusion of pyrite (fool's gold).

Occurrences

Lapis is a rare metamorphic rock produced by the interaction of granitelike magma with marble. In addition to Afghanistan, Chile is a major source of the gemstone. The most productive mine is in the Andes in Coquimbo Province, north of Santiago. It was worked by the Incas in pre-Columbian times and continues in production today. A less important source is near Antofagasta. Chilean lapis usually contains large amounts of calcite, although mining of better-grade material has been reported recently. The Soviet Union produces the stone from mines near Lake Baikal and near Khorog in the Pamir Mountains. In the United States, lapis is produced in Colorado and in California.

Evaluation

The quality, purity, and evenness of color largely determine the value of lapis. The most desired color is deep violet blue. Stones without inclusions of pyrite or calcite are most desirable. Lapis with inclusions of pyrite is more valuable than that with inclusions of calcite. Often the white calcite inclusions are disguised with paraffin treatment, which may include the use of blue dye. (Dye can be detected by rubbing the stone with cotton dipped in acetone or fingernail-polish remover; the blue color comes off on the cotton. In fact, this test works for most dyed gems.) The minerals in lapis have unequal hardnesses and polish differently. Only superior, inclusion-poor lapis can be polished to a smooth, even luster.

Russian lapis lazuli carving decorated with sterling silver, yellow gold, red and yellow enamel, and small rose-cut diamonds. It measures 16 cm (6 1/4 in.) across, and the base is stamped by the famous Fabergé workshop.

Opalized wood, 9 cm (3 1/2 in.) long, from Virgin Valley, Nevada.

The sheer beauty of opal outweighs its disadvantageous physical properties. Rainbow colors pour out of well-lit precious opals, but the gems are easily scratched and so are a poor choice for exposed ring settings. Opals are mechanically fragile and notoriously difficult to set in jewelry; a slight blow or a rapid change in temperature may shatter an opal. The problem is that the gem contains water, which—depending on the opal and its source—may evaporate and leave the opal slightly smaller, stressed, and covered with cracks. Opals need our protection; worn close to the body, they are safe from abrasion, are kept at an even temperature, and receive some body moisture so that they do not lose water.

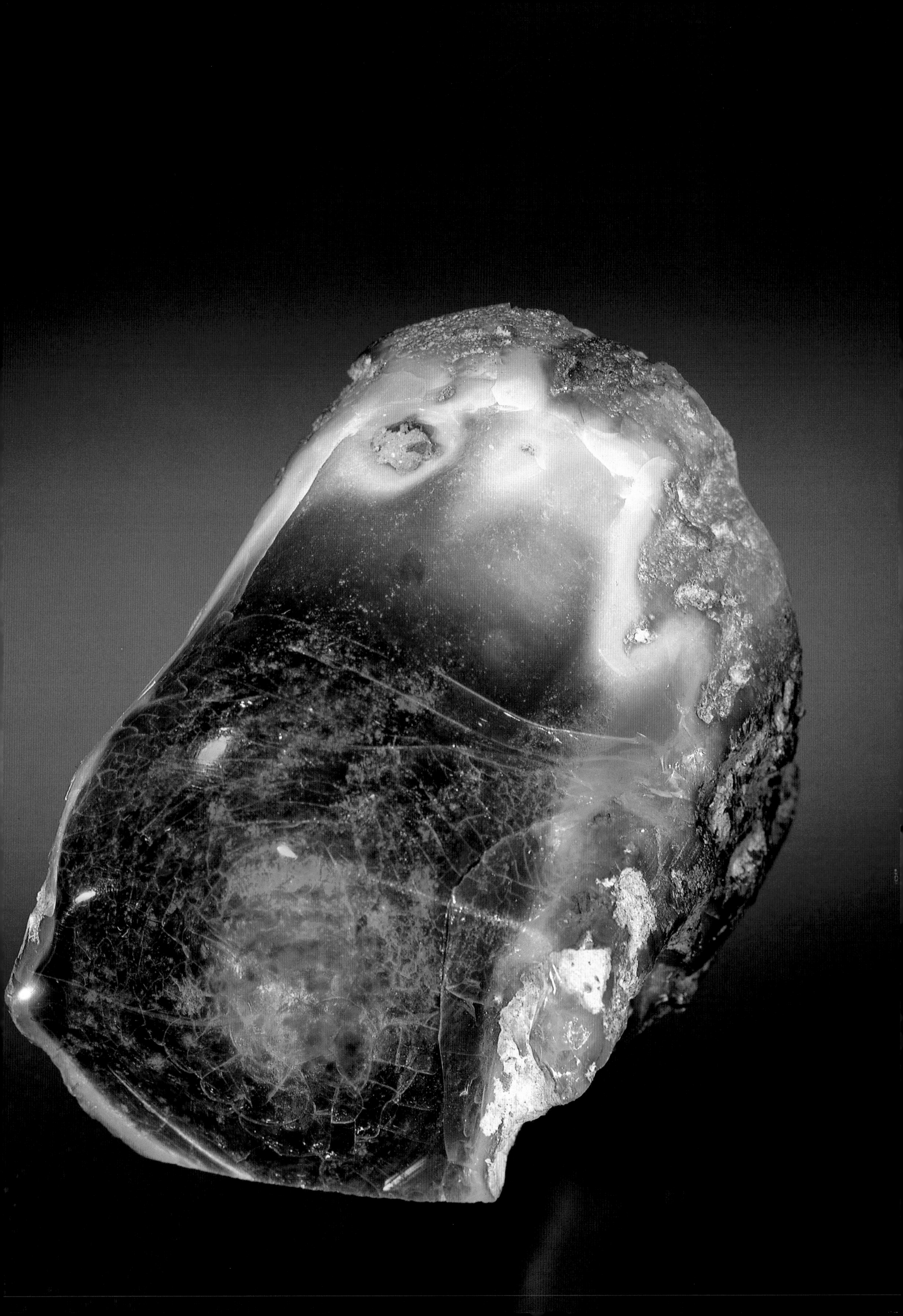

Properties

THE CHARACTERISTIC FEATURE OF GEM OPAL IS PLAY of colors; pure colors can be seen in rapid succession when the gem is moved about. By some standards, opal is not a mineral because it does not have an extended crystal structure. Opal is made up of submicroscopic silica spheres bonded together with more silica and water. The lower the initial amount of water in the opal, the better are its properties. Loss of water or change in temperature causes strain that is relieved by cracking, known as crazing. Also, opal is soft, and its density and refractive index are low.

If the minute spheres in opal are uniform in size and packed into a regular array, they can scatter light in various colors (by diffraction) determined by the size of the spheres and the opal's orientation. Gem or precious opal contains many of these organized zones that display diffraction colors, whereas common opal may be colored but does not show a play of colors. Precious opal is usually cut as a cabochon or carved, but some fire opal is sufficiently transparent to be faceted.

Opals from Mexico ranging in weight from 4.72 to 31.70 cts.

OPAL DATA

Hydrated silica:	$SiO_2 \cdot nH_2O$
Crystal symmetry:	*Largely amorphous*
Cleavage:	*None, but brittle*
Hardness:	*5.5-6.5*
Specific gravity:	*1.98-2.25*
R.I.:	*1.43-1.47 (low)*
Special optical property:	*Diffraction—play of colors*

Pendant 4.5 cm (1 3/4 in.) in length, set with Australian opals, chrysoberyl, sapphires, topaz, demantoid garnets, and pearls.

Precious Opal Body Colors and Their Causes

PRECIOUS OPALS ARE GENERALLY DEFINED BY MANIFESTING A PLAY OF COLORS.

Black:	***A black or dark background in gray, blue, or green—dark inclusions***
White:	***A white background—internal boundaries or fluid inclusions***
Water:	***A transparent or colorless stone—few or no inclusions***
Fire:	***A transparent stone (play of color may or may not appear) or translucent yellow, orange, red, or brown—ochre-colored iron oxide in inclusions***

OPAL CLASSIFICATION BY SIZE AND PATTERN OF COLOR PATCHES

Pinpoint or pinfire:	***Small patches, close together***
Harlequin:	***Larger, more angular patches that resemble the diamonds of a harlequin pattern***
Flame:	***Red streaks and bands that cross the surface like flames***
Flash:	***Flashes of color that appear and disappear when the stone is moved***

Historic Notes

Opal derives from the Sanskrit *upala* and the Latin *opalus*, meaning "precious stone." Pliny described fine opals: "In the opal you will see the refulgent fire of the carbuncle, the glorious purple of amethyst, and the sea green of the emerald, and all these colors glittering together in incredible union."

The oldest opal mine was at Czerwenitza, now in Czechoslovakia (formerly Hungary). Archival evidence indicates that the mine was worked in the fourteenth century, but there are indications that it was in operation much earlier—perhaps the source of opal for Rome. Production of semitranslucent milky-white stones with play of color continued until 1932. Mexican fire opal was known to the Aztecs and was introduced in Europe by the Spanish conquistadors early in the sixteenth century.

Shakespeare referred to opal as "the queen of gems" in *Twelfth Night*. Opal's prominence declined during the nineteenth century, when the stone was associated with bad luck. Many believe that Sir Walter Scott was responsible. In his 1828 novel, *Anne of Geierstein*, the heroine's demonic grandmother died when a drop of holy water touched her enchanted opal and put out its fire.

Black opal was first found in Australia in 1887. Queen Victoria helped popularize it and white opal by giving opal jewelry to all of her children. Opal was a favorite stone of René Lalique (1860-1945), the most talented artist-jeweler of the Art Nouveau movement. He designed opal jewelry for Sarah Bernhardt (1844-1923) among others.

Black and white opals are currently among the most popular gems.

Legends and Lore

The Romans considered opal a stone of love and hope. The Arabs believed it fell from heaven in flashes of lightning. Marbode wrote of an opal talisman of the House of Normandy: "It renders the wearer invisible, enabling him to steal by day without risk of exposure to baneful dews of night." Thus the opal became known as the talisman of thieves and spies. Anselmus de Boot summarized: "Opal possesses the virtues of all gems as it displays their many colors."

According to an Australian legend, the stars are governed by a huge opal that also controls gold in the mines and guides human love. However, the Australian aborigines had a negative slant: opal is a devil, half-serpent, half-human, who lurks in a hole in the ground ready to lure men to destruction with flashes of wicked magic.

(Opposite) The 215.85-ct. Harlequin Prince, found in Australia.

(Above) Opal Indian head carving measuring 4.2 cm (1 4/5 in.) in length. The rough opal is from Myneside, West Queensland, Australia.

Occurrences

OPAL IS FORMED IN CAVITIES AND CRACKS IN NEAR-surface volcanic rocks or as replacements—thus making fossils—of shells, bones, and wood in or near sedimentary volcanic ash by percolating water that dissolves silica and then precipitates opal.

Australia produces 85 percent of the world's opal. Lightning Ridge, New South Wales, is the primary source of black opal. Coober Pedy, discovered in 1915, is the opal capital of the world. The aboriginal name for the location, Kupa Pita, means "white man in a hole." The miners and their families all used to live underground, attempting to escape the intolerable climate; some still do. (Thus the aboriginal legend is made a little clearer.) White Cliffs and Andamooka are also important sources.

The only commercial source of fire opal is Mexico, principally near Querétaro, where mining began in 1870. Mexican fire opal is softer and lighter than opal from other sources because it contains more water. The states of Jalisco, Hidalgo, Guanajuato, and Nayarit also produce opal.

Opal was discovered at Virgin Valley in Nevada in 1908. This material is very beautiful, but once it is exposed to air, it loses some of its water and cracks. Because no effective method has been discovered to prevent cracking, production remains very limited. Other opal occurrences are in Brazil, Honduras, Nicaragua, Guatemala, Japan, and Ireland.

Evaluation

THE PLAY OF COLORS IS MOST IMPORTANT TO THE value of an opal. Fine opals exhibit bright, intense colors. No dead patches should be present. Black opals are more valued than white. Of the types of color patterns, the harlequin is the most valuable. Fire opals without play of color are judged by the beauty of their color and transparency. Red opals are more valued than yellow and brown.

Some white opals from Australia are treated to make them black. After being soaked in sugar solution, they are immersed in sulphuric acid, which carbonizes the sugar, blackening the stone. Some Mexican opals have been turned black by smoke treatment in a mixture of charcoal and cow manure. Plastic and silica-based polymers, with or without dye, have been used to impregnate Brazilian, Mexican, and Idaho opals, significantly improving their appearance.

Opal too thin to be used in jewelry is made into doublets. A thin piece of gem-quality opal is cemented to a piece of common opal, chalcedony, or glass. If an opal doublet is covered with a protective piece of colorless quartz, it becomes a triplet, which is more durable than a doublet.

(Opposite) An opalized clam weighing 69.00 cts. from Coober Pedy, Australia.

(Below) Opal carving of a leaf from rough found in Stuart Range, South Australia, measuring 6.3 cm (2 1/2 in.) in length.

Opal Imitations, Substitutes, and Synthetics

Glass and plastic are used to imitate opal, but do so only poorly. No natural mineral resembles opal, with the possible exception of labradorite feldspar. Synthetic opals were first produced commercially in France by P. Gilson in 1972 and are successful.

Feldspar

Minerals of the feldspar group constitute over half of the Earth's crust, but nature only rarely yields them as gemstones. Remarkable iridescence is the hallmark of the best-known varieties: moonstone and labradorite. The blue shimmer of a fine moonstone forms a subtle but beckoning adornment in soft illumination, whereas laboradorite's color flash, like the peacock's tail feathers, is arresting. Of the latter, Ralph Waldo Emerson said in *Experience:* "A man is like a bit of labrador spar, which has no luster as you turn it in your hand until you come to a particular angle; then it shows deep and beautiful colors."

A polished labradorite slice measuring 7.6 cm (3 in.) long and a 3.2 cm (1 1/4 in.) diameter disk, both from Labrador, Canada.

Properties

IRIDESCENCE AND COLOR ARE THE SOURCE OF THE feldspars' gemstone appeal. They have little brilliance, and crystals are known for their cleavages. The term *feldspar* stems from these cleavages; it means "field spar." The Anglo-Saxon *spar* refers to easily cleaved minerals such as calcite, fluorite, and feldspar.

The iridescence is caused by scattering of light from thin layers in the gemstone; these layers are a second feldspar that develops by internal chemical separation during geologic cooling of an initially single feldspar. Light scattering from the layers results in pure iridescent colors (in labradorite, called "labradorescence" or "schiller") from red to blue or in a broad blue white to yellow white spectrum (in moonstone). Peristerite, intermediate in iridescence, gets its name from the Greek *peristera,* meaning "pigeon stone," in allusion to the color of the bird's neck feathers. Sunstone, also known as aventurine feldspar, reflects internal gold spangles.

Visible intergrowths produce translucency or opacity rather than iridescence in most feldspars, and most are usually not strongly colored. Moonstones are colorless to gray or yellow and semitransparent to translucent. Gem orthoclase is transparent and yellow. Labradorite is usually gray and opaque, but rare transparent crystals are occasionally found. Amazonite is an opaque feldspar with vivid green to blue green color. Translucent feldspars are carved or cut *en cabochon*; the rare transparent material is faceted for collectors.

Historic Notes

AMAZONITE (NAMED AFTER THE AMAZON RIVER) WAS widely used in Egyptian, Sudanese, Mesopotamian, and Indian jewelry; some examples date back to the third millennium B.C. The twenty-seventh chapter of the Egyptian *Book of the Dead* was engraved on this feldspar. A carved scarab and an amazonite-inlaid ring were among the jewels of Tutankhamen (reigned 1361-1352 B.C.). Amazonite was treasured by the Hebrews, and it is generally accepted that the third stone in Moses' breastplate was amazonite. In Central and South America, adornments in pre-Columbian times contained amazonite.

Moonstone appeared in Roman jewelry in about A.D. 100 and even earlier in Oriental adornment. The gemstone was a favorite of Art Nouveau jewelers, and Cartier and Tiffany creations frequently contained the gem.

Labradorite was used in decoration by the Red Paint people of Maine long before the year

FELDSPAR DATA

A mineral group forming two distinct compositional series (solid solutions) of alkali aluminosilicates

Plagioclase Series:	$CaAl_2Si_2O_8$ to $NaAlSi_3O_8$
Alkali Feldspar Series:	$KAlSi_3O_8$ to $NaAlSi_3O_8$
Crystal symmetry:	*Monoclinic or triclinic*
Cleavage:	*Two perfect at right angles; imperfect in a third direction*
Hardness:	*6-6.5*
Specific gravity:	*2.55-2.76*
R.I.:	*1.518-1.588 (low)*
Special optical properties:	*Light scattering and iridescence*

(Left) Labradorites ranging in weight from 2.09 to 3.01 cts., from Clear Lake, Oregon.

(Below) Moonstone intaglios of three siblings carved by noted engraver Ottavio Negri (average, 2 cm high). The moonstone is from Sri Lanka. The central intaglio is enlarged to show details.

Feldspar Gemstones

For varieties, the appropriate mineral name is given in parentheses.

PLAGIOCLASES

Labradorite:	*Middle of series—colorful iridescence, also transparent stones in yellow, orange, red, green*
Sunstone (oligoclase):	*Near Na end—gold spangles from oriented inclusions of hematite*
Peristerite (albite):	*Near Na end—blue white iridescence*

ALKALI FELDSPARS

Orthoclase:	*At K end—transparent gemstone, pale to bright yellow from iron substitution for aluminum*
Amazonite (microcline):	*At K end—yellow green to greenish blue, opaque; color from natural irradiation of microcline containing lead and water impurities*
Moonstone (orthoclase):	*Near K end—blue white to white iridescence*

1000. It was "found" by Moravian missionaries in 1770 in Labrador and named for the locality.

In the late eighteenth and nineteenth centuries, two important deposits of sunstone feldspar were found in Russia; as a result, the gem had extensive use in Russian jewelry. When sunstone deposits were found in Norway in 1850, the gem became increasingly popular in Europe.

Legends and Lore

AMAZONITE WAS A POPULAR AMULET AMONG ANcient Egyptians. According to Pliny, the Assyrians considered it the gem of Belus, their most revered god, and used it in religious rituals.

In India, moonstone was sacred and also had a special significance for lovers; if they placed it in their mouths when the moon was full, they could foresee their future. In Europe, Marbode's eleventh-century lapidary reported that the gem could bring about lovers' reconciliation. Geronimo Cardano wrote in the sixteenth century that moonstone drives sleepiness away.

FELDSPARS CONSTITUTE SUBSTANTIAL PORTIONS OF many igneous and metamorphic rocks. Gem varieties result from the rare geologic conditions that produce clean large grains, particularly in pegmatites and ancient deep crustal rocks.

Important localities for amazonite are in India, Brazil, Quebec in Canada, the Malagasy Republic, the Soviet Union, South Africa, and Colorado and Virginia. The best-quality moonstones came from a dike at Meetiygoda in southern Sri Lanka; this source is now exhausted. Moonstones are found in the gravels of Sri Lanka and Burma as a byproduct of ruby and sapphire mining; Madras, India, produces a less valuable quality—almost opaque and yellowish, reddish brown, or grayish blue. The finest-quality peristerites are found in Ontario and Quebec in Canada and in Kenya. Norway and the Soviet Union continue to be the sources of sunstone, and Labrador and Finland are the principal source of iridescent labradorite. Transparent, facetable labradorite is found in Mexico and in Utah, Oregon, California, and Nevada.

MOONSTONE IS THE MOST VALUABLE FELDSPAR. Finequality moonstone is semitransparent and flawless and exhibits a broad blue sheen. Bright-colored amazonite is the most desirable. To be of fine quality, labradorite must display intense iridescent colors. Spectrolite is a trade name for the Finnish material. Sunstones that are semitransparent and show a pleasing reddish or yellow orange glow are the most desirable. A popular imitation, marketed as goldstone, is glass with copper inclusions.

(Opposite) Amazonite crystal measuring 6 cm (2 3/8 in.) across from the Lake George area in Colorado and cabochons carved with floral designs, with weights ranging from 18.0 to 29.9 cts., from Amelia Court House, Virginia.

GEMSTONES CONFUSED WITH FELDSPAR

Moonstone:	*Quartz, chalcedony, and opal*
Amazonite:	*Jade and turquoise*
Labradorite:	*Opal*

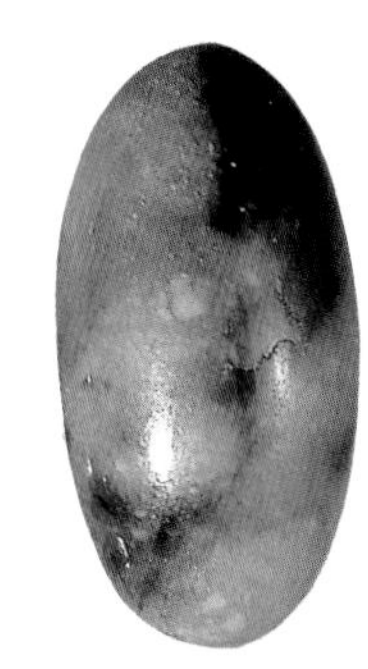

Jade

The term *jade* for the ornamental stone most identified with China is a total misnomer. In the sixteenth century, Spanish conquistadors learned of a stone worn by Mesoamericans as an amulet to cure colic and similar maladies. The Spanish called it *piedra de yjada* (in Latin, *lapis nephrictus*), meaning "stone of the loin," and brought fine examples back to Europe. In translation from Spanish to French, the phrase was misprinted as *pierre le jade.*

In the mid-seventeenth century, the New World sources had disappeared, and Europeans forgot the material but not the name; they applied it to the stone of numerous carvings arriving from China. In 1780, geologist A.G. Werner described the traditional carving material and labeled it *nephrite*, after the Latin term. In 1863, French chemist Augustine Damour chemically analyzed a Chinese carving of Burmese stone and found that it was different from the Chinese nephrite. He labeled the material *jadeite*, derived from *jade*. In 1881, he discovered that Burmese jadeite and the original Mesoamerican material were identical. Nevertheless, the common term persists for both substances; jadeite and nephrite share the common name "jade." To make matters even more complicated, other stones that appear similar or have been used in a similar manner in ancient cultures are also simply called "jade." Such is the confusion with the most important ornamental gemstone.

Vessel with cover and hanging chain, 20.5 cm (8 1/16 in.) high, carved from a single piece of nephrite, mutton-fat jade, during the reign of Ch'ien Lung.

Jadeite cabochons from Burma (nine smaller stones) and Guatemala (two larger stones), with weights ranging from 6.38 to 28.34 cts., displayed on 20-kg (45-lb.) jadeite boulder from Burma.

JADE DATA

Both nephrite and jadeite jade are rocks composed of essentially a single mineral: tremolite or actinolite in nephrite and jadeite in the other. For rocks, crystal symmetry and cleavage are meaningless.

	TREMOLITE/ACTINOLITE (NEPHRITE)	JADEITE (JADEITE JADE)
Composition:	*Calcium magnesium silicate*	*Sodium aluminum silicate*
Formula:	$Ca_2(Mg,Fe)_5(Si_4O_{11})_2(OH)_2$	$NaAlSi_2O_6$
Hardness:	6	6.5-7
Specific gravity:	2.9-3.1	3.1-3.5
R.I.:	1.62 (average)	1.66 (average)

Properties

THE SPECIAL QUALITY THAT NEPHRITE AND JADEITE jade share is exceptional durability; nephrite is one of the toughest known substances. Both rarely yield to a hammer blow—a convenient field identification technique (obviously not suggested for art or artifacts). This property means that jade can be carved into remarkably fine and intricate forms with minimal risk of breaking.

Nephrite owes its exceptional toughness to a solid feltlike structure of intergrown microscopic fibrous crystals. Jadeite, slightly less durable, forms larger prismatic crystals that interlock and create a strong network. Both materials accept a fine polish because of their compactness, though nephrite's polished surface often has many small depressions, like an orange peel.

Colors and patterns are quite variable for both nephrite and jadeite. Veins, clots, zoning, and deformation can produce color variations and juxtapositions. Individual colors are due to substitutions of elements in the major constituent mineral or to "contaminant" minerals in the rock. Boulders of both nephrite and jadeite, particularly the green varieties, frequently have a tan- to ochre-colored rind due to oxidation of their constituent iron.

JADE COLORS AND THEIR CAUSES

NEPHRITE

White:	*Essentially pure tremolite, very little iron; sometimes called "mutton-fat jade."*
Deep green:	*Iron; "spinach green jade" from Siberia has blotches caused by graphite inclusions.*
Creamy brown:	*The color of this stone, sometimes called* **tomb jade,** *was once attributed to the action of heat on lime impurities but research indicates it is the result of reactions between fluids in mummies and jade, both sealed in sarcophagi.*

JADEITE

White:	*Pure jadeite.*
Leafy and blue green:	*Iron.*
Emerald green:	*Chromium; Imperial jade is the finest translucent variety.*
Lavender:	*Manganese and iron.*
Dark blue green and greenish black:	*Iron in omphacite (a calcium-rich jadeitic pyroxene) and aegerine—called "chloromelanite," a term now invalid in mineralogy.*
Deep emerald-green:	*Due to substantial amounts of the mineral kosmochlor, $NaCrSi_2O_6$; this jade variety is called* **mawsitsit** *or* **tawmawite.**

Historic Notes

BOTH JADES SHARE A UTILITARIAN BEGINNING. THE quality that drew primitive people's attention to them was their singular toughness—in this sense, we can consider them the same. Jade was the raw material for celts (choppers), axes, and clubs. One could fashion a thin, strong edge that retained its sharpness. As cultures developed, jade became a substance valued for its beauty. Thus the concept of jade, regardless of the material or name, became comparable around the world.

Jade has had its longest and most continuous history in China, nephrite being the principal jade used. According to folklore, Huang Ti in the twenty-seventh century B.C. had jade weapons and bestowed jade tablets upon officials to confer rank and authority. By the end of the Chou Dynasty (255 B.C.), carving design had reached maturity, and during the reign of Ch'ien Lung (A.D. 1736-1795), technical skill in carving achieved the highest level.

Jade has played a part in almost every aspect of Chinese life—as tools, currency, awards for statesmen, visiting ambassadors, and war heroes. Some of the earliest records of events are inscribed nephrite tablets. Ceremonial libation vessels, incense burners, and marriage bowls were carved from nephrite; and it was the material for innumerable personal ornaments, including beads and pendants inscribed with poetry and worn as talismans.

For the Chinese, jade was not only pleasing to the eye but also to the ear and sense of touch. Musical instruments carved from jade have been used for rituals since ancient times, and rounded and well-polished "buttons" were carried in the sleeves for fingering until the end of the nineteenth century.

Jadeite jade was highly valued by the Olmecs, Mayas, Toltecs, Mixtecs, Zapotecs, and Aztecs. Carbon 14 dating of wood fragments found with jadeite artifacts in Mexico provides a date of use by the Olmecs around 1500 B.C. Jadeite was typically carved in the form of jaguars and decorated celts for ceremonial purposes. The numerous jade artifacts found in tombs include earplugs, diadems, necklaces, pendants, bracelets, masks, and statues of the sun god. The Spanish conquistadors completely destroyed the art of jade carving in America, and soon after the Conquest, the jade sources were lost.

The Maoris of New Zealand used nephrite, starting in about A.D. 1000, first as tools and weapons and later for amulets and decorations.

(Left) Figure of Quan-yin carved from Burmese jadeite in China during the nineteenth century. It is 28 cm (11 in.) high without the stand.

(Opposite) Olmec axe (1200-500 B.C.) of jadeite jade from Oaxaca, Mexico, 27.9 cm (11 in.) in height and weighing 15.5 lbs.

Legends and Lore

SINCE THE BEGINNING OF THEIR HISTORY, THE CHInese have esteemed nephrite more than any other gemstone. From neolithic times until the beginning of the twentieth century, carved pi (flat discs with a central hole) were used to worship heaven. According to a 1596 Chinese encyclopedia, drinking a mixture of jade, rice, and dew water strengthens the muscles, hardens the bones, calms the mind, enriches the flesh, and purifies the blood. One who takes it long enough can endure heat, cold, hunger, and thirst.

Jade was equally important after death. An elaborate burial shroud made of 2,156 jade tablets sewn together with threads of gold covered the princess Tou Wan (second century B.C.). Carved amulets were put in the deceased's mouth, and amulets, insignias of rank, and favorite pieces were placed on different parts of the body and clothing. These "tomb jades" were offerings to the gods, but the durable stone was also believed to protect the body from decay.

In the pre-Columbian civilizations of Mexico and Central America, jade had great talismanic power as well. A piece of jade in the mouth of a dead nobleman was believed to serve as a heart in the afterlife. Powdered and mixed with herbs, jade was used as a treatment for fractured skulls, different fevers, and even reviving the dying.

The Maoris of New Zealand also revered jade as a powerful talisman. Typical are the hei-tiki pendants, which are grotesque human faces or forms that were passed down to the male heirs from generation to generation.

(Left) The largest nephrite boulder ever recorded from Europe weighs 2144 kg (4727 lb.) and has been polished on one side by Tiffany & Co. George F. Kunz, enroute to Russia in 1899, heard that nephrite might be found at Jordansmuhl in Silesia (now Jordanow, Poland). He arrived at the quarry at 6 a.m. and breakfasted with the owner, who provided him with a cart and workers. Kunz found the piece, and the quarry owner gave it to him as the discoverer's right. (On loan from The Metropolitan Museum of Art.)

(Opposite) Nephrite pi disc measuring 31 cm (12 1/4 in.) across from Ming Dynasty (1368-1644). The pi is the symbol of heaven and one of the most important ritual jades. At burials, it was placed under the body of the deceased.

Occurrences

BOTH NEPHRITE AND JADEITE JADE ARE METAMORPHIC rocks that result from chemical reactions between serpentinites (serpentine rock) and adjacent or embedded rock, like granite. Jadeite rock is very rare because its formation also requires conditions of considerable depth of burial (high pressure) that are infrequently preserved geologically. The durability of the jades results in their survival as stream cobbles and boulders, typically the first finds of a deposit.

Canada's British Columbia is the world's major supplier of nephrite. Here, grayish green nephrite is found, mostly along the Fraser River. Taiwan and China now use British Columbian jade.

Alaska, California, and Wyoming are also sources of nephrite. Jade Mountain in Alaska was located in 1886. The deposits are huge, but remoteness and Arctic conditions limit exploitation. South-central Wyoming has the best-quality nephrite in the Western Hemisphere, but the supply is now severely depleted.

In New Zealand, nephrite is found on South Island, both *in situ* and as pebbles and boulders in the streams. The town of Hokitika is the jade center of New Zealand.

Eastern Turkestan (now Xinjiang Province in China) is the oldest known source of nephrite, mentioned in writings from the first century B.C. For millennia, this was the source of China's nephrite. Other occurrences of nephrite are in the Soviet Union, Taiwan, Poland, and India.

Burma is the main source of jadeite and the only significant source of Imperial jade. It may be one of the oldest sources as well, for prehistoric jadeite instruments have been found in the Mogok region, possibly fashioned from pebbles and boulders recovered from the Uru River and other streams. Jadeite *in situ* was not discovered until the 1870s. The mines are nationalized, but miners smuggle much of the better-quality jadeite to Thai border towns. From there, it makes its way to Bangkok and ultimately Hong Kong. Jadeite is newly being smuggled directly to China's Yunnan Province.

Jadeite is also produced commercially in Guatemala and the Soviet Union; some is found in San Benito County in California and near Kotaki in Japan.

Jadite incense burner, 18 cm (7 in.) high, from the reign of Ch'ien Lung, part of a three-piece altar set.

Evaluation

THE DIFFERENCES IN QUALITY AND PRICES OF JADE are great. Color and translucency are the major considerations in evaluating both nephrite and jadeite.

The rarest and most valued color for jadeite is pure, even, and intense emerald green. When this color is combined with maximum translucency and smooth, uniform texture, the stone—known as Imperial jade—commands an extremely high price. Next in value is lavender. A jade piece has a high value if the color is pure, intense, and uniform even if it is almost opaque. Sometimes jade is dyed green or lavender. The dyes are not always permanent, and the green frequently fades.

As a gemstone, jadeite commands substantially higher prices than nephrite. Design, craftsmanship, and antiquity are the major considerations in evaluating carvings.

Jade is imitated with mounted jade triplets, glass, and plastic. Many jade substitutes are on the market, and many other carving materials are readily confused with jade. Serpentine is probably the most common substitute.

JADE SUBSTITUTES AND THEIR TRADE NAMES

Bowenite (gem serpentine):	*Korean or Immature jade*
Amazonite feldspar:	*Amazon and Colorado jade*
Varieties of green grossular garnet:	*Transvaal jade*
Aventurine quartz:	*Indian jade*
Mixture of idocrase and grossular:	*American jade or Californite*
Soapstone:	*Fukien, Manchurian, or Hunan jade*
Green jasper:	*Swiss or Oregon jade*
Green-dyed calcite:	*Mexican jade*
Chrysoprase:	*Australian jade*

Serpentine vase carved from Mongolian material, 16.7 cm (6 9/16 in.). Serpentine is commonly mistaken for jade.

Quartz

Quartz is a very common mineral and is easily recognized because it is so frequently found as transparent, well-formed crystals. It comes in a number of colored varieties—amethyst, citrine, rose quartz—but the colorless variety epitomizes the popular concept of crystal. The beauty and symmetry of the pointed "hexagonal" crystals and their water-clear transparency captivate the eye. It is no wonder that this natural gem has had great significance in many cultures throughout human history. Quartz crystals are among the earliest talismans; beads and seals were the first crystalline objects to be fashioned, and "gazing balls" with mystic significance are virtually synonymous with rock crystal. Whether as a crystal gemstone or in polished form, quartz can be found in the earliest prehistoric grave or the most modern collector's cabinet. The recent "New Age" attention to transcendental perceptions about quartz revives an ancient tradition.

The rock crystal statue of Atlas holding up the world is 12 cm (4 5/8 in.) in height. It was carved during the last century in Russia from a crystal found in the Ural Mountains.

ВЕСТЪ ИНДІЯ
ЦЕНТР.
АМЕРИКА
ЮЖН. АМЕРИКА
АТЛАНТИЧЕСКІЙ ОКЕАНЪ

Properties

THE WIDESPREAD AVAILABILITY (AND THUS MODERATE cost) of large, clear pieces in an array of colors provides quartz's appeal as a gemstone; otherwise, it has low brilliance and fire. Quartz is a remarkably pure mineral, but its coloration does require chemical impurities, although only a little—less than one impurity per thousand silicon atoms. Also, irradiation by natural or artificial means is necessary to produce both amethyst and smoky to black quartz; a large amount of the

(Left) Found in McEarl Mine, Hot Springs, Arkansas, this rock crystal measures 13.5 cm (5 1/3 in.) in height.

(Above) Amethyst scepter crystal 4 cm (1 1/2 in.) across perched on milky quartz, from Hopkinton, Rhode Island.

Quartz Data

Silicon oxide or silica:	*SiO_2*
Crystal symmetry:	*Trigonal*
Cleavage:	*None*
Hardness:	*7*
Specific gravity:	*2.65*
R.I.:	*1.544-1.553 (moderate)*
Dispersion:	*Low*

available smoky quartz is artificially irradiated rock crystal. Similarly, citrine, which is rare in nature, is commercially created by heat-treating natural amethyst.

Quartz has a strong framework crystal structure that makes it hard and free from cleavage—a durable material. It is also a common component of dust, which is the abrasive enemy of all gemstones; this is why quartz's hardness is considered the division between soft and hard gems—those softer or harder than quartz.

Inclusions in quartz are responsible for some very interesting varieties with banded or spangly reflections or just color. Fibers are found in rutilated quartz, cat's eye, and sagenite. In hawk's eye, the blue color comes from blue asbestos veins that have been infiltrated and replaced by quartz; if the asbestos breaks down totally, an iron oxide residue imparts the bronze color of tiger's eye. Small particles, fractures, and fluids are responsible for aventurine, iris quartz, and milky quartz.

Quartz crystals are usually elongate hexagonal-looking prisms capped by a "hexagonal pyramid." However, the crystals only have threefold symmetry. This fact is well demonstrated by the three-bladed pinwheel effect in some amethysts.

Quartz, alone with tourmaline among gemstones, lacks a center of symmetry in its crystal structure; this condition makes it piezoelectric. When pressure is applied across opposing prism faces, they develop opposite charges; relaxation reverses the effect. The property has important application in electronics, but there is no substantiated scientific evidence that humans can directly sense electronic vibrations in quartz.

Gemstone Quartz Varieties, Colors, and Color Sources

Rock crystal:	*Colorless*
Amethyst:	*Purple—iron + aluminum + irradiation*
Citrine:	*Yellow to amber—iron*
Morion:	*Black—aluminum + irradiation*
Smoky quartz or cairngorm:	*Smoky gray to brown—aluminum + irradiation*
Rose quartz:	*Translucent pink—titanium or inclusions*
Green quartz, or praziolite:	*Green—iron + heating*

(Above) Amethyst brooch set with diamonds and a 45.71-ct ametrine (trystine), from Puerto Suarez, Bolivia. Ametrine is bicolor amethyst-citrine, discovered in 1977.

(Left) "Pinwheel" amethyst weighing 41.17 cts., from Brazil. It shows quartz's trigonal symmetry.

(Opposite) An amethyst crystal group measuring 7.5 cm (3 in.) across, from Thunder Bay, Ontario, Canada.

Inclusions in Quartzes

Milky quartz: ***White—fluids, mainly water***

Aventurine: ***Green or brick red—chromian mica or hematite flakes***

Rutilated quartz: ***Golden reflecting—rutile needles***

Iris quartz: ***Iridescence—numerous small cracks***

Sagenite* (or *Venus hair, Thetis hair*):** ***A netlike pattern of needles—rutile, black tourmaline, green actinolite, or epidote

Cat's eye: ***Chatoyancy in several color varieties—fibers of rutile***

Tiger's eye: ***Bronze chatoyancy—brown iron oxides from asbestos weathering***

Hawk's eye: ***Blue chatoyancy—blue asbestos***

Historic Notes

*Q*UARTZ DERIVES FROM THE SLAVIC *KWARDY*, MEANing "hard." The Latinized *quarzum* was first recorded in the sixteenth century by the German scholar Agricola, who made the first scientific classification of minerals. According to him, the term was used by the Bohemian miners of Joachimstal (now Jachimov, Czechoslovakia).

Rock crystal objects have been found with remains of prehistoric man (75,000 B.C.) in France, Switzerland, and Spain. Cylinder seals of rock crystal appeared in the Near East by the fourth millennium B.C. As an amulet and a decorative stone, it found use in Egypt before 3100 B.C. The ancient Greeks and Romans used it extensively in jewelry, valuing its flawless transparency. The Greeks believed that the gemstone was water frozen by the gods to remain forever ice; our term *crystal* derives from the Greek *krystallos*, meaning "ice."

During the Middle Ages and Renaissance in Europe, rock crystal vessels were carved for royal and ecclesiastical use, and for centuries the Japanese and Chinese carved the material, as did the Mayas, pre-Aztecs, Aztecs, and Incas on the other side of the world.

Amethyst was used as a decorative stone before 25,000 B.C. in France and has been found with the remains of neolithic man in different parts of Europe. Before 3100 B.C. in Egypt, beads,

(Opposite) F
crystal carvi
nineteenth c
suring 13.2

Tiger's eye quartz carving of a seated Buddah from South Africa, 8 cm high (3 1/8 in.).

Occurrences

QUARTZ FORMS IN A WIDE VARIETY OF ENVIRONMENTS, but gemstone crystals usually require the openings in rocks, such as veins, cavities, and pockets, to grow to perfection and adequate size. Crystal-lined pockets, called geodes, are a familiar source of quartz. In pegmatites containing quartz and sufficient radioactive minerals to provide the necessary irradiation amethyst and smoky quartz will develop.

There are many commercial sources of gem-quality rock crystal. Brazil is the principal source for all varieties of quartz. Arkansas is important for rock crystal in the United States. The most prolific producers of amethyst are Brazil and Uruguay. Amethyst is found also in Arizona and North Carolina. The major commercial source of citrine is Brazil (Minas Gerais, Goias, Esperito Santo, and Bahia).

Evaluation

AMONG THE MANY VARIETIES OF QUARTZ, AMETHYST IS the most expensive. Intense and uniform purple is the most desired. Any flaws diminish the price considerably. Amethyst may be confused with purple sapphire or spinel. It can be imitated with synthetic sapphire or glass. Synthetic amethyst is manufactured in the Soviet Union.

Clarity is *the* factor in evaluating rock crystal. The price is moderate, except for very large flawless pieces. Herkimer diamond, Arkansas diamond, Arizona diamond, Cape May diamond, Alaska diamond, and Cornish diamond are some of the misnomers used for rock crystal. Synthetic rock crystal is used in industry rather than in jewelry. Poorly formed or coated quartz crystals are polished and sold as natural crystals or "points." This practice is misleading; the quartz is natural, but the faces are not.

For rose quartz, the deeper the color and the more transparent the gem, the greater its value. Rose quartz is sometimes dyed, but the dye fades.

The value of both citrine and smoky quartz is determined by clarity and the attractiveness of their colors. Quartz of yellow and golden brown to orange brown is rare. Citrine is often sold fraudulently as topaz, whose color is richer. However, due to its lack of cleavage, citrine is tougher, wears better, and is less expensive.

Group of three intergrown rutilated quartz crystals measuring 6.7 cm (2 5/8 in.) in height, from Itabira, Minas Gerais, Brazil.

Chalcedony & Jasper

These gemstones were as highly prized by our earliest ancestors as they are by today's lapidary hobbyists. The basis of their appeal comes from the literally hundreds of colorful varieties that can be found. Both are made up of submicroscopic quartz grains—thus are varieties of quartz—and owe their bonanza of colors and patterns to included minute grains of other pigmenting minerals. Chalcedony differs from jasper in that its tiny crystals are parallel fibers rather than sugar-like grains. Distinguishing them requires a microscope, although typically chalcedony is banded and translucent. Listing all the varieties of these gemstones would be daunting to a philologist, let alone a mineralogist. Only the most important and well-known varieties of chalcedony and jasper are discussed here.

Carnelian vase carved in China; it measures 10.5 cm (4 1/8 in.) across.

Properties

THE GEMSTONE PROPERTIES OF CHALCEDONY AND jasper are the properties of quartz—good hardness and durability. As essentially superfine grained rocks, the directional properties of crystals, such as symmetry, are not visible to the naked eye.

In chalcedony the tiny quartz fibers form layers in a velvet-like pile. The layers stack up one upon the other, often producing a banded appearance as seen in the best-known chalcedony variety, agate. The fibrous structure imparts substantial toughness, too. Layers can be translucent to opaque and vary from gray to white when they are free of impurities to almost any color when they are pigmented with an appropriate impurity. Except for white layers, porosity is pronounced; the gemstones are easily dyed. Onyx, a black-and-white variety, is naturally rare but is commercially produced by soaking pale agate in a sugar solution and then carbonizing the sugar in sulfuric acid, rendering the gemstone black and white.

Jasper's granular texture makes it tough and generally more opaque than chalcedony, and jasper lacks the other's banding. While commonly red to ochre from iron oxide pigments, jasper can occur in a multitude of colors. Some materials have mixed textures of both chalcedony and jasper juxtaposed in anywhere from millimeter- to centimeter-scale patches. Gemstone varieties that show either or both textures include bloodstone and chrysoprase.

The layered texture of agates, particularly onyx and sardonyx, have made them very popular materials for carving intaglios and cameos. Cameos are usually carved with the white layer in relief and the colored layer as background. In intaglios, the figure is incised through the dark layer to reveal the white layer—or the reverse.

(Right) A polished agate slab from an unknown locality measures 18 cm (7 in.) across.

(Opposite) Carnelian in an Islamic necklace with tassels; a Chinese belt buckle measuring 6 cm (2 3/8 in.) across; seventh-century Merovingian necklace spanning 28 cm (11 in.), found in France near Soissons; and Native American bracelets.

CHALCEDONY VARIETIES

Agate:	***All forms with parallel to concentric banding, transparent to opaque.***
Bull's eye agate:	*Bands form concentric circles.*
Iris or fire agate:	*Iridescent from thin layers of iron oxide crystals.*
Onyx:	*Bands are black and white—popularly miscontrued to be all black.*
Sardonyx:	*Bands are brown to ochre and white.*
Bloodstone* or *heliotrope:	***Plasma with red hematite or jasper spots and blotches.***
Carnelian:	***Translucent red brown to brick red from hematite.***
Chrysoprase:	***Translucent apple green from nickel serpentine.***
Moss agate:	***Translucent light-colored body with black, brown, or green moss-looking to branchlike (dendritic) inclusions, usually dark oxides. "Mocha stone" is moss agate from a source near Mocha in Yemen.***
Plasma:	***Opaque leek to dark green from various green silicate minerals.***
Prase:	***Translucent leek green from chlorite inclusions.***
Sard:	***Translucent light to chestnut brown from iron oxides and hydroxides.***

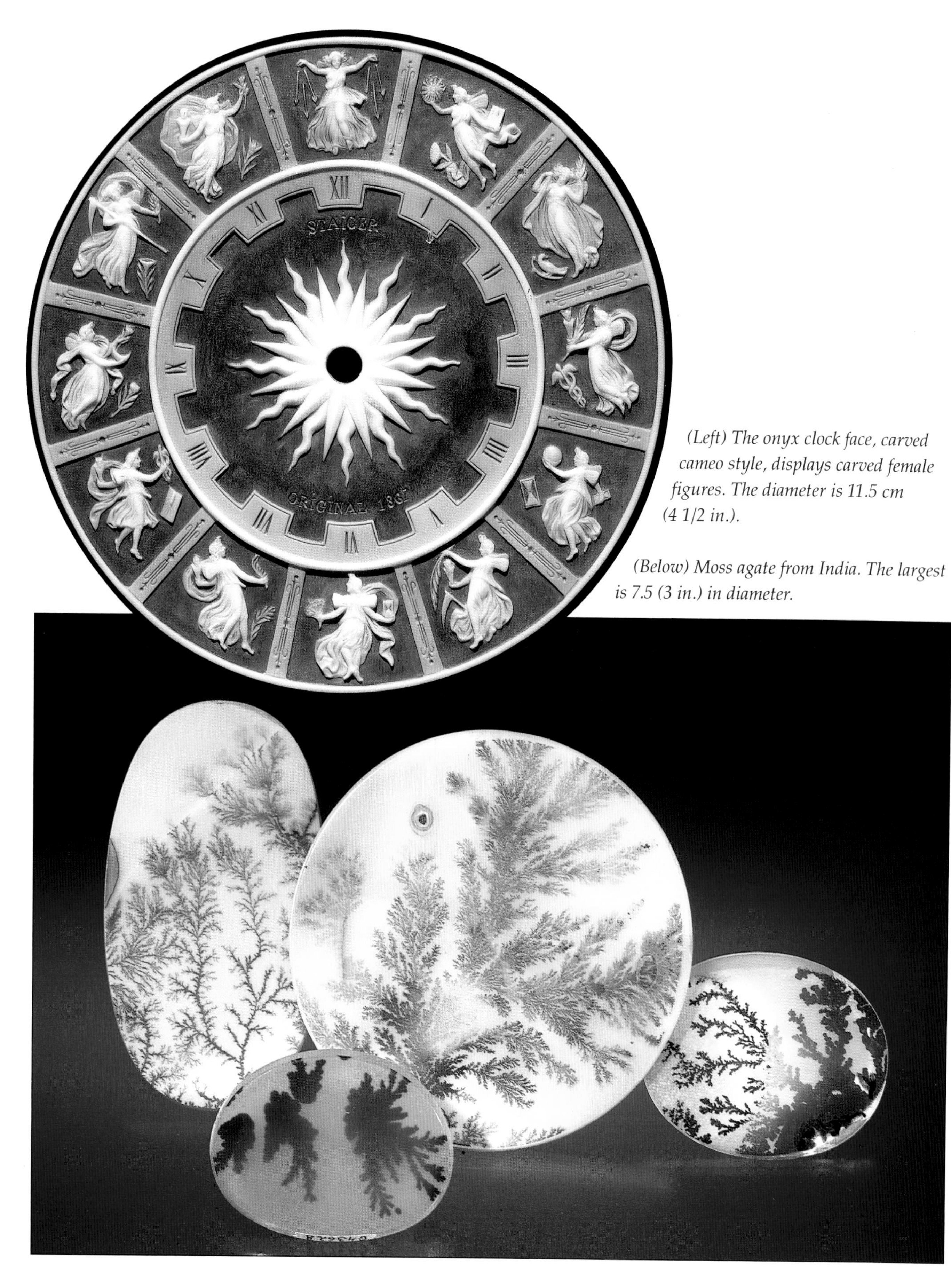

(Left) The onyx clock face, carved cameo style, displays carved female figures. The diameter is 11.5 cm (4 1/2 in.).

(Below) Moss agate from India. The largest is 7.5 (3 in.) in diameter.

Historic Notes

THE TERM *CHALCEDONY* MAY DERIVE FROM THE ancient Greek port of Chalcedon. The terms *chrysoprase* and *prase* come from the Greek *chrysos* and *prase,* meaning "golden" and "leek." Carnelian derives from the Latin *cornum,* meaning "cornel berry" or "cornelian cherry." *Heliotrope* (bloodstone) derives from the Greek *helio,* meaning "sun," and *trepein,* meaning "turning." *Jasper* derives from the Greek *iaspis,* of Oriental origin but unknown significance.

Sard comes from the Greek Sardis, capital of Lydia in Asia Minor. *Agate* is named for the Achates (Drillo) River in Sicily, a major source of the gem, according to Theophrastus. Plasma alone is a use-derived name; the Greek word from which it comes means "something molded" or "something imitated."

The oldest jasper adornments date back to the paleolithic period. Agate has been found with the remains of Stone Age man in France (20,000-16,000 B.C.), and agate, carnelian, and chrysoprase were used by the Egyptians before 3000 B.C. Magnificent agate and jasper jewelry has been found in Harappa, one of the oldest centers of the Indus civilization. Sard was used by the Mycenaeans (1450-1100 B.C.) and the Assyrians (1400-600 B.C.). Carnelian and sard were favorite stones of Roman gem engravers. Carnelian seals have been esteemed by the Muslims; the Prophet Mohammed wore one himself. Prase was used as a gemstone in Greece in around 400 B.C. Mining of agate at around the same time in India has been documented, although the gem was probably used much earlier.

The small German towns of Idar and Oberstein were a source of agate, jasper, and other stones in Roman times, and during the fifteenth century, an agate industry was established there. It flourished until early in the nineteenth century, when the mines were depleted, and many skilled miners and lapidaries went elsewhere. In 1827, German settlers discovered rich chalcedony deposits in Brazil and Uruguay. By 1834, Brazilian agate was being exported to Germany. Although Idar-Oberstein is no longer a supply source, it is renowned for the quality and artistry of its gem craft. Currently, it imports raw material from about 100 countries and employs more than 500 gem polishers and numerous engravers and wholesale gem dealers.

Agate cameo, its rough from Uruguay, measuring 4.7 cm (1 7/8 in.) in length.

Legends and Lore

AS ANCIENT GEMS, THE CHALCEDONIES AND JASPERS have accrued the lore of the ages. Bloodstone preserves an owner's health and protects him or her from deception (Damigeron, first century A.D.). Sard has medicinal virtue for wounds (Epiphanius, fourth-century bishop of Salamis in Cyprus), and protects the possessor from incantations and sorcery (Marbode in the eleventh century). Chrysoprase strengthens the eyesight and relieves internal pain (eleventh-century Byzantine manuscript of Michael Psellius). Carnelian gives an owner courage in battle (Ibnu'l Baitar, botanist of the thirteenth century) and helps timid speakers become both eloquent and bold.

Perhaps the most intriguing virtue of all is noted in Volmar's thirteenth-century *Steinbüch*: a thief, sentenced to death, may escape his executioners immediately—if he puts chrysoprase in his mouth.

Occurrences

CHALCEDONY AND JASPER ARE GEOLOGICALLY COMmon, formed in cavities, cracks, and by replacement where low-temperature silica-rich waters percolate through sediments and rocks, particularly those of volcanic origin. Chalcedonies are common the world around. Brazil, Uruguay, and India produce all varieties of chalcedony and jaspers. With the exception of sard and plasma, all chalcedonies come from diverse localities in the United States.

Additional sources are: chrysoprase—Australia, Zimbabwe and the Soviet Union; carnelian—South Africa and China; agate—Mexico, Namibia, and the Malagasy Republic; jasper—Venezuela, Germany, and the Soviet Union; bloodstone—Australia.

An assortment of jaspers, chalcedonies and other ornamental stones including a heliotrope (bloodstone—green and red cylinder; a banded brown, white and pink jasper cabochon; a carnelian pendant 6 cm (2 3/8 in.) high; a faceted blue chalcedony; an obsidian bowl an onyx intaglio; as well as sodalite, prehnite, turquoise and jadeite cabochons.

Evaluation

THE ATTRACTIVENESS OF COLORS AND PATTERNS determines the value of all varieties. The naturally-colored stones have higher prices than the artifically colored. Chrysoprase is rare and the most valuable variety. Translucency is an important consideration for chrysoprase, carnelian, sard, agate, and prase. (Prase is presently rarely used in jewelry, however.)

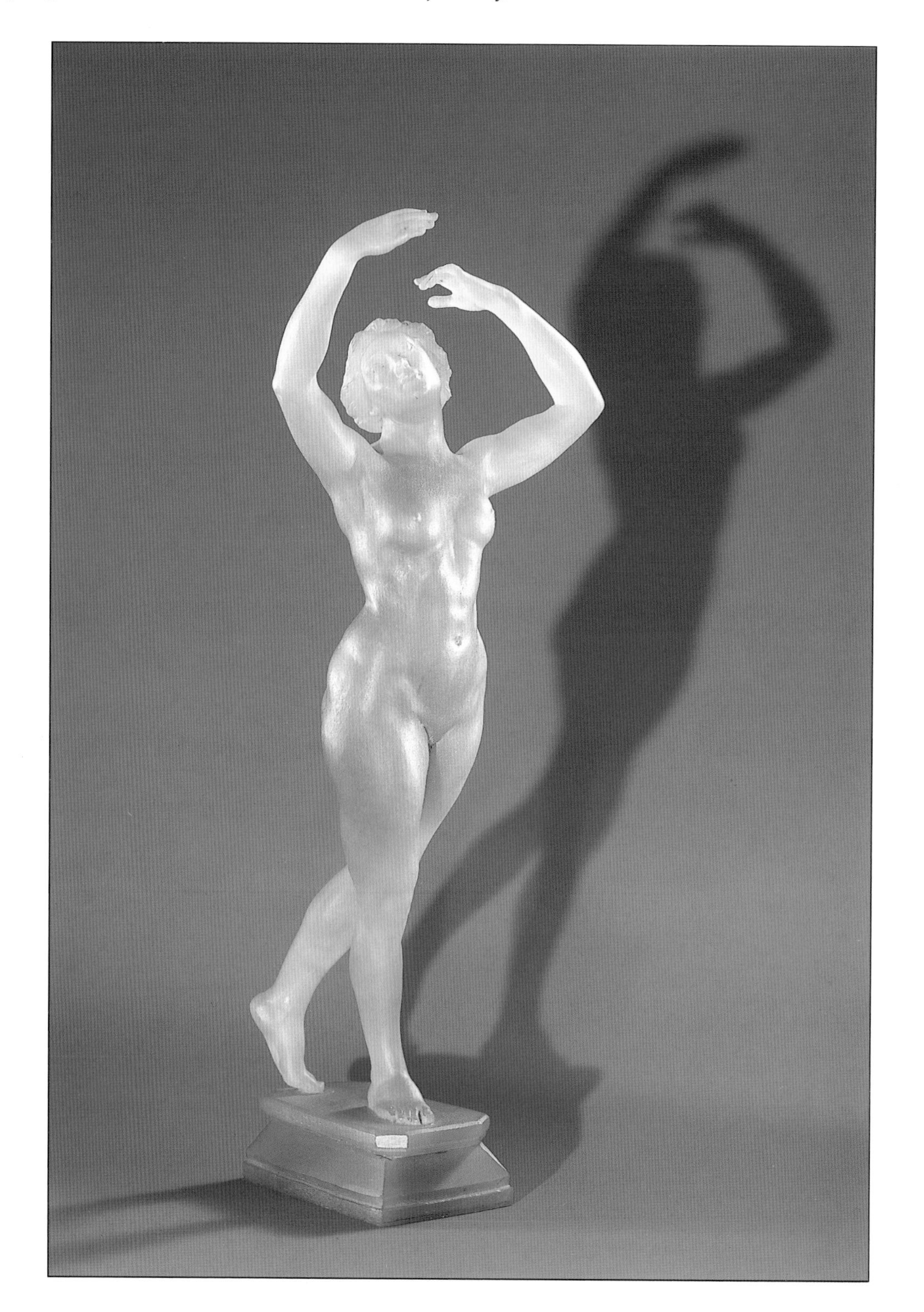

(Right) Pas de Danse, carved by G. Tonnelier, stands 21.5 cm (8 1/2 in.) tall. Its chalcedony rough is from Uruguay.

(Opposite) A chalcedony vase carved in China measures 10.6 cm (4 1/4 in.) high.

Garnet

Garnets are not just red; they come in all colors excepting blues. This is news to many, as may be the fact that new varieties of gem garnet have been discovered in the last few decades. Tsavorite was unearthed in 1968 near Kenya's Tsavo National Park and named for it by Tiffany promoters. It is a beautiful gemstone that rivals emerald but has been in short supply. And about twenty years ago in eastern Africa, a reddish orange garnet was found in the search for the purplish pink rhodolite garnet particularly desired in Japan. Attempts to sell the Japanese on the new garnet were futile, so the stone was called "malaia," a Swahili word meaning "outcast" and "prostitute." To the Africans' surprise, in the late 1970s Americans found the maligned garnet very attractive. But the name has stuck.

Spessartine crystals up to 1.5 cm (9/16 in.) across on quartz from Naugahar Province, Afghanistan, a 28.41-ct. almandine from Tanzania, and an 8.97-ct. round brilliant-cut pyrope from Macon County, North Carolina.

Properties

GARNETS ARE COLORFUL, LIVELY, AND DURABLE—A fine gemstone group, but complex. There are many varieties and many mineral species—like a fruit market with Granny Smith and Delicious apples as well as Concord and green grapes. The colors of the gemstone garnets vary with species as well as with minor substitutions of transition metals into the structure. The iron- and manganese-bearing garnets are intrinsically colored (idiochromatic), whereas those without transition elements are colorless in the pure form (allochromatic). Grossular has the greatest range of colors, and andradite has the highest brilliance and fire—particularly the superb green variety, demantoid. The mistaken concept that garnets are red derives from the predominant use of almandine and pyrope as gems.

Garnet, particularly almandine, can develop fibrous inclusions of several possible minerals in three perpendicular directions; such gemstones can show four- or six-rayed stars when fashioned as cabochons. Garnet's cubic symmetry leads to multifaced equidimensional crystals that can vary greatly in size; some small ones look like natural beads.

GARNET DATA

Garnet is the name of a group of silicate minerals; the gemstone garnets can be described in terms of five limiting members. There is extensive solid solution between minerals listed within each column but not between the columns.

Pyrope:	$Mg_3Al_2(SiO_4)_3$	*Grossular:*	$Ca_3Al_2(SiO_4)_3$
Almandine:	$Fe_3Al_2(SiO_4)_3$	*Andradite:*	$Ca_3Fe_2(SiO_4)_3$
Spessartine:	$Mn_3Al_2(SiO_4)_3$		

Crystal symmetry: *Cubic*
Cleavage: *None*
Hardness: *6.5-7.5*
Specific gravity: *3.5-4.3*
R.I.: *1.714-1.895 (moderate to high)*
Dispersion: *Moderate*

Gemstone Garnet Species, Varieties, Colors, and Sources of Color

SPECIES	VARIETIES	COLORS AND CAUSES
Pyrope		*Colorless, pink to red from iron*
	Chrome pyrope	*Orange red from chromium*
Almandine		*Orangy red to purplish red*
Pyrope-almandine		*Reddish orange to red purple*
	Rhodolite	*Purplish red to red purple*
Spessartine		*Yellowish orange, redder with more iron*
Almandine-spessartine		*Reddish orange to orange red*
Pyrope-spessartine		*Greenish yellow to purple*
	Malaia	*Yellowish to reddish orange to brown*
	Color-change garnet	*Blue green in daylight to purple red in incandescent light due to vanadium and chromium*
Grossular		*Colorless, orange from ferrous iron; also pink, yellow, and brown*
	Tsavorite	*Green to yellowish green from vanadium*
	Hessonite	*Yellow orange to red orange from manganese and iron*
Andradite		*Yellowish green to orangy yellow to black*
	Demantoid	*Green to yellow green from chromium*
	Topazolite	*Yellow to orangy yellow*

(Opposite) A demantoid garnet from Poldenwaja in the Ural Mountains of the Soviet Union, weighing 4.94 cts.

(Right) Spessartine crystal (modified dodecahedron) 1.5 cm (9/16 in.) across on smoky quartz. The specimen was found in Ramona, California.

Tsavorite gem gravel and an 8.16-ct. stone, all from eastern Africa, probably Teita Hills, Kenya.

Historic Notes

Garnet derives from the Latin *granatum*, meaning "pomegranate," and alludes to the crystal's red color and seedlike form. Red garnet gems date back thousands of years. Excavations of lake dwellers' graves in Czechoslovakia have uncovered garnet necklaces and suggest use of the material in the Bronze Age. Other findings indicate widespread use of the gems for beads and inlaid work in Egypt before 3100 B.C., in Sumeria around 2300 B.C., and in Sweden between 2000-1000 B.C. Garnets were the favorite stones in Greece in the fourth and third centuries B.C. and remained popular during Roman times. Garnet-inlaid jewelry has been discovered in southern Russia in graves of the second century A.D. Over 4,000 garnets decorate jewelry that was found when a seventh-century ship burial was excavated in East Anglia in 1939. Aztecs and other Native Americans used garnets in their ornaments in pre-Columbian times.

Pyrope garnets were the basis for a thriving jewelry and cutting center in Bohemia, Czechoslovakia, that started in about 1500. Until the late nineteenth century, the Bohemian deposits were the world's major source of the stone.

Rhodolite garnet from Tanzania, weighing 24.5 cts.

The Names of the Garnets

Pyrope:	*From the Greek* **pyros**, *meaning "fiery" and alluding to the stone's deep red color*
Almandine:	*From Alabanda, an ancient garnet source in Asia Minor (now Turkey)*
Rhodolite:	*Derived from two Greek words that mean "rose stone"*
Spessartine:	*Named for Spessart, the Bavarian district where the gem was first found*
Andradite:	*Named after mineralogist J.B. d'Andrada, who described a variety in 1800*
Topazolite:	*Similar to topaz in color*
Demantoid:	*Derived from the Dutch* **demant**, *meaning "diamond," named for its diamond-like brilliance*
Grossular:	*Derived from the botanical name of the gooseberry,* **R. grossularia**, *alluding to similarity between the colors of the berries and some pale green grossulars*

(Opposite) Nineteenth-century Bohemian jewelry of pyrope garnet. (One bracelet stone is misssing.)

(Above) A hessonite engraved with Christ's head, from the Vatican collection. This piece measures 3.6 cm (1 3/8 in.) in height.

(Right) Engraved almandine garnet bowl from India with a diameter of 5.5 cm (2 1/8 in.).

Legends and Lore

A SINGLE LARGE GARNET PROVIDED THE ONLY LIGHT on Noah's ark, according to the Talmud. During the Middle Ages, garnet was regarded as a gem of faith, truth, and constancy. As late as 1609, Anselmus de Boot contended that garnet drives away melancholy.

Like other red stones, garnet was considered a remedy for hemorrhage and inflammatory diseases and a general protection from wounds, a belief that has been revived among some New Age adherents. In contrast, some Asiatic tribes believed that garnet bullets would be more deadly than those of lead. Accordingly, in 1892, the Hanzas used garnet bullets (some of which have been preserved) against British troops during hostilities on the Kashmir frontier.

Occurrences

GARNETS FORM IN MANY METAMORPHIC AND SOME IGneous rocks. Almandine is the common metamorphic garnet; spessartine is similar, but the purest orangy ones form in pegmatites. Pyrope crystallizes at high pressures. Grossular and andradite form in contact metamorphic zones, particularly next to marble.

Superb rubylike pyropes are found in diamond-bearing kimberlites in South Africa and the Soviet Union. Fine-quality but small pyropes are found in Arizona, New Mexico, and Utah. Other sources are eastern Africa, Australia, Brazil, and Burma.

Major sources of almandine are India, Sri Lanka, and Brazil. Star stones are found in Idaho and India.

Rhodolite, more transparent than pyrope and almandine, was originally found in Lower Creek, North Carolina, in 1882. The major commercial source is Tanzania; rhodolite is also found in India, Sri Lanka, Zimbabwe, and the Malagasy Republic.

Gemstone spessartine's rare occurrences include Brazil, Ramona, California, and Amelia Court House, Virginia, the major source in the late nineteenth century.

Malaia, the new variety, comes from the Umba Valley in Tanzania. Color-change pyrope-spessartine is also found in eastern Africa. Demantoid was first found in about 1851 in placers in the Ural Mountains. A newer deposit is in Chukotka in the Soviet Union. Other occurrences are in Zaire; Korea; and Val Malenco, Italy.

Kenya and Tanzania are the only sources of tsavorite. Other sources of gem grossular are Sri Lanka (yellow, brown, pink, red); Asbestos, Canada (yellow, brown to pinkish); and Chihuahua, Mexico (large crystals, seldom transparent). Green, compact, fine-grained grossular containing small black specks resembles jade and frequently is sold for it. Examples are Transvaal jade, found near Pretoria in South Africa, and material from the Yukon Territory in Canada and California. Massive white grossular from Burma is used for carvings and often sold as jade.

Evaluation

PURITY OF COLOR, CLARITY, AND SIZE ARE THE MOST important considerations. Green garnets are the most highly prized, but the market is plagued by poor availability. Demantoid is the most valuable of the garnets; among all gems, it is prized for its beauty and rarity. Emerald green, transparent, flawless stones are extremely valuable. Brownish red garnets are less valuable than the pure red. The price per carat for a fine-quality garnet increases with size.

Gemstones Confused with Garnets

Pyrope and *almandine:*	*Red spinel and ruby. (Pyrope is occasionally marketed as Arizona ruby, Cape ruby, Elie ruby, and Fashoda ruby; all are misleading names.)*
Rhodolite:	Plum-colored sapphire and tourmaline
Grossular:	Emerald, topaz, zircon, or jade
Demantoid:	Green diamond or zircon
Spessartine:	Zircon and grossular garnet

Etched mass of spessartine garnet sprinkled with pyrite and measuring 6 cm (2 3/8 in.) across and a 98.61-ct. cut stone, from Amelia Court House, Virginia.

Pearls & Other Organic Gems

Pearl, amber, coral, and jet share an organic origin and, in that sense, form a gem group.

PEARL

Pearls are finished gems when found. Their beauty has been valued for centuries, but their popularity has varied over time. In 1916, millionaire Morton Plant wanted to purchase a magnificent rope of pearls at Cartier's for his wife. The price was $1 million! He proposed an exchange—a piece of real estate for the necklace—and Cartier agreed. In 1956, these magnificent pearls were auctioned at Parke-Bernet and brought only $151,000. The real estate is the Fifth Avenue landmark building that Cartier still occupies, one of the most valuable corners in New York City.

Chrysanthemum brooch with American fresh-water pearls and diamonds, 7.5 cm (3 in.) across. A Tiffany creation from around 1900, it was exhibited in "Tiffany: 150 Years of Gems and Jewelry" in 1988. (Courtesy of Mrs. Robert S. Weatherly, Jr.)

Properties

LUSTROUS PEARLS ARE PRODUCED BY MOLLUSCS HAVing a nacreous (mother of pearl) lining, in response to foreign irritants such as parasites or sand grains. Layers upon layers of nacre are deposited on the object, forming pearl's onionlike structure. Conchiolin (a horny substance) binds aragonite microcrystals together about the inclusion. The crystals overlap, producing a slightly irregular surface that feels rough when rubbed across the teeth, a reliable way of distinguishing natural and cultured pearls from imitations. The luster is caused by the scattering and interference of light by the concentric layers. The light interference and diffraction produce an overtone color called "orient." Pearls are semitransparent to opaque and are divided into three color groups: white, black, and colored.

Pearl-bearing molluscs inhabit both salt and fresh water. Salt-water pearls are the more highly prized for jewelry. They come principally from oysters of the *Pinctada* genus. The major source of fresh-water pearls are mussels of the genus *Unio*.

Cultured pearls are produced by artificial stimulation of the natural process. Baby oysters ("spat") are cultured in plastic cages in protected waters. After three years, a bead of mother of pearl and a small piece of oyster mantle tissue are inserted into the body of each oyster. Nacre is secreted around the bead by the foreign tissue. The oysters are returned to the culture cages in the sea. After three years, the oysters are recovered and the pearls removed. The largest Japanese seawater cultured pearls have a diameter of about two-fifths of an inch.

Acids, including those from the human skin, are very damaging to pearl, as are cosmetics, perfume, and hair sprays. Excessive dryness or humidity in the air shortens the life of a pearl. Since pearls are soft, jumbling them together in a box with other gems that scratch them is most destructive. Pearl weight is measured in grains rather than carats: 1 grain = 0.25 carat.

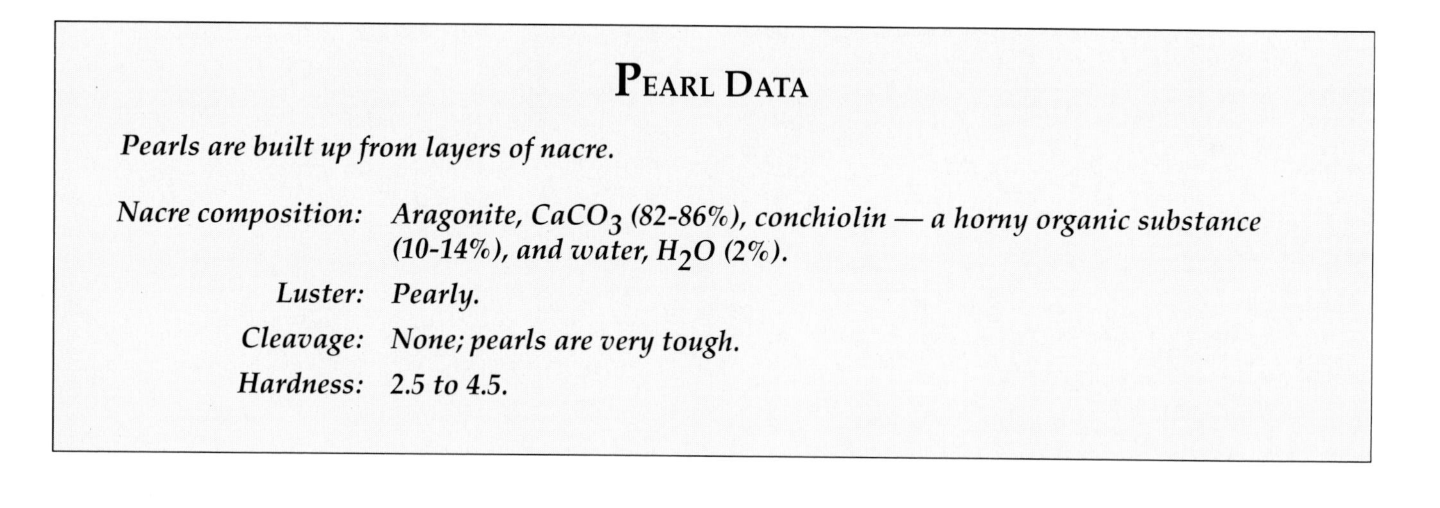

PEARL DATA

Pearls are built up from layers of nacre.

Nacre composition: *Aragonite, $CaCO_3$ (82-86%), conchiolin — a horny organic substance (10-14%), and water, H_2O (2%).*

Luster: *Pearly.*

Cleavage: *None; pearls are very tough.*

Hardness: *2.5 to 4.5.*

Pearl Colors and Shapes

COLOR—*THE SUM OF BODY COLOR PLUS OVERTONE COLOR OR ORIENT.*

White:	*White—white body color, no overtone*
	Cream—cream body color, no overtone
	Light rosé—white body color, pinkish overtone
	Cream rosé—cream body color, deep rose overtone
	Fancy—cream body color, rose and blue overtones
Black:	*Black, gray, bronze, dark blue, blue green, green body colors with or without metallic overtones*
Colored:	*Red, purple, yellow, violet, blue, or green body color—more common in fresh-water pearls*

SHAPES

Round, pear, drop, egg, button (with flat back), baroque (irregularly shaped), blister (attached to shell); seed (unsymmetrical, less than 0.25 grain)

Historic Notes

By 2200 B.C., PEARLS WERE TAX AND TRIBUTE IN China; an Oriental dictionary written before 1000 B.C. specifies pearl as a product of western provinces. Embroidered on costumes and made into ropes and ornaments in ancient Persia (Iran), pearls were the privilege of royalty. In Rome, they were the most admired gems and so frequently worn that the philosopher Seneca (c. 54 B.C.-A.D. 39) criticized women for wearing too many of them.

Throughout the Middle Ages in Europe, pearls were royal gems exclusively, although the Crusaders brought scores of them from the Orient. When Catherine de Medici came to France to marry Henry, Duke of Orleans, in 1532, she brought six strings of fine pearls and twenty-five large single pearls, which she later presented to Mary Stuart (1542-1587). After Mary's death, the Medici pearls were purchased for a trifle by Elizabeth I (1533-1603). In all her portraits, Elizabeth is accoutered with pearls. Men also wore pearls, as demonstrated by royal portraits of the period. In 1612, an edict of the Duke of Saxony stated that the nobility were not to wear dresses embroidered with pearls, and professors and doctors of the universities and their wives were not to use any pearl jewelry. During the late Renaissance, baroque pearl pendants in the forms of dragons, mermaids, and centaurs were popular in Europe. Pearl's prestige continued into the nineteenth century. The gem collection of Dowager Empress Tz'hsi (1835-1908) included thousands of precious gems, many pearls among them, and she owned a cape embroidered with 3,500 pearls, each the size of a canary's egg.

In the early twentieth century, pearl prices remained prohibitive for many, but by the 1920s wealthy American women had ropes of pearls rivaling those of European royalty and Asian potentates. In the 1930s, two events changed pearl's future drastically—the Great Depression and the introduction of cultured pearls.

The spherical cultured pearl was first produced in Japan in about 1907. The creator of the industry is Kokichi Mikimoto (d. 1955), known as "The Pearl King." Fine jewelry stores rejected cultured pearls initially; Tiffany & Co. did not sell them until 1956. Today, although natural pearls may be up to ten times more expensive than equivalent cultured pearls, the natural ones amount to less than 10 percent of the total pearl trade in value.

Legends and Lore

PEARLS, IN INDIAN MYTHOLOGY, WERE HEAVENLY dewdrops that fell into the sea and were caught by shellfish under the first rays of the rising sun during a period of full moon, a belief adopted by Europeans. According to Hebrew legend, pearls are the tears shed by Eve when she was banished from Eden. For the ancient Chinese, pearls represented wealth, honor, and longevity. Pearls were widely used as medicine in Europe until the seventeenth century. The lowest-grade pearls are still ground and used as medicine in the Orient.

Occurrences

COMMERCIAL FISHING FOR NATURAL PEARLS IS essentially nonexistent, but in the past it was important in the Persian Gulf and the Gulf of Mannar between India and Sri Lanka. In addition to Japan, cultured salt-water pearls are produced in Australia and a few equatorial islands in the Pacific, where warmer waters and a larger mollusc permit the growth of larger pearls. The largest and finest pearls, famous for their creamy color and fine pink orient, are produced in Burma's pearl farms.

There are over 100 fresh-water pearl farms in Lake Biwa, Honshu, Japan. Biwa pearls are usually white with fine luster and baroque shape. Fresh-water cultured pearls are also produced in China, Australia, and, most recently, Tennessee.

Fresh-water United States pearls and a shell measuring 9.5 cm (3 3/4 in.) wide and containing a blister pearl and a long hinge pearl.

West Indian emperor helmet shell cameo, "Chariot of the Muses," a late nineteenth-century Italian work.

Evaluation

DISTINCTION BETWEEN NATURAL AND FINE CULTURED pearls can be accomplished conclusively only by X-radiography. Important factors for both are size, shape, color, and orient. The most valued shape is a perfect sphere, followed by symmetrical drop, pear, and button. A perfect pearl is a semitranslucent sphere with even color, fine orient, deep luster, and fine texture. Highly valued colors are white and cream with pink overtones and black with iridescent green orient. Most cultured pearls are bleached. Many are tinted pink and sometimes dyed to imitate naturally-colored black pearls. Occasionally, pearls are irradiated to produce gray, gray blue, and black colors. The colors are permanent.

Pearls in a necklace should be matched in color, luster, and translucency and strung with knots between them to prevent rubbing—also ensuring that only one pearl will be lost if the string breaks.

The thickness of the nacreous layers is important for cultured pearls. Pearls with a coating of less than one-fiftieth inch are considered low quality. Lacquer is often applied to prevent cracking and wear. Cheap imitation pearls are made of plastic or glass beads with a thin coating of synthetic pearl essence. Finer imitations consist of opalescent glass beads dipped many times into a solution of guanine (manufactured from fish scales), then polished and coated with lacquer to prevent discoloration. Majorcan imitation pearls are known for their good quality.

Small lead Buddah figures were implanted in a live fresh-water mussel and became covered with mother of pearl. The shell measures 11.3 cm (5 in.) across.

AMBER

Greek philosopher Thales (sixth century B.C.) noted that, after it has been rubbed, amber attracts lightweight objects. The Greeks termed the substance *elektron*, a word associated with the sun. Thus the Greek name for amber is the word from which words like *electron* and *electric* derive. *Amber* derives from the Arabic *ambar*, meaning "ambergris," a substance obtained from the sperm whale and used in making perfumes.

(Left) Chinese carving of amber 10.9 cm (4 1/4 in.) high, a string of 108 beads from the Baltic Coast, and irregular polished piece from Sicily, 11.5 cm (4 1/2 in.) long.

(Opposite) Amber dress ornament carved in China and of Burmese origin, 7 cm (2 3/4 in.) across.

AMBER DATA

Chemical formula:	*A mixture of hydrocarbons*
Cleavage:	*None, but sometimes brittle*
Hardness:	*2-2.5*
Specific gravity:	*1.05-1.096*
R.I.:	*1.54*
Luster:	*Resinous*
Colors:	*Yellow, brown, whitish or red; occasionally green and blue caused by fluorescence or interference of light by included air bubbles*

Properties

AMBER IS COMPOSED OF FOSSILIZED NATURAL BOTANIC resins of various sorts. Amber is transparent to translucent, often found in sizable pieces, and often contains interesting inclusions—flora and small arthropods trapped by the once-fluid resins. Such fossils date to as early as the Cretaceous Period, 120 million years ago. Amber is soft but relatively tough, capable of being drilled and carved. Its specific gravity is so low that it floats in a saturated salt solution—a quality that distinguishes it from substitutes, which sink.

Some reserve the term "true amber," sometimes called "succinite," for amber from the Baltic region. Baltic amber is derived from various coniferous trees that lived 30 to 60 million years ago. Dominican amber is somewhat younger than Baltic amber and probably derives from a leguminous plant. Dominican amber is also softer than Baltic amber.

Amber 1.3 cm (1/2 in.) wide from Kinkora, New Jersey, containing the oldest known bee from the Cretaceous Period, about 80 million years old.

Classification of Baltic Amber

Clear amber:	*Transparent*
Fatty amber:	*Full of small air bubbles, resembling goose fat*
Bastard amber:	*Clouded because of the presence of many bubbles*
Bone amber:	*White or brown, more opaque than bastard amber*
Foamy or frothy amber:	*Opaque, with a chalky appearance*

Historic Notes

AMBER PENDANTS, BEADS, AND BUTTONS DATING TO 3700 B.C. have been found in Estonia, and amber treasures found in Egypt date as early as 2600 B.C. Amber beads from 2000 B.C. have been found in Crete and Mycenae, and graduated beads are of a similar age in England. In 1000 B.C., the Phoenicians were trading Baltic amber in the Mediterranean region. In Etruria (west central Italy) amber was used in fashioning inlays, beads, scarabs, and small-figure pendants. Amber has been burned as incense since early Christian times.

During the Middle Ages in Europe, the demand for use as rosary beads consumed the available amber. As the supply increased, so did amber's popularity. The skill of amber carving reached a peak in the sixteenth and seventeenth centuries; examples of carved objects include chalices, candlesticks and chandeliers, religious sculpture, and jewelry. During the nineteenth century, amber jewelry was very popular; but attention focused on the intrinsic value of the gem rather than on workmanship. Today, most amber is simply polished to display the gem's natural beauty and warm glow.

Legends and Lore

In Greek mythology, amber was formed when Phaeton, son of Helios, the sun god, was killed by lightning. Grief turned his sisters to poplar trees; their tears were drops of amber.

Occurrences

Ninety percent of the world's gem-quality amber is found along the southeastern shores of the Baltic Sea. Floating "sea" amber from these deposits is dispersed around the Baltic's shores. Most Baltic "pit" amber is mined from blue glauconite sand, called "blue earth." The largest deposits in this area are in the Samland Peninsula near Kaliningrad in the Soviet Union and around Gdansk in Poland. The second most important source is the Dominican Republic. Other occurrences are in Sicily (simetite), Burma (burmite), and Romania (romanite).

Evaluation

The best-quality amber is clear, transparent, and flawless. The most valuable colors are greens, blues, and reds. Of the common colors, yellow is the most highly prized. Pressed amber or amberoid is made by heating small pieces of amber and hydraulically compressing them into blocks. Amberoid is distinguished from true amber by its flow structure and the elongation of air bubbles. Imitation amber is made with plastics, modern natural resins, or glass.

Fossilized ammonite, marketed as ammolite or korite, from northern Alberta in Canada, an 18.3-cm (7 1/8 in.) wide piece and two cabochons of 9.67 cts. and 41.98 cts.

CORAL

The orange-to-red gem often seen as Italian "horn" good-luck charms was long thought to be a sea plant with flowers but no leaves or roots. In 1723, the French biologist G.A. Peyssonel identified coral as the exoskeleton of colonial polyps, small animals that create their dendritic forms from calcite dissolved in sea water. Although coral is a potentially renewable resource, reckless exploitation has placed the corals in jeopardy of extinction. Conservation efforts, initiated in the 1970s, aim at selective harvesting to preserve the gem corals.

Coral Data

Coral is formed primarily of either calcite, $CaCO_3$, or conchiolin, a horny organic substance.

Cleavage:	*None*
Hardness:	*3.5-4*
Specific gravity:	*2.6-2.7 (red calcite coral); 1-3 (black, gold, blue conchiolin coral)*
R.I.:	*Not measurable*

Nineteenth-century Chinese coral carvings, 35.5 cm (14 in.) high.

Properties

RED COLOR IS THE DISTINCTIVE ATTRIBUTE OF THE traditional gem coral, although it ranges from red through orange to pink and even white. Red is due to iron and organic pigments, and fine skeletal structure renders coral opaque. The most valued gem coral is created by the coelenterate species *Corallium rubrum*, a hard coral. Black coral, known as "akbar" or "king's coral," golden coral, and the rare gray blue "akori" are soft corals. The lengthwise striped or patterned skeletal structure to their branches distinguishes all corals from imitations.

Historic Notes

CORAL HAS BEEN FOUND WITH PALEOLITHIC REMAINS in Wildscheuer Cave, north of Wiesbaden, Germany. It was depicted on a Sumerian vase of about 3000 B.C. Coral was popular with the ancient Greeks and Romans. Pliny, writing in the first century A.D., mentions coral trade between the Mediterranean countries and India. In the thirteenth century, Marco Polo noted coral in jewelry and adorning idols in Tibetan temples. Chinese mandarins wore coral buttons of office. The Spaniards introduced coral to Mesoamerica during the sixteenth century, and the Navajo and Pueblos used it extensively in jewelry.

The Victorians favored coral, and it was a favorite of Art Deco jewelers. Today, coral enjoys great popularity, but the supply will probably decrease or even cease if conservation efforts are unsuccessful.

Legends and Lore

CORAL IN GREEK MYTHOLOGY ORIGINATED WITH Medusa's death at the hands of Perseus; the drops of her blood became red coral. Coral amulets were thought to protect children from danger during Roman times. In "Metamorphosis," Roman poet Ovid (c. 43 B.C.-A.D. 17) praised it as a cure for scorpion and serpent bites. Coral has promoted good humor, according to Arab physician Avicenna (A.D. 980-1037). A twelfth-century English manuscript recommends coral engraved with a gorgon or serpent as protection against all enemies and wounds. A medieval English belief was that a coral necklace helps in childbirth. In Italy to this day, coral is worn as protection against the "evil eye."

Occurrences

CORAL GROWS IN CLEAR, SHALLOW WARM WATER AT depths of about 10 to 45 feet. The sources of the finest coral from early times have been the Mediterranean and Red Seas, and the center of the coral industry, Torre del Greco, south of Naples, Italy, is also known as the City of Coral.

Two species of black coral were discovered in 1957 off Maui; this coral is also found off Australia and the West Indies. And a pink coral, previously known from other regions, was discovered in 1966 off Oahu. Gold coral is the rarest of the Hawaiian corals and varies in color from gold, brownish gold, bamboo beige, or brown to dark olive green. Coral is also found in the coastal waters of Japan, Malaysia, Australia, Ireland, and Mauritius.

Evaluation

COLOR, SIZE, AND POLISH DETERMINE CORAL'S VALUE. Pale rose ("angel skin") and deep red ("ox blood") are the most valued, and coral is sometimes stained to produce a more valuable shade. Large pieces are rare, and large, fine carvings command high prices. Necklaces are evaluated by matching color and evenness of the beads.

Coral is imitated with conch pearl, conch shell, and powdered marble compacted under pressure. Plastics, wood, and sealing wax are also used. The Gilson coral, produced in orange to red colors, is an excellent recent imitation.

JET

Though jet has an ancient tradition and was a charm of good fortune, not long ago it was associated with mourning jewelry. Queen Victoria wore it for forty years following Prince Albert's death in 1861, raising jet's popularity to its pinnacle. But by the early twentieth century, fashion changed, and the jet jewelry industry vanished. Today, with black in vogue, jet's popularity has revived.

JET DATA

Composition:	*Carbon plus various hydrocarbon compounds*
Cleavage:	*None, but brittle*
Hardness:	*3-4*
Specific gravity:	*1.3-1.35*
R.I.:	*About 1.66*

Properties

JET IS A DARK BROWN TO BLACK VARIETY OF LIGNITE (derived from the Latin *lignum,* meaning "wood"), a low-grade coal. Jet will burn. It takes a high polish but scratches and abrades easily. It is sufficiently tough to be carved and faceted; a softer, brittle and less "workable" variety is called "bastard jet." Rubbed vigorously on wool or silk, jet develops an electric charge and attracts small pieces of straw or paper. This similarity to amber earned it the name "black amber."

Historic Notes

JEWELRY FROM BRITAIN DATES FROM THE MIDDLE OF the second millennium B.C., and the area around Whitby on the northeast coast of England has been the major source of the finest jet in the world. During Roman times, jet mining in Britannia was active, and a significant amount of jet jewelry made in Eburacum (York) was shipped to Rome. According to Pliny, the material was named for the town and river Gagas in Lycia (Turkey), where it or a similar substance was found.

Jet carving flourished in Spain during the fourteenth and fifteenth centuries, when jet was used in talismans and during periods of mourning. Pre-Columbian Mayas, Aztecs, Pueblos, and Native Alaskans used jet as decoration. The eighteenth and nineteenth centuries saw jet's extensive use in rosaries, crosses, carvings, and jewelry.

A polished jet slab 9.5 cm (3 3/4 in.) long, a faceted jet stone of 2.92 cts., an oval cabochon of 5.26 cts. All from unknown localities. The jet and turquoise frog from Chaco Canyon in New Mexico, measuring 8.1 cm (3 3/16 in.) in height, from the Department of Anthropology at the Museum.

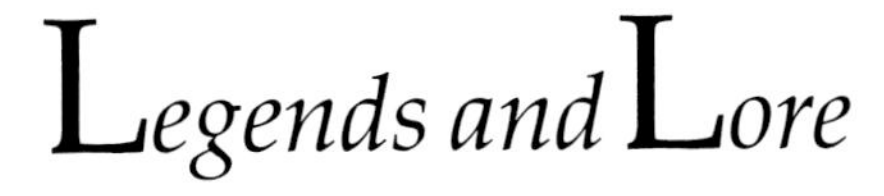

Legends and Lore

JET HAS BEEN CONSIDERED PROTECTIVE, A GEM FOR seafarers. It drives away venomous beasts, according to a book written in 1213 by Ibnu'l Baitar, an Arabian botanist. Since the tenth century, Spanish jet *hijas,* a hand-shaped talisman, has been worn as protection from the evil eye.

Occurrences

JET IS FOUND IN LENTICULAR MASSES EMBEDDED IN hard black bituminous shale, known as "jet rock," where it formed by the lithification of submerged driftwood in sea-floor mud. Jet is found in Germany, Spain, France, Poland, the United States, the Soviet Union, and India, as well as in England.

Evaluation

UNIFORM COLOR AND TEXTURE ARE THE MAJOR FACtors to be considered. The compact homogeneous hard types take better polish and are considered the finest quality. Jet is moderately priced.

GEMSTONES CONFUSED WITH JET, JET SUBSTITUTES, AND IMITATIONS

Obsidian, dyed chalcedony, and black tourmaline can be confused with jet. Scotch cannel coal and Pennsylvania anthracite have been used as substitutes. Imitation jet is made with glass, plastics, and vulcanite (hard vulcanized rubber). Black glass stones are often known as "Paris jet."

Rare & Unusual Gemstones & Ornamental Material

Many minerals—for lack of sufficient abundance, uniformly good properties, or a popular tradition—do not rank among the better known gems. Some are beautiful but not suited for use in jewelry and are mainly of interest to collectors. In this chapter, gem minerals have been segregated from the carving materials; this separates the facetable crystals from the rocks or "ornamental material." These gem minerals, often called "the rare and unusual gemstones," have been sorted into three categories: (1) minerals that have excellent properties but are too rare or are not so rare but have only adequate properties—usually lacking sufficient color or brilliance; (2) gem crystals that are too soft or fragile to be anything other than collectors' stones or part of a museum exhibition; and (3) opaque metallic minerals that have been faceted and fashioned for use in jewelry. The ornamental materials are listed last.

Tanzanite crystal from Tanzania measuring 4.2 cm (1 5/8 in.) in height.

A kunzite crystal 9.8 cm (3 13/16 in.) long and two cut stones of 121.48 and 191.84 cts. from Pala, California.

A 3.57-ct. benitoite gem and a crystal measuring 2 cm (3/4 in.) across, with neptunite crystals in natrolite matrix. Both are from San Benito County, California.

The entries in each group are ranked in descending order of "gem quality," a somewhat subjective evaluation. (See table, pages 200–201, for specific data on each mineral; some rare ones are also included only in this table.) We fully acknowledge our own biases in this ranking.

Very Rare and Moderately Good Gemstones

Zoisite was first described in 1905, and the pink variety, thulite, particularly from Norway, was used as an ornamental stone for cabochons and carvings. In 1967, a new, magnificent intense blue variety from Tanzania was named *tanzanite* by Henry B. Platt, vice-president of Tiffany & Co., the firm that created a market for this gem. The crystals are transparent, sapphire blue to amethyst violet, and very strongly pleochroic. Some tanzanite is heat-treated to eliminate yellow or brown tinges and deepen the blue color. Because of its magnificent color and beauty, tanzanite has become popular as a faceted gem.

Benitoite was discovered in San Benito County, California, in 1907 and recently established as that state's gem, since it is found nowhere else. This rare gem has the color of blue sapphire and the dispersion of diamond.

Spodumene varies in color from colorless to

yellow, yellow green, pink, violet, pale to deep green, and pale green blue. *Kunzite* and *hiddenite* are the two most popular gem varieties. Kunzite is pink, lilac, or violet and was named after the famous gemologist George F. Kunz. Some kunzites fade on prolonged exposure to sunlight. Major sources are California, Brazil, and Afghanistan. Hiddenite is a rare gem, restricted in occurrence almost exclusively to Hiddenite, North Carolina.

Kornerupine can be colorless, brown, and yellow, but green is the most valued color. Cat's eye and star kornerupine are very rare. Gem-quality material was first found in the Malagasy Republic in 1911; eastern Africa is an important source.

Sinhalite was identified in 1952 and named after Sinhala, the ancient Sanskrit name for Sri Lanka, where the gem was found. It also occurs in Burma and Tanzania (pink to brownish pink). Previously, sinhalite was considered as brown peridot. Its typical colors are yellowish brown, greenish, and very dark brown.

Rutile is a common titanium-rich mineral. Its color is usually dark red or reddish brown to black. It is characterized by high refractive indexes and very strong dispersion. Its fire exceeds by six times that of diamond but is usually masked by dark colors. The rarity of transparent material and the darkness of the colors restricts its use to collectors.

Euclase is usually transparent and varies from colorless to assorted colors; blue is the most prized. The name derives from the Greek *eu*, meaning "easy," and *klasis*, meaning "fracture," in allusion to the mineral's perfect cleavage. If the supply were not so limited, it could be a popular gem; Minas Gerais, Brazil, is a major source for gem-quality euclase.

(Above) A kunzite crystal from Nuristan, Afghanistan, 14.5 cm (5 3/4 in.) high.

(Left) A group of color-zoned euclase crystals, the largest of which is 5.5 cm (2 1/8 in.) long, from Zimbabwe; and two cut gems of 7.94 cts. and 8.64 cts., from Minas Gerais, Brazil.

Titanite: a 10.07-ct. stone from Switzerland and a twinned crystal from Austria, 5.5 cm (2 1/8 in.) long.

Titanite, also known as *sphene,* is yellow, brown, or green and in gem quality is transparent. It has high refractive indexes and strong dispersion, giving the well-cut stones high brilliance and fire. It would be an important gem mineral if it were harder and less brittle. Gem titanite is found in the Malagasy Republic, Brazil, and Mexico.

Diopside is a member of the pyroxene group of rock-forming minerals, seldom found in gem quality. Occasionally, it is cut into faceted stones, cat's eyes, and four-rayed star stones. It most commonly occurs in different shades of green. India, China, and New York are important sources of gem diopside.

Obsidian is the most important of the natural glasses for use in jewelry. Obsidian is a transparent to opaque volcanic glass; it is usually black but may also be brown, green, yellow, red, or blue. Occasionally, it exhibits a golden or silver iridescent sheen caused by reflection from tiny inclusions. *Snowflake,* or *flowering,* obsidian is a black variety with white inclusions. *Mahogany obsidian* is a banded black and red variety. *Apache tears* are small rounded pebble-like pieces, usually translucent and light to dark gray in color, found in the American West. Major occurrences are worldwide.

Scapolite is actually a mineral group and may be colorless, pink, violet, yellow, or gray. It makes attractive cat's eyes and faceted stones. Gem-quality scapolite was first found in 1913 in Mogok, Burma, but the Malagasy Republic and Brazil are also sources.

Amblygonite of gem quality is usually yellow, greenish yellow, or lilac. The relative rarity and pale colors of amblygonite restrict its use in jewelry. The major sources of gem amblygonite are Brazil, Burma, and Maine in the United States.

Soft and Fragile Gemstones

Calcite is the most common carbonate mineral and is very abundant. Aragonite is chemically identical to calcite but has a different crystal structure. Faceting calcite is difficult because it has perfect cleavage in three directions. Calcite may be colorless, white, gray, red, pink, green, yellow, brown, or blue. *Iceland spar* is the transparent colorless variety of calcite. Gem-quality calcite is found in many localities, particularly Mexico. *Marble* is a metamorphic rock consisting predominantly of calcite and used for statues and decorative objects. *Onyx marble* is banded calcite and/or aragonite. Its major source is Baja California in Mexico, and hence it is occasionally called "Mexican onyx" or, if dyed green, "Mexican jade."

Fluorite is fragile but occurs in a wide variety of colors. *Fluorescence* derives its name from this mineral, which displays the property vividly. Because of its attractive colors, it is occasionally faceted for collectors. The variety banded in white and blue, violet, or purple—known as Blue John or Derbyshirespar—has been used since Roman times for ornamental objects. Southern Illinois is a major source of fluorite.

Rhodochrosite has been used commercially for decorative objects, beads, and cabochons since a beautifully banded massive material was discovered at San Luis in Argentina before World War II. Rhodochrosite from Argentina is occasionally referred to as "Inca Rose" because the Inca worked the same deposits. Rhodochrosite is also found as deep red crystals, which are occasionally faceted for collectors, particularly those from Hotazel, South Africa.

(Above) A 59.65-ct. rhodochrosite gem from Kuruman, South Africa, the largest on record, and a group of crystals that is 7.0 cm (2 3/4 in.) wide.

(Opposite) Amblygonite cut stone of 34.00 cts. and an irregular crystal of gem-quality amblygonite 8 cm (3 1/8 in.) high from Minas Gerais, Brazil.

Metallic Opaque Gemstones

Pyrite, known as "fool's gold," is the most common sulfide mineral and occurs throughout the world. When used in jewelry, it has been called "marcasite." This is a misnomer; marcasite, though chemically identical, is a mineral with a different crystal structure. Pyrite was used by the ancient Greeks and Romans, Mayas, Aztecs, and Incas. Its popularity revived in the late 1980s. Pyrite is opaque with a brass-yellow color and bright metallic luster.

Hematite, one of the most important ores of iron, is black to dark gray with a metallic luster. It is fashioned into intaglios, cameos, and occasionally beads imitating black pearls or faceted stones often sold as black diamonds. There are many sources; Alaska is now a major one.

Calcite crystal 7.3 cm (2 7/8 in.) long and a 99.6-ct. gem from Gallatin County, Montana.

A 4 1/2-ton block of azurite-malachite 5 feet tall from the Copper Queen Mine in Bisbee, Arizona.

(Above) A polished slice of intergrown azurite and malachite from Bisbee, Arizona. It is 7.5 cm (3 in.) in diameter.

(Opposite) Chinese malachite vase, 20 cm (7 7/8 in.) high.

Ornamental Material—Carvings, Beads, Inlays

Gypsum has three varieties that have been used as ornamental stones since ancient times. *Alabaster* is the massive, fine-grained, translucent variety. *Satin spar* is the fibrous variety with a pearly luster. *Selenite* is the transparent colorless crystal form. Gypsum is very soft and can be scratched with a fingernail. It is usually white, but it may also be yellowish, brownish, reddish, or greenish. The massive variety is porous and is easily dyed. The most important sources of alabaster are Tuscany, Italy, and Derbyshire and Staffordshire, England.

Talc, when free from admixture, is silvery white but with impurities becomes gray, green, reddish, brown, or yellow. It is the softest gem mineral. *Steatite,* a popular material for carvings, is massive talc containing impurities that increase its hardness. It has a greasy, soapy feel and is also known as soapstone. The translucent material has a higher value than the opaque. *Agalmatolite* is a brownish variety of steatite. Steatite occurs at many locations.

Pyrophyllite is a rare mineral occasionally used for cabochons and more frequently for carvings. It is soft and usually opaque, has a pearly to greasy luster and comes in colors varying from white to gray, pale blue, and brown. Translucent material is the most valued. A considerable part of the so-called *agalmatolite,* commonly used in Chinese carvings, is compact pyrophyllite. An important source is China.

Malachite is a vivid green copper mineral widely used for cabochons, beads, carvings, and inlaid work. It was known in Egypt as early as 3000 B.C. and was used for amulets, jewelry, and, as powder, for eye shadow. In the early nineteenth century, the famous Ural mines were highly productive and supplied malachite to Europe. It was worn in Italy as an amulet against the evil eye. Malachite seldom occurs in visible crystals but is usually massive or fibrous. The massive variety, banded in two green hues, is the most attractive for jewelry. Malachite is soft, brittle, and sensitive to heat, acids, and ammonia—not sufficiently durable for ring stones. Currently, the major producer of malachite is Zaire.

Azurite is an intense azure blue, soft and opaque gem. Most often cabochons, beads, or decorative objects are fashioned from massive azurite. Important sources are Tsumeb, Namibia and China.

Rhodonite in its massive form is popular for cabochons, beads, vases, boxes, goblets, and other decorative objects. It was first used in Russia in the late eighteenth century. Rhodonite has an attractive rose red color, usually with black veins of manganese oxides. Major sources are in the Ural Mountains of the Soviet Union and Japan.

RARE AND UNUSUAL GEMS

SPECIES VARIETY	CHEMICAL FORMULA	HARDNESS	SPECIFIC GRAVITY	REFRACTIVE INDEX	COMMENT
VERY RARE GEMSTONES					
Zoisite	$Ca_2Al_3(SiO_4)_3(OH)$	6–6.5	3.15–3.38	1.685–1.725	
Tanzanite—a beautiful blue pleochroic variety rare					
Thulite—a massive pink ornamental stone					
Benitoite	$BaTiSi_3O_9$	6–6.5	3.64–3.7	1.757–1.804	sapphire-blue, strong fire, very rare
Spodumene	$LiAlSi_2O_6$	6.5–7	3.0–3.20	1.660–1.676	good cleavage in two directions
Kunzite—lilac colored, some popularity					
Hiddenite—emerald green, exceedingly rare					
Kornerupine	$Mg_3Al_6(Si,Al,B)_5O_{21}(OH)$	6–7	3.28–3.35	1.661–1.699	green to brown and rare
Rutile	TiO_2	6–6.5	4.2–4.3	2.62–2.90	dark and somewhat soft
Sinhalite	$MgAlBO_4$	6.5–7	3.47–3.5	1.665–1.712	mistaken for brown peridot
Euclase	$BeAlSiO_4(OH)$	6.5–7.5	3.05–3.1	1.650–1.676	one perfect cleavage; rare
Titanite	$CaTiSiO_5$	5–5.5	3.44–3.55	1.843–2.11	fine brilliance and fire but soft
Diopside	$CaMgSi_2O_6$	5–6	3.2–3.3	1.664–1.721	green pyroxene, rare in multi-carat sizes
Obsidian	natural glass	5–5.5	2.40	1.48–1.51	usually dark, brittle
Scapolite	$(Ca,Na)_4Al_3(Al,Si)_3$-$Si_6O_{24}(Cl,CO_3,SO_4)$	5–6	2.50–2.74	1.539–1.579	in various colors and cat's eyes
Amblygonite	$(Li,Na)AlPO_4(F,OH)$	5.5–6	3.0–3.1	1.578–1.619	various pale colors and rare
SOME COLLECTOR GEMSTONES					
Calcite	$CaCO_3$	3	2.70	1.486–1.658	high birefringence, soft; perfect cleavage in 3 directions
Iceland Spar—optically flawless colorless calcite					
Fluorite	CaF_2	4	3.18	1.434	many colors, soft, octahedral cleavage
Rhodochrosite	$MnCO_3$	3.5–4.5	3.45–3.6	1.97–1.817	soft with 3 cleavages—massive form ornamental stone
Barite	$BaSO_4$	3–3.5	4.3–4.6	1.636–1.648	various colors but soft and 1 perfect cleavage

RARE AND UNUSUAL GEMS

SPECIES VARIETY	CHEMICAL FORMULA	HARDNESS	SPECIFIC GRAVITY	REFRACTIVE INDEX	COMMENT
METALLIC GEMSTONES					
Pyrite	FeS_2	6–6.5	5.02		brassy, called macasite in marketplace
Hematite	Fe_2O_3	5.5–6.5	5.26		steely black, called black diamonds
ORNAMENTAL MATERIALS					
Gypsum	$CaSO_4{\cdot}2(H_2O)$	2	2.3	1.520–1.530	soft and abundant
Alabaster—massive fine grained rock					
Satin spar—fibrous with pearly luster					
Selenite—transparent colorless crystal					
Talc	$Mg_3Si_4O_{10}(OH)_2$	1	2.2–2.8	1.54	soft and greasy feeling
Steatite or Soapstone—massive fine grained rock, often apple green					
Pyrophyllite	$Al_2Si_4O_{10}(OH)_2$	1–2	2.65–2.90	1.58	soft, resembles talc
Agalmatolite—creamy white to brown massive form					
Malachite	$Cu_2CO_3(OH)_2$	3.5–4.5	3.6–4.1	1.85	light to dark green, banded and massive
Azurite	$Cu_3(CO_3)_2(OH)_2$	3.5–4	3.77	1.730–1.836	dark azure blue, alters to malachite
Rhodonite	$MnSiO_3$	5.5–6.5	3.57–3.76	1.73	rose, pink to brownish red, usually massive
Variscite	$AlPO_4{\cdot}2H_2O$	3.5–4.5	2.2–2.57	1.56	massive blue-green mistaken for turquoise
Serpentine	$Mg_3Si_2O_5(OH)_4$	2.5	2.44–2.62	1.56	commonly in soft green rock serpentinite
Bowenite—green jadelike rock (Hardness = 4–6)					

Glossary

Allochromatic Pertaining to color resultant from a mineral impurity, such as minor chemical substitutions or radiation damage.

Alluvial deposit See Placer deposit

Amulet See Talisman

Asterism Chatoyancy in two or more directions giving a starlike appearance in illumination.

Axis (or crystal axis) A reference direction in a crystal that is parallel to symmetry directions or the intersection of faces.

Birefringence is the magnitude of the difference in the R.I.s of birefringent minerals.

Birefringent (or doubly refractive) Having two or three refractive indices (R.I.s) a characteristic of minerals not possessing cubic symmetry.

Brilliance Degree to which faceted gem sparkles and returns light from within; dependent upon cut and refractive index. Synonyms: life, liveliness.

Cabochon A gem cut style distinguished by its smooth convex top and no facets.

Carat The standard unit of gem weight (mass); 1 ct. = 0.2 grams.

Chatoyancy Cat's eye appearance when stone is illuminated. Caused by parallel arrangement of tiny needles within a crystal

Cleavage The tendency of a mineral to break along a plane due to a direction of weakness in the crystal.

Cryptocrystalline Constituted of submicroscopic crystals.

Crystal A solid body having a regularly repeating arrangement of its atomic constituents; the external expression may be bounded by natural planar surfaces called "faces."

Crystalline Having the properties of a crystal: a regular internal arrangement in three dimensions of constituent atoms.

Cubic (crystal system) Defined by three mutually perpendicular axes of equal length—the highest symmetry class.

Dichroism Pleochroism in two directions.

Dispersion The systematic variation of refractive index with color in a substance; colors separate during refraction of white light. It leads to fire in a gem.

Fire Division of colors in a colorless transparent gem such as diamond; due to dispersion.

Gem A mineral (gemstone) that has been fashioned to enhance its natural beauty.

Gemstone A substance that has beauty, durability and rarity and that can be fashioned into personal adornment.

Group (mineral group) A set of minerals that share the same crystal structure.

Habit A characteristic shape of a mineral, either a crystal shape or the shape and style of polycrystalline intergrowths.

Hardness Resistance to scratching; measured from 1 to 10 on the Mohs scale.

Hexagonal (crystal system) Defined by three equal axes lying in a plane and intersecting at 120° angles and a fourth perpendicular axis that is a six-fold rotation.

Idiochromatic Color is inherent and due to some aspect of chemical composition and crystal structure.

Imitation A substance that simulates a genuine gem, although typically applied to glass, plastic and other non-crystalline materials.

Intergrowth A composite of crystals in intimate contact.

Iridescence Color produced by light interference, as in labradorite feldspar.

Luster The manner in which a substance reflects light from its surface; it is affected by the surface's smoothness and the substance's reflectivity.

Magma Mobile molten rock material from which igneous rocks form by solidification.

Mineral A naturally-occurring substance (usually inorganic) that is crystalline and has a composition that can be defined by a simple chemical formula.

Monoclinic (crystal system) Defined by three nonparallel axes where there are only two right angles between the axes and no high-order rotation axes.

Orthorhombic (crystal system) Defined by three unequal mutually perpendicular axes.

Pegmatite (gem) An igneous rock with conspicuously large mineral grains and often enriched with volatile elements in minerals such as beryl (Be), spodumene (Li), topaz (F) and tourmaline (B).

Piezoelectric Capable of producing a surface electric charge when deformed elastically; a property of some minerals without a center of symmetry.

Placer deposit An accumulation of dense mineral grains at the bottom of a sediment pile by the weathering action of a moving fluid such as water (alluvial deposit) or wind.

Play of colors A range of colors seen in a gemstone such as opal when it is viewed from different angles. The phenomenon is due to optical diffraction.

Pleochroism The phenomenon whereby the color intensity or the actual color is different depending on the orientation in which a crystalline substance is observed.

Pseudochromatic Coloring due to physical causes such as dispersion or foreign included particles and internal boundaries

Pyroelectric Capable of producing a surface electric charge when temperature changes; a property of some minerals that do not have a center of symmetry.

Refraction The bending of light (or any wave phenomenon) when it moves between media with different conductive velocities

Refractive index (R.I.) A mathematical constant equal to the ratio of the velocity of light in a vacuum to that in the substance; it determines the angle at which light bends when it enters a substance obliquely.

Rock A consolidated assemblage of grains of one or more minerals.

Rough The raw gemstone.

Schist A metamorphic rock having a subparallel alignment of the principal constituent mica or micalike (platy) minerals.

Simulant A substance used to simulate a gemstone, usually a synthetic material with a similar appearance to the simulated gemstone.

Sixling A twin intergrowth of six crystals that appears to have hexagonal symmetry; a common habit for chrysoberyl.

Specific gravity A dimensionless measure of density (numerically equivalent to the value in grams per cubic centimeter).

Symmetry The correspondence in shape or length of elements of a body; as repeated by a mirror, rotation about an axis, or inversion through a point (center of symmetry).

Synthetic A man-made substance that is identical to a natural one.

Talisman An object, sometimes fashioned and engraved with a symbol, that is believed to provide magical, medicinal, or protective power. Synonym: Amulet.

Tetragonal (crystal system) Defined by three mutually perpendicular axes, two of which are of equal length.

Trigonal (crystal system) Defined by three equal axes lying in a plane and intersecting at 120° angles and a fourth perpendicular axis that is a three-fold rotation axis.

Trilling A twin intergrowth of three crystals that appears to have trigonal symmetry.

Twin (twinned crystal) A nonparallel intergrowth of separate crystals related by symmetry not possessed by the substance.

Variety (gemstone) A named specific color or other quality of a gemstone species, such as ruby for red corundum.

Volatiles (components) In a magma, those materials that readily form a gas and are the last to enter into and crystallize as minerals during solidification.

GEMS, CRYSTALS, *and* MINERALS

Part Two

MINERALS

MINERALS

An Illustrated Exploration of the Dynamic World of Minerals and Their Properties

George W. Robinson, Ph.D.
Earth Sciences Division
Canadian Museum of Nature

Photography by
Jeffrey A. Scovil

Mineral specimens from
the collection of the
Canadian Museum of Nature

■

A Peter N. Nevraumont Book

SIMON & SCHUSTER

New York London Toronto Sydney Tokyo Singapore

SIMON & SCHUSTER
Rockefeller Center
1230 Avenue of the Americas
New York, New York 10020

10 9 8 7 6 5 4 3 2 1

Printed in Hong Kong by Everbest Printing Company through Four Colour Imports.

Library of Congress Cataloging in Publication Data

Robinson, George W. (George Willard), 1946-
Minerals / by George W. Robinson : photography by Jeffrey A. Scovil.
p. cm.
ISBN 0-671-88002-0
1. Minerals. I. Scovil, Jeffrey A. II. Title.
QE372.2R63 1994
549—dc20 94-6344
CIP

This book was created and produced by
Nevraumont Publishing Company
New York, New York

President: Ann J. Perrini

Book design: José Conde
Figure illustrations: Jane Axamethy

The Earth

PART I

and Its Minerals

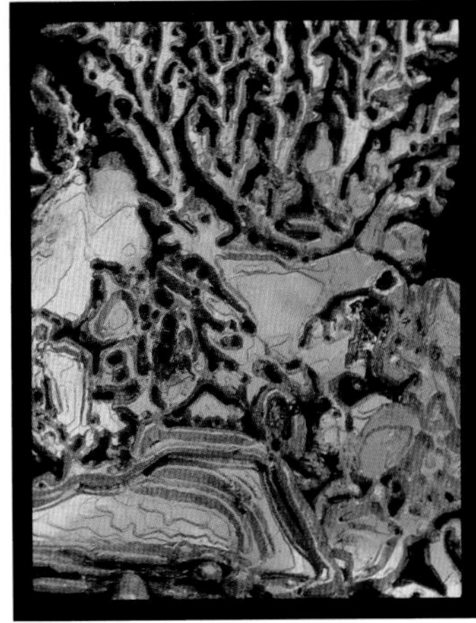

chapter one

Thinking like a Geologist

EVEN in childhood I was fascinated by the natural beauty of minerals and intrigued by their seemingly endless diversity of colors and shapes. Finding and studying these treasures of nature became a lifelong obsession. Later I learned that minerals, along with the familiar landscape of mountains, valleys, plains, and oceans that we all too often take for granted, are the natural consequence of billions of years of interaction between dynamic global forces and the materials that make up the planet. What are these forces and materials, and how do they interact? These are the questions that have enticed and will continue to motivate me and other earth scientists for generations to come. ℭ The devastating blasts of Krakatoa or Mount Saint Helens, cataclysmic earthquakes in Italy or California, seasonal avalanches in British Columbia or the Alps, or periodic floods by the Mississippi or Yellow Rivers are continual reminders of the powerful geological forces at work around us. While we mourn our losses, we also perplexedly fail to learn from them and continue to build our homes and cities on major faults and floodplains in defiance of nature. The geological record provides overwhelming evidence that these processes have been and will continue to be operative for millions of years. ℭ The record also shows that numerous dynamic relationships exist between these forces and the materials that make up the Earth. Study has taught me that the same geological forces that cause earthquakes in San Francisco, volcanos in Iceland, or black smokers along the Juan de Fuca ridge beneath the Pacific Ocean, also provide the heat, pressure, and other requirements for making minerals. How do I know what these requirements and materials are? How can I tell that one particular rock or mineral must have formed in an ancient sea and another in an ancient lava flow? By observation and experimentation. Many geological processes can be observed directly. We can watch mountain streams tumble pebbles as they flow downhill to the plains. We can watch a thin crust of salt form along the edge of a mud puddle and cracks develop as the mud dries up. We can watch a volcano erupt. If I encounter a field full of rounded pebbles at the base of a mountain, I don't suspect that the site is an ancient volcano or mud puddle! Common sense and the law of **uniformitarianism** tell me otherwise.

Simply stated, the law of uniformitarianism says that *the present is the key to the past*. Because streams carry pebbles downhill today, they probably did so yesterday, last week, last year, and even a million years ago. Thus, even if there is no flowing water within a hundred kilometers of the pile of pebbles I came upon, logically I can assume that water once flowed here. Although now it may seem merely to state the obvious, when first proposed by James Hutton in 1785 and systematized by Sir Charles Lyell in the 1830s, the law of uniformitarianism not only revolutionized geology but established it as a hypothesis-testing science, based on observation and reasoning. Seeing is believing. Witnessing lava cool before our eyes to form solid rock provides the direct evidence we need to infer a similar origin for ancient lava flows or cinder cones we may encounter elsewhere.

Some things, such as atoms, gamma rays, or the center of the Earth, we can't see, so we must rely on indirect means to visualize their properties. A bat can't see an insect flying past him in the night sky, yet it is able to locate the insect and track its position with sound waves. Similarly, I can't see atoms, yet by diffracting X rays from a tiny crystal of a particular mineral, I am able to learn how its atoms are arranged internally. I can't see, smell, hear, feel, or taste gamma rays, so when I go to the field in search of radioactive minerals, I bring a Geiger counter that can detect them. The fact that many of the Earth's properties and processes, including some of those that characterize and make minerals, cannot be detected or observed directly by human senses does not preclude their existence.

Even though I can't watch most minerals forming, I can predict how and where they might originate. This is not magic. If I know that a certain kind of cake requires baking at 175°C for 45 minutes, then I also know what kind of oven I'll need to make it, and that a refrigerator won't be required. Likewise, if I know the physical and chemical conditions that a mineral needs to form, then I need only look for possible places in the Earth that provide those conditions. If a temperature of 1,500°C is required to synthesize a particular mineral in the laboratory, then looking for that mineral in rocks that form at lower temperatures would make little sense. Sound, straightforward reasoning such as this, taken a step at a time and tested in the field, is the modus operandi of the geologist.

During the last half of the twentieth century, geophysical data from the study of earthquakes and gravity measurements combined with detailed geological mapping of the Earth's surface and ocean floors has provided

PLATE 1

Nickel-Iron meteorite.
Canyon Diablo, Arizona. 12 x 13 cm. (Geological Survey of Canada specimen)

By studying meteorites such as this one, geologists are able to learn about the composition of the Earth and its history.

PLATE 2

Olivine, variety peridot.
Mogok, Burma (faceted gem, 44.73 ct); San Carlos Indian Reservation, Arizona (necklace and specimen, 7 x 8 cm).

Olivine is a common mineral in the Earth's mantle and in some meteorites.

geologists with a whole new concept of the Earth's structure. We now know that the Earth is layered, as an egg is. It has a relatively thin, light, outer **crust** or "shell," which is underlain by a denser **mantle** (the white of the egg), with an even denser "yolk," or **core.** The solid crust on which we live averages 30 to 50 kilometers thick beneath the continents, but only 6 to 8 kilometers thick beneath the oceans. The mantle, which is also solid, is approximately 2,900 kilometers thick, but the core, with a radius of 3,400 kilometers, has a liquid outer portion. Naturally we know more about the crust than we do about any of the other layers, since we can directly sample and analyze the crust. Some direct sampling of mantle rocks also has been possible because pieces of them are sometimes transported to the Earth's surface by volcanic eruptions, but most of what we know about the mantle and core we have learned indirectly from geophysical measurements and the study of meteorites [PLATE 1].

You are probably now asking yourself, What can we possibly learn about the inside of the Earth by studying meteorites from outer space? To answer that question, I will pose another: How did the Earth form in the first place? Of course, no one knows for sure, but the theory upheld by most geologists today suggests that all the planets in our solar system formed by the coalescence and accretion of matter by gravitational attraction. The first particles to combine were probably very small, but as more accumulated, the increase in gravitational attraction due to increased mass attracted larger pieces of matter, such as meteorites and small would-be planets forming at the same time. In fact, the moon probably formed when one such "planet" collided with the Earth, breaking off a section that eventually became the moon.

Ancient meteor-impact craters on the surfaces of both the moon and the Earth are well documented, and meteorites are continually being found all over the Earth. Is it mere coincidence that two of the main minerals found in mantle rocks, olivine [PLATE 2] and pyroxene, are also those found in stony meteorites, or that the observed velocity of seismic waves traveling through the Earth's outer core matches that expected for a molten nickel-iron meteorite containing some silicon? Is it also mere coincidence that the age of the Earth based on the radioactive decay of lead isotopes is the same as that for nickel-iron meteorites: 4.5 billion years?

Much has been learned about the Earth by studying **isotopes.** Isotopes are atoms of the same element with different numbers of neutrons in their nuclei and thus different atomic masses. Some isotopes are unstable and tend to "decay" into more stable forms. To do so they emit energy in the form of radiation at a constant rate. If this rate is known, as it is for many isotopes, then by comparing the amounts of initial isotope and decay products present, we can calculate how much time has passed since the decay began. This process is the basis for the radiometric dating of rocks. In addition, as radioactive isotopes of potassium, uranium, and thorium in the Earth's interior decay into more-stable products, the energy they emit is converted into heat. Inside the Earth, temperature increases with depth, possibly reaching 5,000°C in the core. This increase in temperature with depth is known as the **geothermal**

gradient and averages about 30°C per kilometer in the Earth's crust.

Isotopes are useful geological indicators in other ways as well. Because they have different masses, different isotopes of the same element may partition differently in certain reactions or occur in different ratios in rocks from different parts of the Earth. Increasing knowledge of how various isotopes are distributed throughout the Earth and how they selectively partition from one another during certain chemical reactions is helping geologists interpret many phenomena.

Because heat always flows from hotter areas to cooler ones, within the Earth heat flows from its hot interior to its cooler surface. Much of this heat transfer is accomplished by **convection.** What is convection and how does it work? Here's a simple experiment to try. Prepare an ordinary cup of hot coffee with cream, milk, or whitener and wait a minute or two for motion caused by the stirring to stop. Now, taking care not to cause any motion in the cup, slowly lower a thin, clear ice cube into the coffee. Peer through the ice cube as if it were a glass-bottomed boat and watch what happens. In a few seconds you should see a swirling motion develop beneath the ice cube, which will start to move about, seemingly under its own power. You are observing convection at work. As heat is transferred from the coffee to the ice, it causes the ice to melt. Because the ice water is denser, it sinks toward the bottom of the cup and is replaced by rising, less-dense, hot coffee, which melts more ice, continuing the process. The relative motion set up in the cup by the rising and sinking liquid creates **convection currents** that cause the turbulent flow you see beneath the ice cube. As the convective motion increases, the resulting current is usually strong enough to move the ice cube across the surface of the cup.

A similar situation exists within the Earth. Heat flowing from the core through the mantle to the crust causes convection cells to develop. The crust responds by breaking apart and, like our ice cube, moving about. Of course, the crust moves much more slowly than our ice cube—only a few centimeters a year. This simple concept is the driving force behind what is probably the most important geological theory proposed in modern time: **plate tectonics**●

chapter two

Continents Adrift

THE **plate tectonic** theory proposes that the Earth's crust consists of large **plates** that "float" about its surface because of convective motion through the mantle beneath them. If this sounds a bit far-fetched, don't worry; I was not a believer at first either. After all, for convection to work there must be a flow of materials, and I told you earlier that both the Earth's crust and the mantle beneath it are *solid*. How can a solid flow? The answer lies in *time*, that unique, all-important parameter that separates geology from the other sciences. As humans, we have difficulty believing what we can't see happening. When we toss a stick into a river, we can watch it flow downstream because fluid flow in a river is easily observable within our human time frame and points of reference. A stick thrown onto a glacier, however, even if we stare at it for a month, does not seem to move. But the question remains, did the stick really not move, or did it move so little or so slowly that we couldn't see it happening? ◖ Glaciologists have provided an unequivocal answer. Suspecting that solid ice does indeed flow, glaciologists have proven the point by driving a straight row of posts across the surface of a glacier and marking reference points on land in line with the row at each of its ends. After several months, when the line of posts is surveyed, it is no longer straight, but is bowed outward near its midpoint. With more time, the bow becomes pronounced and visually obvious. The solid ice is flowing, but too slowly for human perception. Given enough time and a slow, steady application of force (in this case, gravity), the solid ice "bends" rather than breaks. Sealing wax, which is also solid, will shatter if given a quick blow by a hammer but will bend, or flow, with a slow, steady application of force, especially if warmed. The same is true of most solid metals. Might not a solid (especially one at 1,200 to 2,500°C), such as the Earth's mantle, also be capable of maintaining a similar slow rate of convective flow?

If you are still a little uncertain about the plausibility of plate tectonics, there is a good deal more tangible evidence supporting the theory. For the plates to "float" they would have to be less dense than the rock beneath them. This is exactly the case. Although the composition of crustal rocks varies widely, in general we find that the continents are **granitic** in composition, while the ocean basins are **basaltic.** Granites contain more lightweight elements, like sodium, aluminum, silicon, and potassium, than do basalts, which contain more of the denser elements calcium, iron, and magnesium. Therefore, the average density of granite is less than that of basalt. Peridotite, which is one of the most common rocks in the upper mantle, has a density greater than that of either granite or basalt. Thus, considering only differences in density, the observed positioning of the continents on top of the oceanic crust, which is on top of an even denser mantle, makes sense.

From deep-sea submersibles scientists have been able to watch submarine volcanic eruptions produce oceanic crustal basalts along midoceanic ridges. Studies of these rocks show that bands of basalts with identical magnetic properties and ages are symmetrically distributed at equal distances on either side of the ridge, with progressively older basalts as one travels farther from the ridge. This compelling evidence for a spreading seafloor is corroborated by calculations indicating that North America and Europe are moving apart at a rate of about one centimeter per year. A quick look at a world map shows that the shapes of opposing continental margins do appear to fit together. Fitting continents back together reveals much more evidence suggesting that they were once joined. Rocks of similar age and structure on opposite sides of the oceans match up, as do rocks containing similar fossil assemblages, glacial features, and magnetic alignment of iron-bearing minerals.

All the evidence points to the conclusion that the continents were formerly joined together and have drifted to their present positions like ships on a sea of moving basaltic plates. But what happens to the continents and plates as they move about? As FIGURE A shows, there are three possibilities: (1) they spread apart, (2) they slide past one another, or (3) one sinks beneath the other. The first scenario produces midoceanic ridges and volcanos, such as those in Iceland. The second produces long fractures along which there is movement, such as the San Andreas **fault** that parallels the California coastline. The frequent earthquakes experienced in California result from the sudden release of energy and fracturing of rock as the plates slide by one another. The third case is exemplified off the east coast of Japan, where a deep-sea trench marks a zone of **subduction.** There a westward-moving oceanic plate **subsides,** or sinks beneath the Asian continent, carrying along with it a good deal of sediment piled on the ocean floor. The deeper the sediment and oceanic crust sink beneath the continent, the hotter they get, until they finally melt. Surrounded by much-denser rock, the less-dense,

FIGURE A

Plate Boundaries.

When two plates meet, they spread apart (e.g., East Pacific Rise), slide past one another (e.g., San Andreas Fault), or pass beneath one another (e.g., west coast of South America).

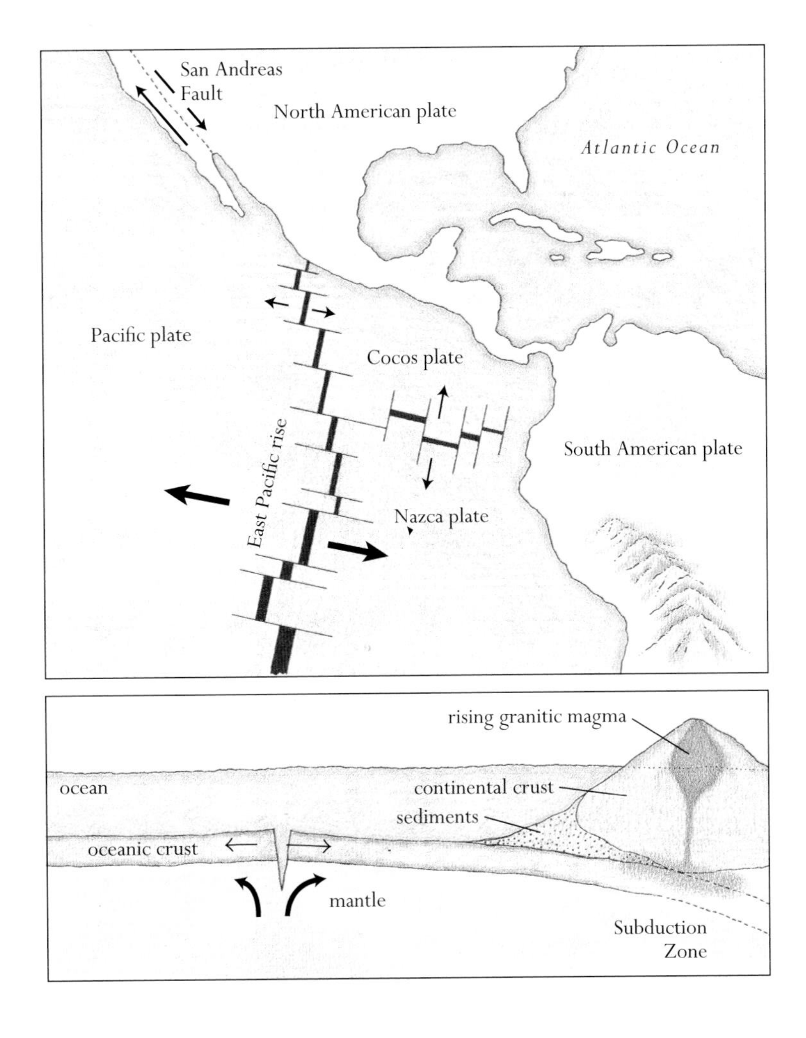

molten rock, known as **magma,** follows zones of weakness back toward the surface, where it eventually may erupt as one of Japan's many volcanos.

Because the Earth's crustal plates ferry the continents about its surface, continental collisions also occur. When two continents collide, huge mountain ranges are formed, as crustal rocks are either folded like an accordion or scraped off one continent and piled atop those of the other. The Himalayas formed by such a collision between India and Asia, the Alps by the collision of Africa with Europe.

Why did I say that the plate tectonic theory was probably the greatest geological discovery in modern time? Because it accounts for and ties together so many different geological phenomena. It explains the cause and global distribution of volcanos and earthquakes, the shape of the continents, why we find fossils of ancient marine organisms high and dry on mountaintops, and even why we find certain plants and animals on one continent but not another. It also provides a rational explanation for the evolution and distribution of many minerals, but to understand why, we must first understand what a mineral is and what is needed to make it●

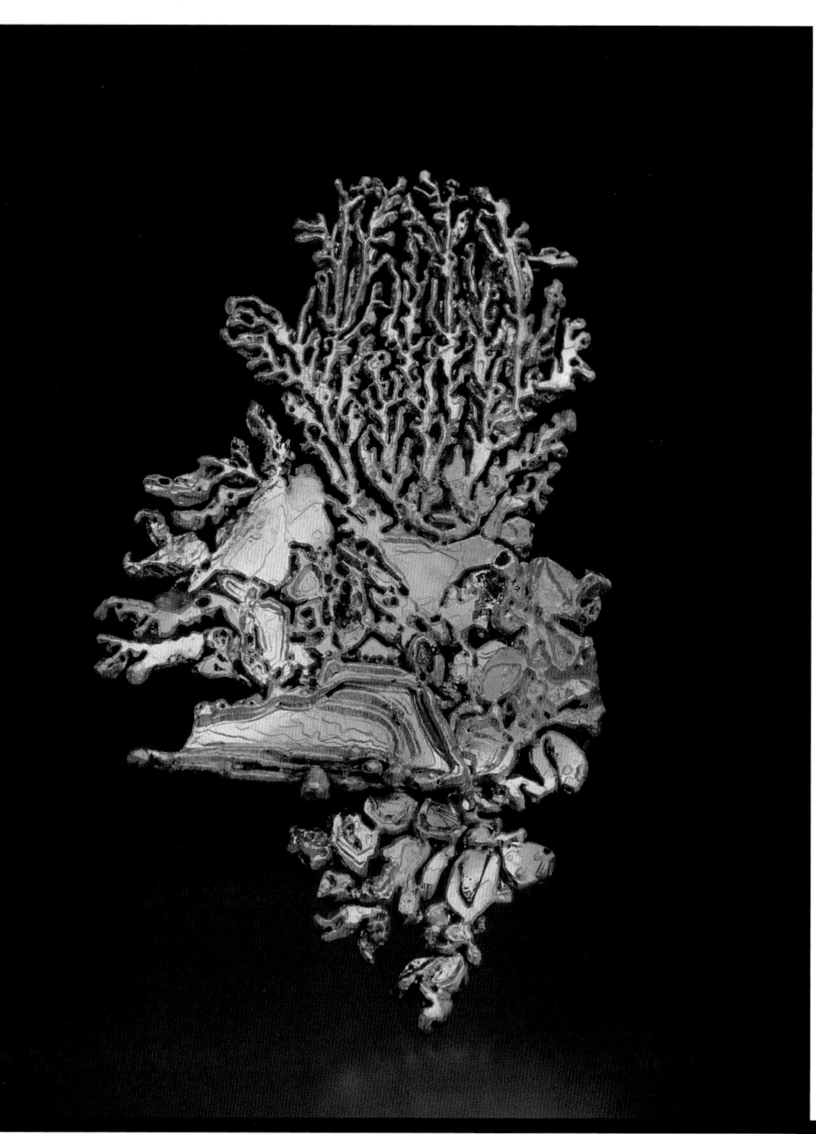

PLATE 3

Gold.
Grass Valley, California. 2 x 3.5 cm.

Gold is one of the few minerals that occurs as a native element. Most minerals are compounds, composed of two or more elements.

PLATE 4

Calcite.
Herja, Transylvania, Romania.
5 x 7 cm.

Calcite is stable over a broad range of temperature and pressure. These spherical aggregates of calcite crystals are only one of its many forms.

chapter three

Minerals

MOST **minerals** are *naturally occurring inorganic solids with defined chemical compositions and crystal structures.* I usually think of them as nature's inorganic chemicals and the Earth as nature's laboratory. A few minerals, such as gold, sulfur, or diamond [PLATES 3, 12, 63, and 135], exist as individual **native elements,** but most are **compounds,** composed of two or more elements chemically bonded together. Like all chemicals, minerals have specific ranges of stability with respect to heat, pressure, acidity, oxidation, and numerous other parameters. If one or more parameters change, some minerals may no longer be stable and will change to form different minerals that are stable in the existing conditions. In that regard minerals are like people. They like to be "comfortable" in their surroundings. No one working in a foundry wants to wear a heavy overcoat, even on a cold winter day, until, of course, it comes time to go outside and walk home. We respond to changes in our environment, and so do minerals. ℭ The plate tectonic model described in Chapter 2 provides many combinations of parameters and the means to change them. For example, at midoceanic ridges there is high heat flow but low pressure; submarine trenches provide little heat but enormous pressure; the collision of continents places dissimilar rocks in contact with one another and provides great lateral pressure; a subducting plate exposes huge areas of rock and sediment to increasing heat and pressure while a rising body of magma created when they melt provides a localized source of heat but not much pressure; and so on. *The wide range of physical and chemical parameters provided by natural geological environments and the availability of specific chemical elements ultimately dictate which minerals will be produced.* ℭ It doesn't matter to a growing crystal of calcite [PLATES 4, 40, 49 and 107] where it forms, as long as its surroundings provide its essential

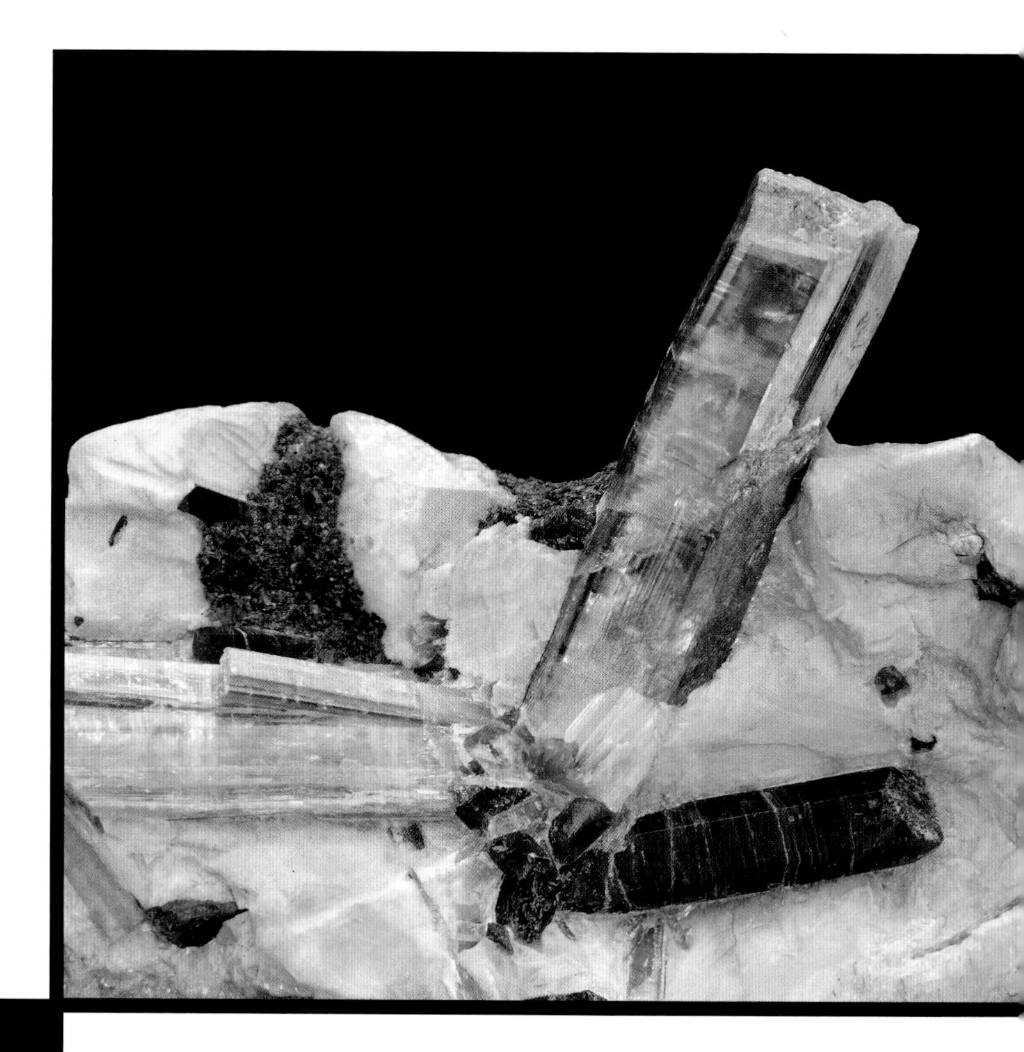

PLATE 5

Kyanite (blue) **and Staurolite** (brown) **in Mica Schist.** Monte Campione, Ticino, Switzerland. 6 x 8 cm.

Because they form exclusively under conditions of high heat and pressure, kyanite and staurolite are used by geologists to interpret the history of metamorphic rocks.

PLATE 6

Mesolite on Apophyllite. Poona, Maharashtra, India. 4 x 6.5 cm.

Sprays of delicate crystals like this mesolite can form only by unimpeded growth into open spaces, such as gas bubble cavities in basaltic lava flows.

ingredients and a stable environment for its growth. Calcite may form equally well in an igneous rock, such as a carbonatite at several hundred degrees Celsius, in a vein in a metamorphic rock like slate at only 200°C, or in a cavity in limestone, a sedimentary rock, at only 20°C. (Igneous, metamorphic, and sedimentary rocks are discussed in Chapter 4.) Calcite has a broad range of stability with respect to heat and pressure. The physicochemical conditions in each of these extremely different geological environments overlap the stability field for calcite, so calcite may occur in any of them. Therefore, by itself, the discovery of a calcite crystal in a rock tells us very little about the geological history of the rock.

Some minerals, however, have more-restricted limits of stability and therefore occupy only a few geological niches. The conditions of high heat and pressure required to form diamonds, for instance, exist only in the Earth's mantle. Stishovite, a rare form of silica (SiO_2), forms only under extremely high pressures, such as realized in rocks that have sustained a meteor impact. Andalusite, sillimanite, and kyanite [PLATES 5 and 112], all of which have the chemical formula Al_2SiO_5, are stable only at specific ranges of pressure and temperature and thus are extremely useful to geologists in deciphering the geological record of rocks that contain them.

Minerals are all around us; the Earth's rocks are composed of them. If minerals are so common, though, why do well-formed crystals seem so rare? Why is it, in spite of hiking over miles of rocks, you have probably never encountered spectacular crystals like those pictured in this book? To understand why this is so, we must look carefully at the underlying mechanism that forms all minerals: **crystallization.** Simply stated, crystallization is the organization of atoms from random into ordered, symmetric arrangements. It is the crystal structure that determines all the properties of a mineral: its shape, color, hardness, and even how it breaks. No two mineral species have the same kind of atoms arranged in the same way.

Most of the magnificent crystals housed in museums and illustrated in this book crystallized in a fluid medium in open spaces. Nonrestrictive surroundings are essential for the growth of large, perfect crystals [PLATE 6]. Time and the availability of essential chemical constituents are also crucial. All other factors equal, large crystals require more time to grow than small ones, but if supplied with the proper ingredients for too long a time, continued growth results in the crystals heading each other off to form a mass of interlocking crystals without well-developed forms. Given all these requirements, it is apparent why large, perfect crystals are scarce in nature: in most geological settings their essential constituents are seldom supplied in the right amount, at the right rate, and at a suitable temperature and pressure with sufficient time or space for them to develop ●

PLATE 7

Azurite (blue) **and Malachite** (green).
Bisbee, Arizona. 6 x 8 cm.

Azurite and malachite form as a result of the oxidation of earlier-formed copper-bearing minerals such as chalcopyrite [see PLATE 8].

PLATE 8

Chalcopyrite and Calcite.
Groundhog mine, Vanadium, New Mexico. 4 x 5 cm.

These crystals formed in open space by precipitation from an aqueous solution.

chapter four

The Rock Cycle

ALTHOUGH in general we can say that the genesis of minerals is controlled by crystallization, this statement does not explain the distribution of minerals within the Earth. Such an explanation can be obtained only by considering crystallization in context with geological processes. Only then does it become clear why certain minerals occur where they do. A dynamic balance exists between constructive and destructive forces within the Earth. Vulcanism and continental collisions build up mountains; weathering and erosion tear them down. In the process rocks, and the minerals that comprise them, are recycled. Even a hard, durable rock like granite is not immune to these forces. ℂ The very moment granite is exposed to the Earth's surface, physical and chemical weathering begin to take their toll. Repeated heating and cooling by the sun causes the outermost surface of the rock to expand and contract, physically weakening it until it splits off (of course, the granite could always fall victim to a quarrying operation, which would greatly accelerate the process). As rain falls on the granite, it chemically breaks down its **feldspars,** the most abundant mineral group in the Earth's crust, into clays, releasing grains of quartz and other minerals. These smaller particles are carried from the mountains to the plains by wind, water, or ice, and eventually by rivers to the sea, where they accumulate as layers of sediment. With continued accumulation, sediments compact under pressure to form **sedimentary** rocks, which are eventually carried back toward the continent by the moving plates. ℂ When these rocks reach the continent, some may be carried downward with the plate as it plunges beneath the continent. As they continue their downward journey, increased heat and pressure recrystallizes both the sediments and the

FIGURE B

The Rock Cycle.

Rocks are continually recycled by geological processes. Magma crystallizes to make igneous rocks, which are broken down and carried as sediments to the sea by weathering and erosion. The sediments compact to form sedimentary rocks, which recrystallize under heat and pressure, forming metamorphic rocks. Continued heating causes melting, which forms a magma, completing the cycle. Water may be driven from the rocks at various points along the way, producing aqueous solutions.

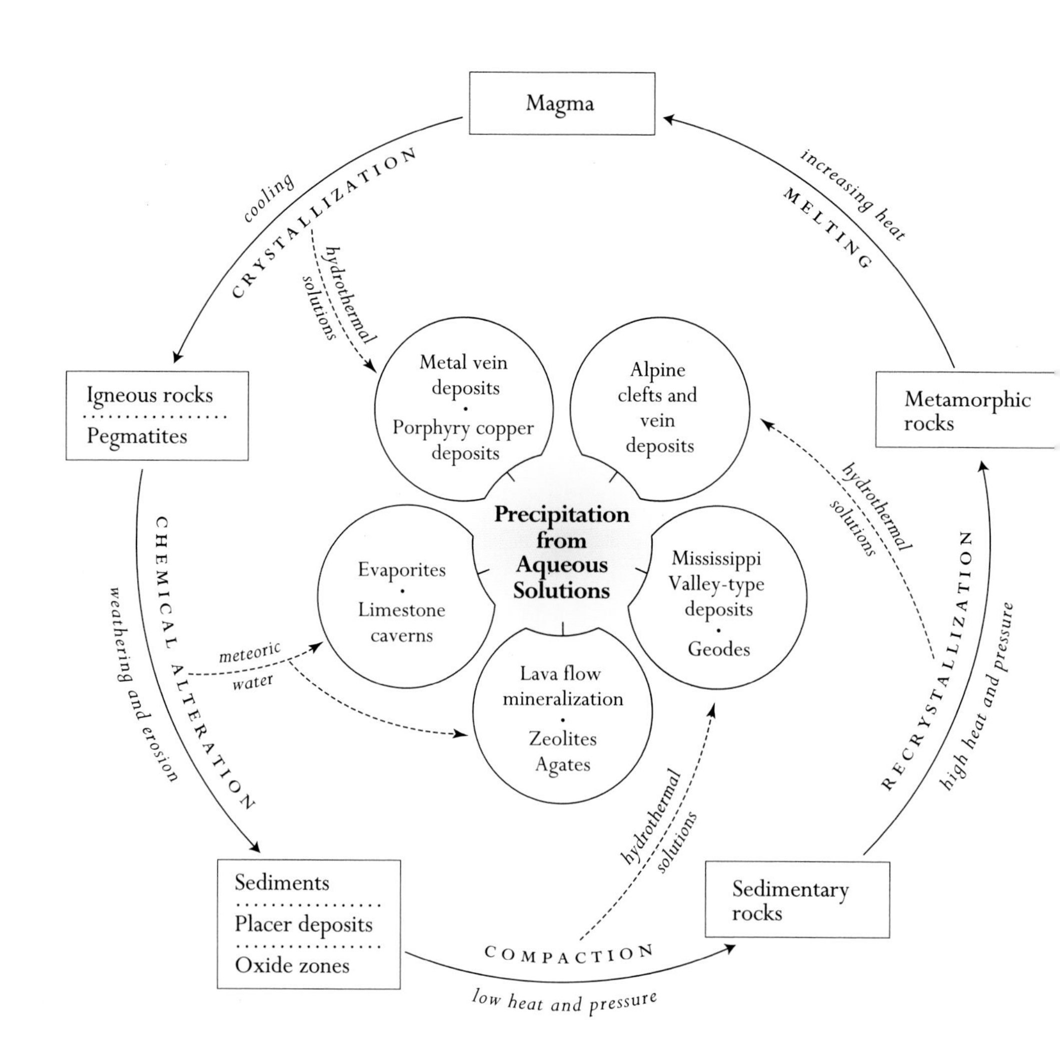

oceanic crust into **metamorphic** rocks. As the rocks descend even deeper, the rising temperature causes them to melt, forming a **magma.** The magma, which is under pressure and is less dense than the surrounding rock, rises upward to form an igneous rock—maybe a granite—as it cools and completes the cycle [FIGURE B].

Of course, there are many shortcuts and side roads that can be taken along the way. For example, igneous rocks may be recrystallized into metamorphic rocks or remelted to form magma without ever having been eroded. Igneous rocks are not the only ones subject to erosion. Sedimentary and metamorphic rocks are equally susceptible and may be recycled directly back to the sea as sediments. Life, too, enters the geological picture. Consider for a moment where each of the calcium atoms in your bones or teeth may have been in the geological past. Perhaps they were once dissolved in an ancient sea, where they were removed by a mollusk and incorporated into its shell as aragonite ($CaCO_3$). When the mollusk died, its shell would have fallen to the sea floor, where it could have been buried for millions of years in the sediment. The aragonite would probably recrystallize into calcite (also $CaCO_3$), while the imprint of the shell remained as a fossil record of former life. As more sediment accumulated, the increased heat and pressure due to burial would have formed limestone. Millions of years later, the processes of uplift and erosion could expose the limestone to the Earth's surface. There it could be discovered and quarried by humans, who might blast it apart and haul it off in a truck to a kiln, where it would be roasted to make lime (CaO) for use as a soil conditioner. Vegetables grown in the conditioned soil could take up some of the calcium, which your body might assimilate when you ate them.

Although these hypothetical calcium atoms in the aragonite, calcite, limestone, and lime represent only a miniscule fraction of the materials that make up the Earth, they illustrate an important point: They may be constituents of very different things depending on the time at which they are observed within an ongoing, natural process of recycling. Eventually, given enough time, the rock cycle is usually completed.

Completing the rock cycle requires that some minerals be created and others destroyed. As FIGURE B shows, the cycle involves four major processes: (1) **crystallization from molten rock,** (2) **precipitation from aqueous solutions,** (3) **chemical alteration of preexisting minerals,** and (4) **recrystallization by heat and pressure.** In the first of these, atoms of different elements are naturally sorted and recombined to form specific minerals as molten rocks cool to form igneous rocks. The minerals that form depend on the composition of the magma and what happens to it as it cools. Minerals such as diamond [see PLATE 12] form deep within the Earth at high temperatures and pressures and therefore are early to crystallize from a magma. Other minerals, such as beryl or topaz [see PLATES 20 and 23], form much later, at lower temperatures and pressures.

The second process centers on the role of water in dissolving, transporting, and recombining various chemical elements to make minerals as different as gold, halite (rock salt), or opals [see PLATES 3, 37, and 41], depending on the source of the water and the kinds of rocks through which it flows. The third process involves specific chemical processes, such as oxidation or changes in acidity, that result in the creation of new minerals like azurite from old ones like chalcopyrite, formed previously by other processes [PLATES 7 and 8]. The fourth process embodies the roles of heat and pressure in restructuring the atoms of previously formed minerals into new arrangements to make various minerals in metamorphic rocks [see PLATE 5]. Some minerals, such as emeralds [see PLATES 20, 120, and 133], may form by more than one process or require combinations of several processes to form. Finally, living organisms also play a role in the creation of some minerals, such as sulfur [see PLATE 135].

The remainder of this book has been organized around the mineral-forming processes that I just described. It is imperative to remember, however, that this scheme of classification is simply one of convenience; it is not inviolable. No definite boundaries have been established between these processes. As with many natural phenomena, often explanations of the genesis of minerals offer more shades of gray than black or white. Many questions have no clear-cut answers. For example, at exactly what point does the heat and pressure

PLATE 9

Goethite.
Bennington, Vermont. 15 x 18 cm.

The formation of these stalactites of goethite required at least two mineral-forming processes: chemical alteration and precipitation from an aqueous solution.

required to compress sediments into sedimentary rocks become enough to call them metamorphic rocks instead? At what point does recrystallization by heat and pressure in a semisolid, plastic state leave off and partial melting to form a magma begin? These points are all arbitrary, and my judgment may differ from that of other geologists. Where the line is drawn is not important. What is important is understanding how each process works and its relationships to the others.

Often more than one process is involved in the formation of a particular mineral. For example, consider the formation of the simple iron-oxide mineral goethite [PLATE 9], which forms the familiar brown rust stains commonly seen on rocks of all types. The first requirement is a source of iron, such as the iron sulfide pyrite [see PLATES 65, 78, and 109]. When pyrite comes in contact with groundwater containing dissolved oxygen, it oxidizes into iron sulfate, which further oxidizes in solution and precipitates goethite (i.e., goethite separates from the solution, since it is insoluble). Operative are at least two processes: chemical alteration and precipitation from aqueous solution. Which process is more important to form goethite? The question is irrelevant because both are essential.

As human beings we tend to see all things relative to our own brief lifetimes. Slow change appears to us as no change and therefore goes unnoticed. Most rocks and minerals change at an imperceptibly slow rate, but they do change! Thus, the specimens illustrated in this book are a snapshot in time, subject to transformation by geological processes. Although they may currently reside in human hands, they did not always, nor will they likely forever remain. To explain the genesis of minerals, a geologist must work like a detective, interpreting clues left behind by the processes that formed them. He or she must look for a motive, opportunity, and means to solve the mystery. In the case of minerals, the motive is their inherent tendency toward chemical and physical equilibrium with their surroundings. Opportunity is provided by the diverse geological environments created by plate tectonics and other geological processes; the means are the four main mineral-forming processes: crystallization (see Part II), precipitation from aqueous solutions (Part III), chemical alteration (Part IV), and recrystallization (Part V) ●

Minerals

PART II

from Molten Rock

chapter six

Gems from the Deep

NOW that we have explored generally how minerals crystallize from molten rock, it's time to turn our attention to some specific minerals that formed that way, one of the most popular of which is diamond [PLATE 12]. (So far, only Superman has succeeded in making it from coal!) Traditionally, primary diamond occurrences have been limited to the relatively uncommon, though widely distributed, igneous rock, kimberlite (named for Kimberley, South Africa, where diamonds have been mined for over a century). More recently, diamonds have been discovered in another kind of igneous rock, lamproite, and they are currently being mined from at least one such deposit at Argyle, in Western Australia. Both kimberlite and lamproite are thought to have formed 150 to 200 kilometers beneath the Earth's surface, in the upper mantle, where the extremely high temperature and pressure required to produce diamonds exist. For years it was assumed that diamonds must somehow form in kimberlitic or lamproitic magmas. Recent studies, however, have produced overwhelming evidence to the contrary, proving that the diamonds did not form in either of these rock types! ◖ Although we cannot directly determine the age of diamonds themselves with present technology, we can date other minerals (e.g., garnet) that crystallize from the magma at the same time and are trapped as inclusions in the diamonds. Relatively new microchemical techniques capable of detecting minute quantities of certain radioactive isotopes that are sometimes present in garnets have shown that garnet inclusions in diamonds from Kimberley, South Africa (and by analogy, the diamonds themselves), are as much as 3.2 billion years older than the kimberlite in which they occur. No way could these diamonds have crystallized from the same magma as the

PLATE 12

Diamond crystal in Kimberlite.
Mir Pipe, Yakut, Russia. 6 x 8 mm.

Although embedded in kimberlite, this diamond crystal did not originally form in this rock.

PLATE 13

Some gemstones that crystallize from magmas.
Top row *(left to right)*:
Peridot. Arizona, 3.5 ct;
Zircon. Cambodia, 37.7 ct;
Peridot. Burma, 16.46 ct.
Bottom row *(left to right)*:
Chromian Diopside. Russia, 3.25 ct;
Diamond. Locality unknown, 1.02 ct;
Ruby. Thailand, 1.61 ct.

Most of these gems contain appreciable amounts of magnesium, chromium, or zirconium, elements common to early-formed magmatic minerals.

kimberlite! An intrusion of kimberlite (or lamproite) simply provides a convenient shuttle service to transport the diamonds, which it incorporates as inclusions along the way, to the Earth's surface. Diamonds probably form in the mantle, as huge slabs of basaltic oceanic crust are carried down zones of subduction at continental margins in accordance with the model of plate tectonics. As the basalt slabs are forced beneath the continental crust, they are subjected to a tremendous change in temperature and pressure, which transforms them into a rock known as eclogite. The wide variation in carbon isotopes of diamonds from eclogites suggests that the carbon that constitutes them may be derived from carbonate-rich sediments and/or organic material dragged downward along with the oceanic crust. Maybe Superman is onto something after all!

Placer deposits are another well-known source of diamonds, although the diamonds clearly do not form in them. Placer mining employs water to separate gems or precious metals from gravels; perhaps the most familiar example is gold panning. Placer deposits result from weathering and erosion. (The diamond deposits around Diamantina, Minas Gerais, Brazil, near the Orange River in South Africa, and in the famous Golconda mines in Maharashtra, India, are all of this type.) At the Earth's surface, peridotites and kimberlites are broken down into soils by weathering. In the process, diamonds and other resistant minerals, such as pyrope garnet and chrominum-rich minerals, that may accompany them are freed. Because these minerals are relatively hard and similar in density to diamonds, they are transported with diamonds by erosion, and are useful as prospecting indicators. The diamonds that have been recovered from placer deposits are some of the world's finest, which according to Darwin's theory of natural selection we might predict, since only the fittest survive the harsh processes of weathering and erosion.

Many other gem minerals form by crystallization from magma. Like diamond, most have high melting points and relatively simple chemical compositions. A few of the more common gems that form this way are some rubies and sapphires, spinel, zircon, pyrope garnet, olivine (peridot), and chrome diopside [PLATE 13]. Also like diamonds, many of these minerals occur as distinct inclusions in volcanic rocks, and many have subsequently been concentrated or relocated in placer deposits by weathering and erosion ●

chapter seven

Carbonatites: Unusual Rocks with Unusual Minerals

CARBONATITES, a group of unusual igneous rocks, often host rare minerals. No one knows exactly how carbonatites form, but most geologists believe they are derived from the mantle. As their name implies, carbonatites are composed chiefly of carbonate minerals, most commonly calcite. This composition sets carbonatites apart from all other igneous rocks, which are composed predominantly of silicate minerals. Another distinguishing feature of carbonatites is that they normally contain relatively high concentrations of strontium, niobium, thorium, and rare earth elements. These elements may combine to form such exotic species as pyrochlore [PLATE 14] and other minerals of interest to both scientists and collectors because of their unusual compositions and rarity. Many are also of economic importance. Niobium is used in manufacturing alloys with great strength and resistance to high temperatures, and both niobium and the rare earth elements are becoming increasingly important in the superconductor industry. Other more-common elements, such as iron, titanium, and phosphorus, have also been mined from carbonatites in the form of magnetite, rutile, and apatite. ❨ Minerals that form in carbonatite magmas share similar crystallization trends with those that form in other magmas. Species with high melting points, such as magnetite and rutile, form before those with lower ones, such as amphiboles or micas. Because carbonatite magmas also may differentiate, certain minerals crystallizing from them may show compositional changes with time, just as pyroxene and olivine do in basalts [see Chapter 5]. Although the actors may have changed, the story line has not. ❨ One of the most mineralogically interesting carbonatite occurrences I have ever visited was exposed by the Francon quarry on Montreal

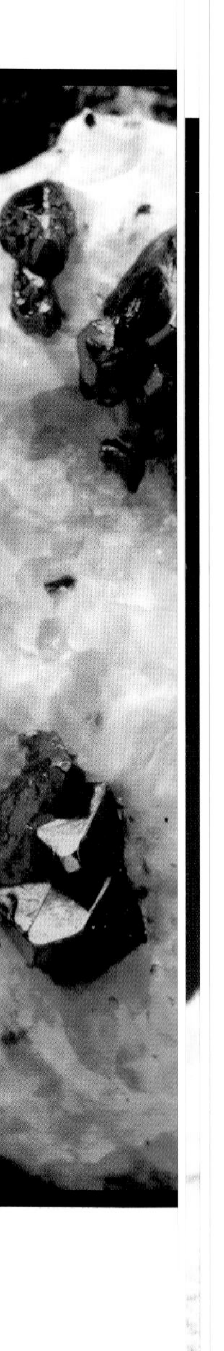

PLATE 23

Topaz.
Kleine Spitzkoppe, Namibia (crystal, 4 x 4 cm); Mason County, Texas (gem, 48.35 ct).

Topaz is one of many gem minerals mined from granitic pegmatites throughout the world.

PLATE 24

Almandine.
Hedgehog Hill, Peru, Maine.
6 x 9 cm.

Perhaps the most common species of garnet in granitic rocks, almandine crystals such as this one have been found in many of New England's pegmatites.

minerals crystallized from it because of the rapid decrease in pressure encountered at the Earth's surface.

Not all rhyolites or granitic pegmatites contain exotic minerals. Like granites, which consist largely of quartz, feldspars, and mica, some granitic pegmatites are mineralogically simple. The type of pegmatite that forms depends largely on two factors: the chemical makeup of the original magma and the depth at which it forms. In general, pegmatites that are derived from granites already enriched in boron, lithium, and rare elements and that form at medium to shallow depths in the Earth's crust show the most complex and diverse assemblages of minerals.

As a genetic group, no other type of deposit is known to host a more varied array of important gem minerals than do the granitic pegmatites. They are the primary source of aquamarine, tourmaline, chrysoberyl, and topaz, as well as significant producers of almandine and spessartine garnets, some emeralds, and a long list of less familiar gems [PLATES 20 to 24]. The range of colors, clarity, superior hardness, and relatively high refractive indices of these minerals make them ideal gemstones. Some are varieties of the same mineral species. For example, aquamarine, emerald, heliodore, and morganite are each a differently colored variety of the species beryl [see PLATE 20]. Similarly, kunzite and hiddenite are the lavender and green varieties of the species spodumene.

Individual gems or crystals of some of these minerals may show two or more zones of different colors. In the tourmaline species elbaite [see PLATE 21], one color variant occurs with sufficient regularity to have earned the name "watermelon" tourmaline because of its pink interior and green rim. The ability of certain elements to induce color in minerals has earned them the name **chromophores.** The color in all these gem minerals is caused by extremely *minor* amounts of contaminant chromophores such as iron, manganese, chromium, vanadium, or copper, but it is their *major* element constituents that define the species. The dramatic color zonation exhibited by multicolored and watermelon tourmaline reflects changes in the availability of chromophores to the growing crystal. Color zonation may be due to crystal structural preferences for specific chromophores or may indicate an abrupt change in fluid composition, as might occur during pocket rupture.

Granitic pegmatites have become so well known for their exotic and gem minerals that we too often forget that they also produce some of the world's finest examples of their primary constituents: feldspars, quartz, and micas. Were it not for the industrial demands for these more common minerals, most pegmatite deposits would never have been worked! Most granitic pegmatites host two different kinds of feldspars: the potassium-rich species, orthoclase or microcline [PLATE 25], and the sodium-rich species, albite [PLATE 26].

Every known color variety of quartz has been found in granitic pegmatites. One of the most common is smoky quartz [PLATE 27]. Granites in general, and their pegmatites in particular, contain above-average concentrations of the

PLATE 25

Microcline, variety amazonite.
Pikes Peak, Colorado. 7 x 12 cm.

The Pikes Peak area of Colorado is famous for its crystals of amazonite, which form in open voids in granite, known as miarolitic cavities.

PLATE 26

Albite (white) **and Microcline** (beige).
Minas Gerais, Brazil. 15 x 18 cm.

Together these two feldspars constitute much of the volume of all granitic pegmatites.

PLATE 27 *(opposite)*

Smoky Quartz and Albite.
Middle Moat Mountain, North Conway, New Hampshire. 4 x 4 cm.

These crystals of smoky quartz and albite were collected from a miarolitic cavity in granite by the author, circa 1966.

PLATE 28

Rose Quartz.
Minas Gerais, Brazil. 8 x 14 cm.

Unlike other varieties of quartz, rose quartz crystals such as these are found exclusively in granitic pegmatites.

PLATE 29

Muscovite.
Conselheiro Penã, Minas Gerais, Brazil. 7 x 13 cm.

Because of the abundance of potassium, aluminum, and silicon in granitic magmas, muscovite is one of the most common micas produced when the magma cools to form a granitic pegmatite.

PLATE 30 *(opposite)*

Molybdenite.
Malartic, Quebec. 1.5 x 2 cm.

Sulfide minerals such as this crystal of molybdenite are usually found in hydrothermal vein deposits rather than in pegmatites.

PLATE 31

Triphylite.
Smith mine, Newport, New Hampshire. 2.5-cm crystal.

Well-formed crystals of triphylite such as this one are rarely preserved in granitic pegmatites because they are chemically attacked by late-stage, residual fluids near the end of a pegmatite's crystallization sequence.

PLATE 32

Monazite.
Joaquim Felicio, Minas Gerais, Brazil. 2.5 x 3 cm.

Monazite, a phosphate of cerium, often contains uranium, thorium and other rare elements typical of granitic pegmatites.

radioactive elements uranium and thorium, as well as some radioactive potassium in their feldspars and micas. With time, the radiation emitted from these elements causes quartz that contains trace amounts of aluminum substituting for some of its silicon atoms to appear a dark, smoky color. Rose quartz [PLATE 28] is also common and sometimes extremely abundant in granitic pegmatites. After much investigation the prevailing theory is that its pink color is due to the presence of titanium or manganese substituting for silicon. Less commonly, citrine quartz and amethyst are found in some pegmatites.

Only three species of mica are common in granitic pegmatites: muscovite [PLATE 29], biotite, and to a lesser extent lepidolite, which is restricted to lithium-bearing pegmatites, and often accompanied by the lithium-rich tourmaline, elbaite. Biotite and muscovite are more widespread. Unlike muscovite, biotite seldom forms pocket-type crystals because in most pegmatites biotite crystallizes before the pocket-forming stage begins.

Many other minerals are found in individual granitic pegmatites. Some are common, others exceedingly rare. Occasionally unexpected minerals appear in granitic pegmatites. For example, excellent crystals of molybdenite occur in a pegmatite near Malartic, Quebec [PLATE 30], and well-formed silicic edenite crystals occur in the McLear pegmatite near DeKalb Junction, New York. The chemical constituents required to form such minerals can usually be traced to the source rocks of the magma or to assimilated rocks. Some pegmatites host significant quantities of phosphate minerals. The most common is fluorapatite, although some pegmatites contain large pods of lithium-, manganese-, or iron-rich phosphates such as amblygonite, lithiophilite, or triphylite [PLATE 31]. Less-common phosphates, such as xenotime or monazite [PLATE 32], may be locally abundant in pegmatites, as are simple oxide minerals such as rutile or cassiterite (tin oxide). Some pegmatites in Western Australia have even been commercially worked for tin●

chapter nine

Agpaitic Pegmatites

NAMED for Agpat, a locality in southern Greenland, **agpaites** are feldspar-rich igneous rocks with unusually high concentrations of alkali metals, especially sodium, and low concentrations of aluminum and silica. Agpaites are often found near carbonatites. In addition to feldspars, the common rock-forming minerals in agpaites reflect their unusual chemistry and often include sodium-rich species such as nepheline and sodalite. Differentiation of agpaitic magma generally parallels that of other magmas, with pyroxenes crystallizing before amphiboles, which crystallize before micas. Because of the high alkali metal content of the magma, however, the specific minerals formed are different. The main pyroxene mineral is the sodium-rich species aegirine [PLATE 33], and the sodium-rich amphibole tends to be arfvedsonite. ◖ By the time an agpaitic magma differentiates to a pegmatite stage, it has usually become enriched with zirconium, as well as titanium, niobium, thorium, strontium, barium, beryllium, and rare earth elements because these elements do not fit easily into the structures of the early-formed minerals. Because of their unusual chemistry, agpaitic pegmatites host a long list of uncommon to exceedingly rare minerals, some of which (e.g., eudialyte) may be potentially useful as a source of zirconium. Other rare species in these rocks constitute a veritable treasure trove for scientists and collectors. Intrusive complexes of agpaitic rocks are known around the world, but the most famous are those in Russia's Kola Peninsula, in southern Greenland, in Norway, and at Mont Saint-Hilaire in Quebec. ◖ Conditions of formation for several species common in agpaitic pegmatites must be similar, for their mutual presence and relative orders of formation in the crystallization sequence appear similar even in widely

PLATE 33

Aegirine.
Mont Saint-Hilaire, Quebec.
4 x 4.5 cm.

This species of pyroxene contains abundant sodium, as might be expected for a mineral commonly found in alkalic rocks.

separated occurrences. For example, when collecting at the famous locality of Narssârssuk in southern Greenland, I was struck by similarities between many of the minerals I found there and those I had seen at Mont Saint-Hilaire, Quebec, which I had visited only a few weeks earlier. At both localities well-formed crystals of similar minerals occur in miarolitic cavities in a gray, nepheline-bearing rock. Frequently the rock coarsens and displays a pegmatitic texture around the cavities. Since crystallization in these cavities must proceed from the outside in, the earliest minerals form at the outer margins of the pockets and the last to crystallize fill the pocket interiors. The major constituents of the pegmatite band surrounding the pocket at both localities are typically potassium feldspar and aegirine, showing that these were the first minerals to crystallize.

Well-formed crystals of eudialyte are also among the early-formed minerals, and the main plagioclase feldspar at each locality is the sodium-dominant species albite, most of which crystallized after the potassium feldspar. I was surprised to find transparent, pale pink albite forming overgrowths on the potassium feldspar at Narssârssuk, since I had previously seen that association on only a few specimens from Mont Saint-Hilaire. These minerals were succeeded by a series of uncommon minerals, such as elpidite [PLATE 36], that are rich in zirconium and other elements whose atoms are of unusual sizes or charges. Among the last minerals to form in these pockets were natrolite, quartz, and various carbonate minerals, such as synchysite [PLATE 35] and calcite. Although my observation of these trends in crystallization at both Narssârssuk and Mont Saint-Hilaire is

purely empirical, their similarity leads me to speculate that such a crystallization sequence may be characteristic of agpaitic pegmatites.

Of all known occurrences of agpaitic pegmatites, no single one has produced as many mineral species (nearly 300 at the time of this writing, 28 of which are new to science) or as many large, well-crystallized specimens as Mont Saint-Hilaire, Quebec, has. No other locality has produced equally splendid rosettes of catapleiite crystals (up to 20 cm) or similarly fabulous crystals of orange serandite [PLATE 34]. Mont Saint-Hilaire is one of the world's most important mineral localities, but why? What unique set of circumstances brought about its mineralogical importance?

Commercial quarrying operations there have removed a large volume of rock, providing collecting opportunities for hundreds of people, and many keenly interested private collectors have brought their finds to the attention of professionals who could identify them properly. Nature has had significant input, too. Three episodes of magmatic intrusion are evident at Mont Saint-Hilaire. Partial assimilation of rock inclusions from earlier intrusions by later intrusions, as well as assimilation of overlying sedimentary rocks diversified the compositions of the magmas. Of key significance, however, is that before it was emplaced, the last intrusion of magma encountered and mixed with a brine, which explains both the enrichment of sodium, manganese, chlorine, and bromine in these rocks and the high volatile content, which was required to produce the explosively fractured rocks that characterize them. Of great importance, too, is that this final magma evolved to a pegmatite stage.

The formation of pegmatites was no doubt facilitated by the addition of volatiles from at least three sources: the original magma, the brine, and the overlying sediments. Finally, the pegmatite stage was able to express itself in more than one way at Mont Saint-Hilaire: as dikes, miarolitic cavities, and infillings in fractured rocks, depending on the quantity of fluids involved, the concentration of volatiles, and the confining pressure and mobility of fluids within the enclosing rock. All these factors combined to produce many small, unique pegmatites, each of which differentiated by the "law of constant rejection" to form the myriad exotic, rare minerals for which Mont Saint-Hilaire is famous ●

chapter ten

Summary of Crystallization in Magmas

WE have examined one important process by which minerals are formed: the crystallization of magma, or molten rock. Most magmas are created when heat from the Earth's interior melts rocks in the upper mantle or crust. As a magma cools, crystallization organizes randomly arranged atoms or groups of atoms in the liquid magma into orderly, symmetric arrangements to form crystals of individual minerals. Those with the highest melting points are the first to form, followed by minerals with successively lower melting points. As different minerals crystallize, they remove specific chemical elements from the melt, changing its composition. The elements removed are those whose atoms are of a size and charge suited to fit the available spaces in the structures of the forming crystals. The continual compositional change in the magma through the selective removal of specific elements by crystallization is known as differentiation. ◖ Atoms of unusual size or charge do not fit into the forming crystals and thus accumulate in the liquid fraction of the melt along with water and other volatile components. The resulting late-stage magmas often produce the most interesting and diverse minerals. The specific minerals that form depend largely on two factors: the initial composition of the magma, and its cooling history. Contamination by assimilation or diffusion, as well as segregation of immiscible components, can alter a magma's chemistry, and too-rapid cooling can halt differentiation. Generally, rocks that host more-interesting and diverse minerals have cooled more slowly, taking more time to differentiate and growing larger crystals. Some of the largest crystals known are found in granitic pegmatites, where their growth probably was promoted by slow cooling, the fluxing action of boron or other elements, and favorable conditions for diffusion due to the presence of volatiles ●

PART III

Minerals and Water

chapter eleven

Dissolution and Precipitation

WATER is perhaps the most important substance on the Earth. It is a unique compound that makes soil out of mountains, drives chemical reactions, and supports life as we know it. Without water, not only would no life exist on the Earth, but probably there would not be as many well-crystallized minerals. As we learned in Part I, minerals are nature's inorganic chemicals, and like chemicals stored in jars in a laboratory, most minerals require water to dissolve them and enable them to react with one another. For example, salt, the mineral halite [PLATE 37], is a compound composed of two elements, sodium and chlorine. In a crystal of halite, the atoms of these two elements exist as **ions.** Ions are atoms that have gained or lost one or more of their electrons and, as a result, have an electric charge. An ion with a positive charge (i.e., an atom that has lost electrons) is known as a **cation.** A negatively charged ion (i.e., an atom that has gained electrons) is called an **anion.** In halite sodium cations are bonded to chloride anions. ◖ Because of their unique shape, water molecules (H_2O) have opposite charges on the ends. Thus, the reason I can dissolve a spoonful of salt in a glass of water is that the water molecules pull the sodium and chloride ions apart. The sodium cations are attracted to the negatively charged oxygen ends of the water molecules, and the chloride anions to the positively charged hydrogen ends. This process of separating the sodium and chloride ions, thereby placing them into solution, is called **dissociation.** Because the solvent is water, the resulting solution may be described as **aqueous.** ◖ The degree to which a substance will dissolve in water—its **solubility**—depends on how strongly its constituent atoms are bonded together. If their own attractive force is greater than that of the

PLATE 37

Halite.
Neuhof, Hessen, Germany.
11 x 14 cm.

As water containing dissolved sodium and chloride ions evaporates, these two elements bond together to form crystals of halite such as these.

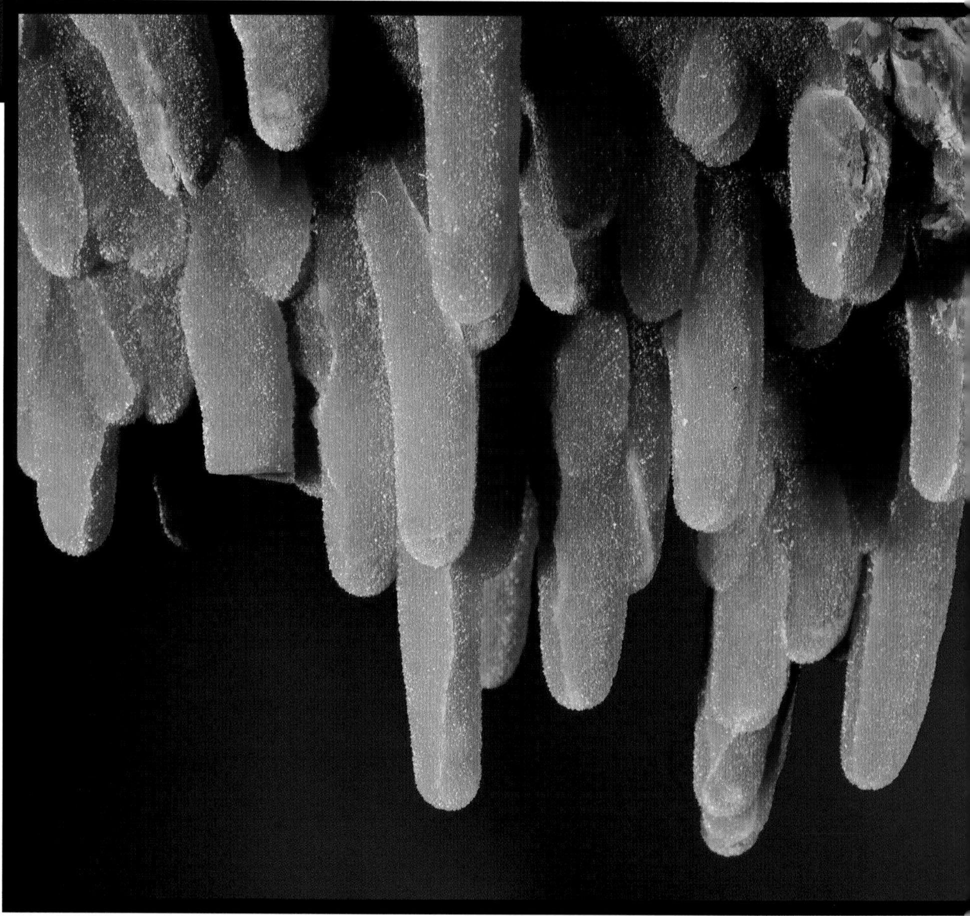

PLATE 39

Colemanite.
Boron, California. 3.5 x 7 cm.

The source of the boron required to make colemanite and other borates in many seasonal lakes (playas) is probably linked to volcanic activity and supplied by nearby hot springs that once flowed into the basins.

PLATE 40

Calcite.
Carter County, Montana. 8 x 10 cm.

The iciclelike form of these calcite stalactites is graphic evidence of formation by precipitation from dripping water.

PLATE 41 *(opposite)*

Opal.
Barcoo River area, Queensland, Australia. 4 x 5 cm.

The play of color in opal results from the diffraction of light by an orderly arrangement of similarly sized spheres of silica.

There are numerous other places where minerals form by dissolution and precipitation. The eerie, cool, dark silence of Carlsbad Caverns, a fairyland of magnificent cave formations, seems about as far removed from the clear, warm, sunlit pools of water at Yellowstone National Park's Mammoth Hot Springs as one can imagine. Yet the same geological process is occurring at each place: the dissolution and precipitation of calcium carbonate. Among the more alluring of geological wonders, limestone caverns are also the product of dissolution by and precipitation from aqueous solutions. As rainwater falls through the air, it dissolves minute amounts of carbon dioxide, producing a weak solution of carbonic acid. Less commonly, acidic water derived from other sources, such as a magma, also may be present. Following fractures and openings in the rock, the acidic water slowly dissolves calcium carbonate (calcite) from limestone, thereby widening its path as it goes. With sufficient time, large openings develop into an intricate system of underground caverns.

In open areas above the water table, contact with air causes some of the water to evaporate. When this happens, the dissolved carbon dioxide is lost from the solution, and calcium carbonate precipitates because it is not soluble in the purer water. When the calcium-charged water drips from an isolated point on the ceiling of the cave, the beautiful iciclelike formations we know as **stalactites** are born [PLATE 40]. Like annual rings of a tree, the concentric bands nearly always visible in cross sections of stalactites attest to their mechanism of growth [see PLATE 139]. At Mammoth Hot Springs the groundwater that forms the pools flows through a large area of limestone before emerging as springs at the Earth's surface. As it trickles over the precipice facing the valley below, the water evaporates, losing its dissolved CO_2 and causing calcium carbonate to precipitate in the form of an elegant terrace resembling a petrified, white waterfall.

Like calcite, quartz (or silica, SiO_2) is nearly insoluble in pure water at normal surface temperatures but becomes more soluble in warmer and more-alkaline solutions. Silica makes up much of the porous, white mineral matter that accumulates around geysers and hot springs and is the major constituent of most petrified wood, but perhaps the most intriguing and valuable form of silica is opal [PLATE 41]. Chemically, opal is $SiO_2 \cdot nH_2O$, where *n* denotes a variable amount of water that can range up to about 20 percent by weight. The most important deposits of precious opal, such as those in Australia, form when silica is dissolved from SiO_2-rich sediments by groundwater as a result of chemical weathering. The groundwater trickles downward until it reaches an impermeable layer, where it accumulates in cracks and cavities. In dry seasons the supply of water is cut off, and the silica separates as submicroscopic spheres that accrue in a gel-like suspension. As long as the rate of evaporation is fairly constant, all the silica spheres reach about the same size. As they pile up into layers, the spheres form an orderly array, as would equal-sized glass marbles if dropped into a bathtub of water. A small amount of water is usually trapped in the spaces between the spheres, but as evaporation continues, the whole mass eventually solidifies to form opal.

PLATE 42

Barite.
Frizington, Cumbria, England.
7 x 16 cm.

Barite, together with calcite, aragonite, and other minerals, is often found in the iron deposits in western Cumbria, England, since all these minerals form by the same process as the iron deposits: precipitation from aqueous solutions.

PLATE 43

Manganese Oxide dendrites on Limestone.
Solnhofen, Bavaria, Germany.
12 x 14 cm.

Like a miniature river delta, groundwater containing dissolved manganese randomly spread along a flat, planar surface in this limestone, leaving behind a trail of manganese oxide when it dried up.

PLATE 44

Rhodochrosite.
Hotazel, Cape Province, South Africa. 2 x 2.5 cm.

The introduction of carbon dioxide to water containing dissolved manganese was probably responsible for the formation of these crystals of rhodochrosite on manganese-oxide ore.

PLATE 45

Wavellite.
Avant, Arkansas. 4 x 6 cm.

Wavellite is a phosphate mineral that can form when water containing dissolved phosphorus encounters aluminum-bearing minerals in a rock.

PLATE 46

Variscite.
Fairfield, Utah. 12 x 15 cm.

The yellowish parts of this specimen of variscite are made of crandallite, wardite, and other phosphate minerals that formed as solutions penetrated cracks in the variscite, liberating some of its phosphorus and replacing it in the process.

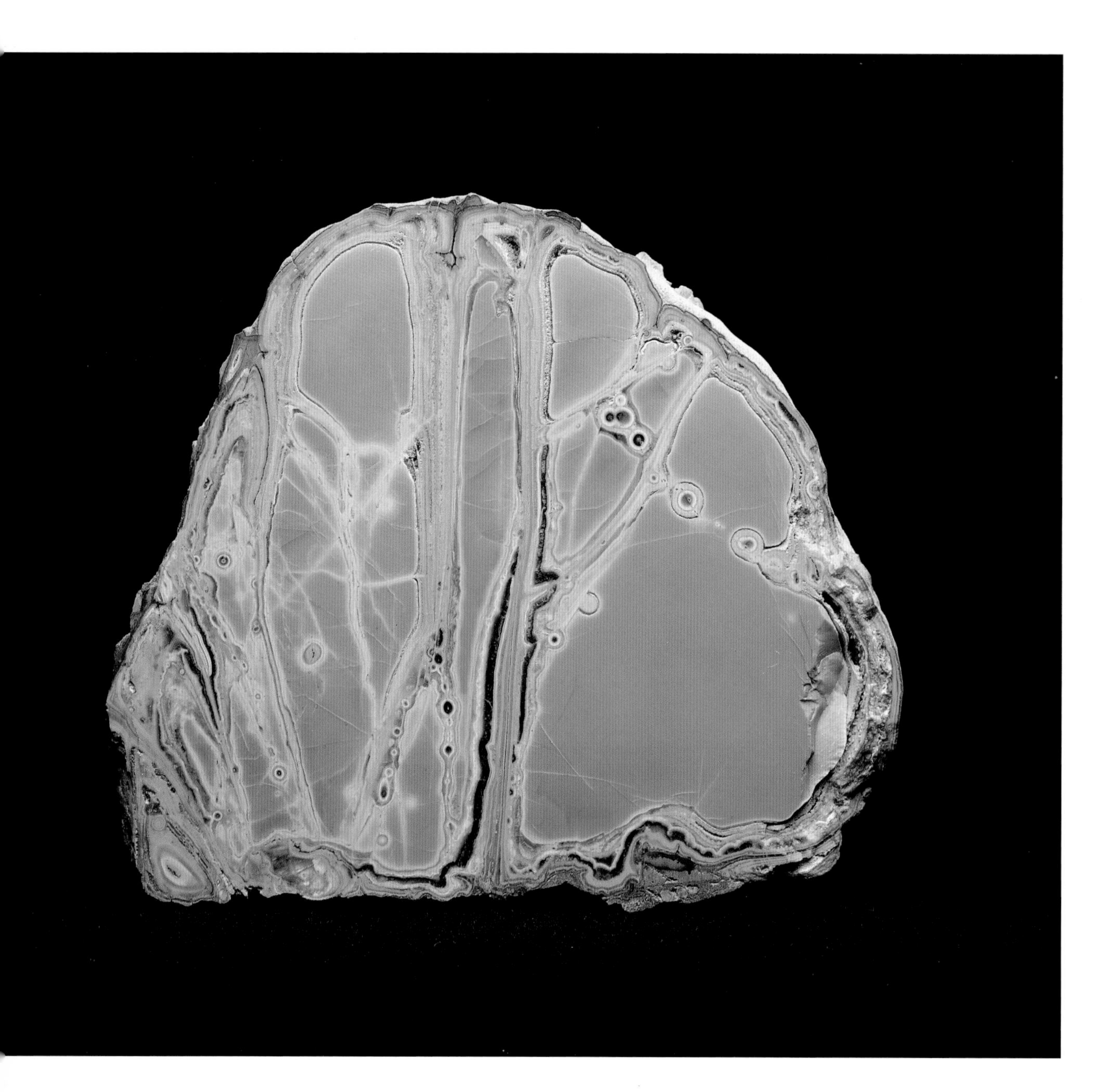

PLATE 47

Strengite (pink) **on Rockbridgeite** (green).
Svappavaara, Norrbotten, Sweden.
3 x 4 cm.

The presence of both ferrous (Fe^{2+}) and ferric (Fe^{3+}) iron in rockbridgeite, but only ferric iron in strengite, indicates that increasingly oxidizing conditions prevailed in the solution from which these minerals precipitated.

In the mineral world examples of dissolution and precipitation of the same compound are relatively uncommon, since most natural aqueous solutions contain other dissolved elements that interfere and cause the formation of additional minerals. One of the most common of these elements is iron, which may combine with other elements to form minerals such as hematite (Fe_2O_3), goethite [α-FeO(OH)], siderite ($FeCO_3$), or pyrite or marcasite (both FeS_2) when it precipitates. Which mineral forms depends largely on how acidic or basic the solution is and on the availability of dissolved oxygen, sulfur, and carbon dioxide, which determines whether oxides, sulfides, or carbonates can form. Stalactitic and **botryoidal** (resembling a bunch of grapes) growths of goethite [see PLATE 9] and hematite commonly fill voids in sedimentary iron deposits and are indicative of precipitation from solution.

Other minerals, such as calcite, aragonite, and barite [plate 42], often occur with the iron minerals in these deposits, as do a variety of manganese oxide minerals (e.g., pyrolusite and manganite), which normally precipitate after the iron minerals because they are more soluble and remain in solution longer. Occasionally the beautiful red manganese carbonate mineral rhodochrosite is encountered [PLATE 44]. Other manganese oxide minerals, such as romanechite or hollandite, commonly precipitate on fracture surfaces in rocks to form fernlike growths known as **dendrites** [PLATE 43]. Groundwater circulating through phosphorus-rich rocks may dissolve some of the phosphate from them and deposit it elsewhere in the form of various phosphate minerals, depending on what other kinds of rocks or minerals the groundwater encounters. Reaction with clays or aluminum-containing rocks brings about the precipitation of aluminum-bearing phosphates, such as wavellite or variscite [PLATES 45 and 46]. Reaction with iron-rich rocks forms iron phosphates, such as rockbridgeite, strengite, or vivianite [PLATES 47 and 67]. The presence or absence of any of these minerals at a given locality depends on whether their constituent ions were present in sufficient quantity to exceed their limits of solubility ●

chapter twelve

Hydrothermal Solutions

IN nature's laboratory, aqueous solutions may vary from a few drops to an ocean in volume, from near freezing to several hundred degrees in temperature, and may be at the Earth's surface or several kilometers beneath it. Similarly, the compositions of these solutions may range from nearly pure water, to water saturated with many dissolved materials. Geologists broadly categorize these solutions as either **hydrothermal** or **meteoric,** based on their temperature and mode of occurrence. Generally, meteoric water is derived from the atmosphere, such as rainwater, and makes up most of the water we see on the Earth's surface or a short distance beneath it, such as the groundwater in a well or sinkhole. Meteoric water is the type responsible for the minerals we have discussed up to now. Hydrothermal water, as its name implies, is much hotter. Its evolution is more complex than that of meteoric water [FIGURE E], and it is usually hidden well beneath the surface of the Earth. ℂ Sometimes meteoric and hydrothermal water are difficult to distinguish because one may mix with or contribute to the other. Certainly the water in some hydrothermal solutions originally was meteoric. Hot springs and geysers are familiar examples. These geological wonders typically occur in areas of recent volcanism, where a source of magma lies relatively close to the Earth's surface. As groundwater seeps downward through channels and cracks in the rock in these areas, it comes closer to the magma and is warmed by it. Even without a near-surface body of magma, as we descend into the Earth's crust beyond the first 1 to 2 kilometers the temperature increases. Therefore, any meteoric water that sinks into the crust or is buried along with sediments accumulating in a geological basin, inevitably will be warmed. The temperature it must reach to be considered hydrothermal water is a matter of opinion.

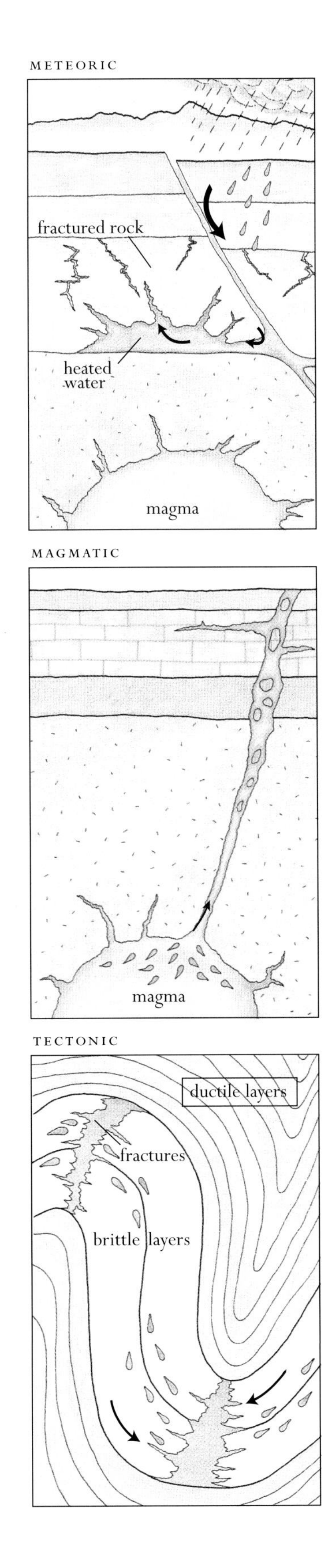

FIGURE E

Sources of Hydrothermal Solutions.

There is no single, universal source for hydrothermal solutions. Some are simply hot meteoric water. Others may evolve from magma or be driven from the rocks by heat and pressure during burial or tectonic events.

Hydrothermal solutions also may be generated by magmatic or tectonic processes [see FIGURE E], and the vein-type deposits they produce historically have supplied a significant percentage of the world's gold, silver, copper, and other metals. The idea that mineral deposits might form by deposition from hot, watery solutions is not new. As early as 1546 the "father of mineralogy," Georgius Agricola, proposed that veins of ore minerals form by rainwater circulating deep into the Earth's crust and depositing dissolved minerals in open channels. A century later the French philosopher René Descartes postulated that these mineralizing solutions originate from molten rock deep in the interior of the Earth. Who was correct? They both were. In Part II we saw how differentiation concentrates certain metals and volatile components, such as water, in the leftover, residual magmatic fluids. These hot, aqueous fluids are one type of hydrothermal solution, and the metal vein-type deposits they form are found in and around large bodies of granitic rocks throughout the world. We also find similar metal-bearing hydrothermal vein deposits in folded mountain belts and other zones of high tectonic activity, such as along plate boundaries, where there are no apparent magmatic or meteoric sources available. In these cases the tectonic activity itself provides the heat and pressure required to "squeeze" out and force the water through the rocks.

Mineralization by precipitation from aqueous solutions, whether hydrothermal or meteoric, follows a common pattern. Water travels through open spaces or permeable sections of the Earth's crust, such as along faults or through porous rocks, dissolving soluble minerals as it goes. If the water is hot, it dissolves even more minerals. As we will see in Part IV, the solution may chemically alter the rocks through which it passes and itself be changed in composition by these reactions. Whenever one of the components dissolved in the solution exceeds its solubility limit, it precipitates. Any number of factors, such as evaporation of the water, falling temperature or pressure, reaction with the wall rock, or mixing with other solutions, can cause the solubility limits of one or more dissolved components to be exceeded.

This process is the general way that minerals crystallize from aqueous solutions. Remember, though, that the specific chemical reactions involved in mineral dissolution, transport, and deposition are governed by complex physical and chemical mechanisms, many of which are still poorly understood. Not the least of these is the process of crystal growth itself. Because they typically grow unimpeded in open spaces, crystals precipitated from aqueous solutions are frequently very well formed and are among the most spectacular examples of the mineral world●

chapter thirteen

Warming the Rain

AT the Earth's surface weathering and erosion are constantly at work, wearing down mountains into soil. Running water, probably the single most important agent of erosion, carries the soil from the mountains to the plains or sea, where it accumulates as sediment. It is not surprising, then, that sediments commonly contain up to 20 percent water trapped in pore spaces. On the continents sediments tend to collect in broad, low areas known as **basins.** The basins themselves often contain marine sediments, indicating their former existence as seafloors that were uplifted by tectonic movements as the continents evolved. As sediments accumulate in basins, they are compacted under their own weight and warmed by the Earth's heat as they are buried deeply under more sediments or occasional lava flows. Such burial may result in temperatures as high as 250 to 300°C, causing organic materials and hydrated minerals like gypsum or clays to give up more water. Of course, additional water may always flow into the basin from above or from along its margins following porous layers. As the water is warmed by burial, its ability to dissolve the minerals it contacts increases. Evaporite minerals, such as halite and gypsum, are common in sediments and are a source of soluble chlorides and sulfates, which form a **brine** when they dissolve. The addition of these ions to the solution greatly increases its capacity to dissolve metals such as copper, lead, and zinc by forming complex ions that are more stable (i.e., less likely to precipitate) at lower temperatures. ◖ Driven by differences in pressure, temperature, and elevation, solutions may travel great distances through basins, following porous sediments or the open spaces of fractured rocks. Carbonate rocks, such as limestones or dolostones (limestones in which calcite is replaced by

PLATE 48

Calcite and Chalcopyrite.
Sweetwater mine, Reynolds County, Missouri. 22-cm crystal.

Resembling a spire on a Gothic cathedral, such large crystals of calcite are relatively common in MVT deposits.

PLATE 49

Galena on Dolomite and Sphalerite.
Joplin, Missouri. 2-cm crystal.

The repeated association of these minerals in MVT deposits around the world suggests that their origins are closely linked.

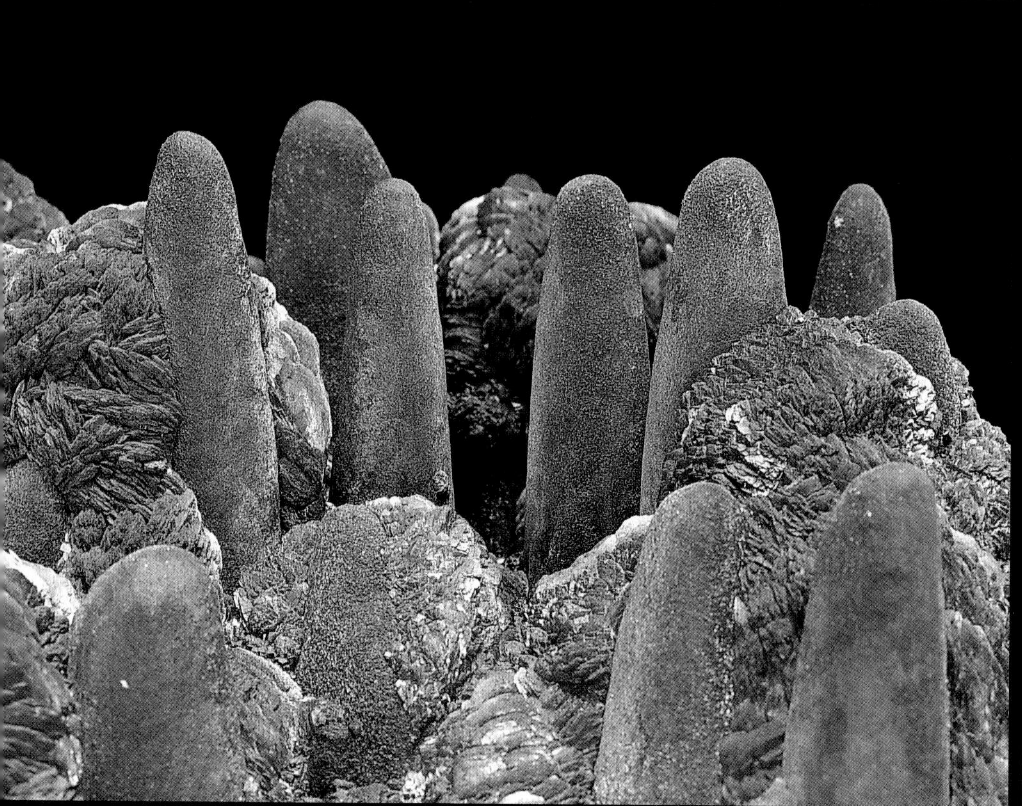

PLATE 50

Marcasite.
Shullsburg, Wisconsin. 15 x 20 cm.

The presence of stalactites of marcasite and other minerals in MVT deposits underscores the importance of precipitation from aqueous solutions in their formation, since stalactites require dripping water to form.

PLATE 51

Fluorite and Sphalerite.
Denton mine, Hardin County, Illinois. 12 x 19 cm.

The occurrence of sphalerite and fluorite together on a limestone matrix is an association typical of specimens from MVT deposits.

dolomite), are especially susceptible to invasion by these solutions because they are highly soluble and are porous. As the solutions enter open spaces, they are cooled by a reduction in pressure. They also may react with the rocks or other solutions that they contact or be cooled by upward transport. Any one of these events can cause dissolved minerals to precipitate as their saturation limits are exceeded. Perhaps the most famous hydrothermal deposits formed in this manner are called **Mississippi Valley Type (MVT)** deposits, named for the region of the central United States where they have been studied in detail. Some of the world's finest examples of galena, sphalerite, calcite, marcasite, and fluorite have come from these deposits [PLATES 48 to 51].

Some of my most memorable mineral collecting was in the Mid Continent mine near Treece, Kansas, which exploited one of many MVT deposits in the tristate district of Missouri, Kansas, and Oklahoma. Stepping into the infamous "iron maiden," as the local mine operators had christened the oversize tin can being used as an elevator, I was lowered 100 meters below the Earth's surface into an entirely different world. As I walked through the underground maze of tunnels, the light from my lamp was reflected by numerous small fracture fillings of galena and sphalerite, the two main ore minerals at the mine, and here and there a natural black asphalt oozed from cracks in the dolostone. Eventually I came to an area where the rock was much more fractured; in fact, all that held it together were well-formed crystals of sphalerite and calcite. Farther along I encountered even larger openings, some extending nearly 15 meters, completely lined top and bottom with golden-yellow calcite crystals up to 30 centimeters long. Elsewhere I found galena crystals as much as 15 centimeters across and a 2-meter wide fracture zone entirely filled with slabs of rock covered with pink dolomite crystals. Perched upon the dolomite were small, brassy-yellow crystals of chalcopyrite and larger individual crystals of sphalerite, galena, and calcite. It was a mineral collector's dream!

A mineral collector's dream, however, can be a geochemist's nightmare! Explaining the origin of MVT deposits is difficult. Since the two predominant sulfide minerals in MVT deposits, galena (PbS) and sphalerite (ZnS), are nearly insoluble in water, how can they travel in solution to their site of deposition? If lead or zinc ions encounter sulfide ions, as they must to form galena or sphalerite, precipitation is nearly instantaneous. These properties of galena and sphalerite imply that crystals as large as those I found at the Mid Continent mine should never have formed, since rapid precipitation nearly always produces very small crystals. How then do such large crystals of galena and sphalerite form in an MVT deposit? This is the question that plagues the geochemist.

There is no single, simple answer. What is needed for these large crystals to form is a mechanism that slowly introduces sulfide ions to the solution at the site of metal precipitation. Analyses show that most subsurface brines contain very low amounts of sulfur, which is nearly always in the form of sulfate (SO_4^{2-}) rather than sulfide (S^{2-}) ions. From experimental studies we know that

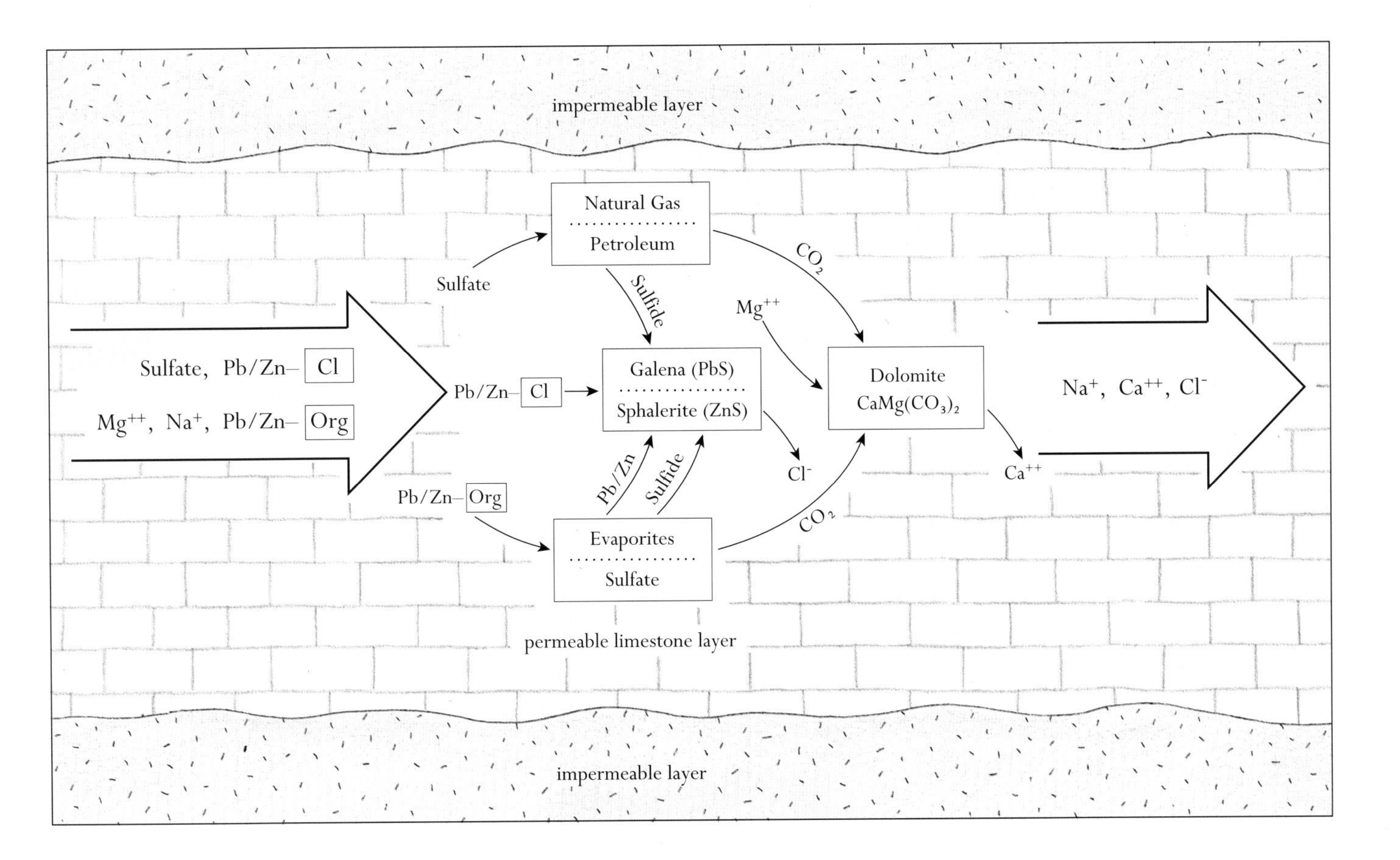

FIGURE F

Mineralization in MVT Deposits.

Because the lead and zinc sulfides galena and sphalerite are nearly insoluble in water, these metals probably travel in solution as complex ions with chlorides or organic molecules. Reactions between organic molecules and sulfate minerals may provide the sulfide necessary to form the galena and sphalerite, to liberate chlorides, and to cause dolomite to precipitate. Fluid inclusions in minerals from MVT deposits commonly contain chlorides, sulfates, and organic molecules, and dolomite is nearly always present in the rocks.

PLATE 52

Barite on Calcite.
Pennington County, South Dakota.
5-cm crystal.

The occurrence of this barite crystal in a type of geode called a septarium suggests that it was formed by precipitation from groundwater percolating through sediments.

both lead and zinc travel in brine solutions as chlorides and may do so with certain organic compounds as well. We also know that certain organic compounds can convert sulfate to sulfide and that when sulfate is removed from solutions containing dissolved calcium, magnesium, and carbonate ions, dolomite [$CaMg(CO_3)_2$] readily precipitates.

Interestingly, fluid inclusions in minerals from MVT deposits typically contain abundant chlorides as well as organic compounds. Was all the asphalt and pink dolomite I saw in the Mid Continent mine there just by chance, or does each play a definite role in the creation of an MVT deposit? Do lead and zinc travel as chlorides along with dissolved calcium, magnesium, carbonate, and sulfate ions until the solution encounters organic compounds that convert sulfate to sulfide, causing the precipitation of dolomite, sphalerite, and galena? Or is sulfide added to the solution by thermal decomposition of petroleum, bacterial reduction of sulfates, or another mechanism? All are possible, and all have been suggested as probable mechanisms for the development of MVT deposits [FIGURE F].

There are many other examples of hydrothermal mineralization by solutions derived from sediments. In the midwestern United States, such mineralization in beds of limestone has led to the formation of hollow, crystal-lined nodules called **geodes.** The area around Keokuk, Iowa, is particularly well known for the number of fine specimens it has produced. Resembling a cantaloupe on the outside, when broken open the geodes are found to be lined with inwardly projecting crystals of quartz, calcite, barite [PLATE 52], celestine, or other minerals. Similar hollow structures, called **septaria,**

PLATE 53

Mordenite on Cristobalite in Basalt.
Poona, Maharashtra, India.
6 x 10 cm.

Once thought to have originated from basaltic magmas, mordenite and most other zeolite minerals found in basaltic lava flows are now believed to have crystallized from meteoric solutions.

PLATE 54

Stilbite.
Poona, Maharashtra, India. 5 x 7 cm.

Stilbite, here in the form of a bow tie, is one of the most common zeolite minerals in basaltic lava flows.

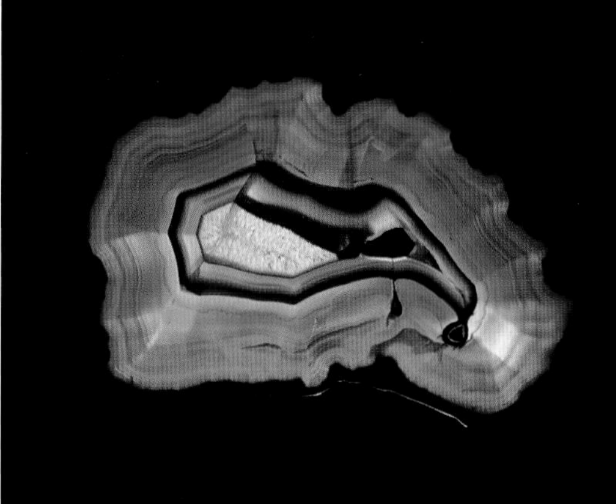

PLATE 55

Agate.
Rio Grande do Sul, Brazil.
14 x 23 cm.

The variously colored bands in this agate are due to impurities such as iron and manganese oxides.

PLATE 56

Amethyst.
Artigas, Uruguay. 6 x 10.5 cm.

The stalactitic form of this amethyst specimen suggests that it formed in a cavity by precipitation from an aqueous solution.

PLATE 57

Native Copper.
Keweenaw Peninsula, Michigan.
10 cm.

Like zeolites and agates, the occurrence of native copper crystals such as these in basaltic lava flows for years led geologists to believe they precipitated from magmatic rather than meteoric solutions.

often contain similar minerals. How and why geodes form is not completely understood. They tend to form around fossils, which probably provide the initial opening for mineral-bearing solutions to enter. As the organic remains in the fossil decompose, microchemical environments with characteristics different from those of the enclosing sediment may be established, causing various minerals to precipitate.

Sometimes things are not as they seem. Because of the widespread occurrence of zeolites, agates, amethyst, and native copper in basaltic lava flows [PLATES 6 and 53 to 57], formed by the crystallization of magma, one might conclude that these minerals also form by magmatic rather than hydrothermal processes, especially any involving meteoric water. But this is not the case. As molten rock spreads across the Earth's surface during a volcanic eruption, two obvious things happen: first, the lava warms the area beneath it, and second, in doing so the molten rock is cooled along its lower contact, like hot fudge poured onto a cold table. If the lava encounters water (rain, snow, a lake, or ocean), it will cool even faster and form a glass, since there is not enough time for crystallization to occur. Rising bubbles of gas often become "frozen" in place, forming layers of cavities as the lava cools, and continued flow can fracture already solidified portions. If a larger body of water is encountered, bulbous **pillow** structures (named for their shape) may develop.

Gas bubbles, pillows, and areas of fractured rock all provide open spaces through which water can flow. As the lava cools, water preferentially reacts

FIGURE G

Hydrothermal Mineralization in Lava Flows.

Minerals such as zeolites, amethyst, agate, and native copper that are common in basaltic lava flows form by precipitation from aqueous solutions that permeate open spaces such as fractures or gas bubble cavities in the basaltic lava.

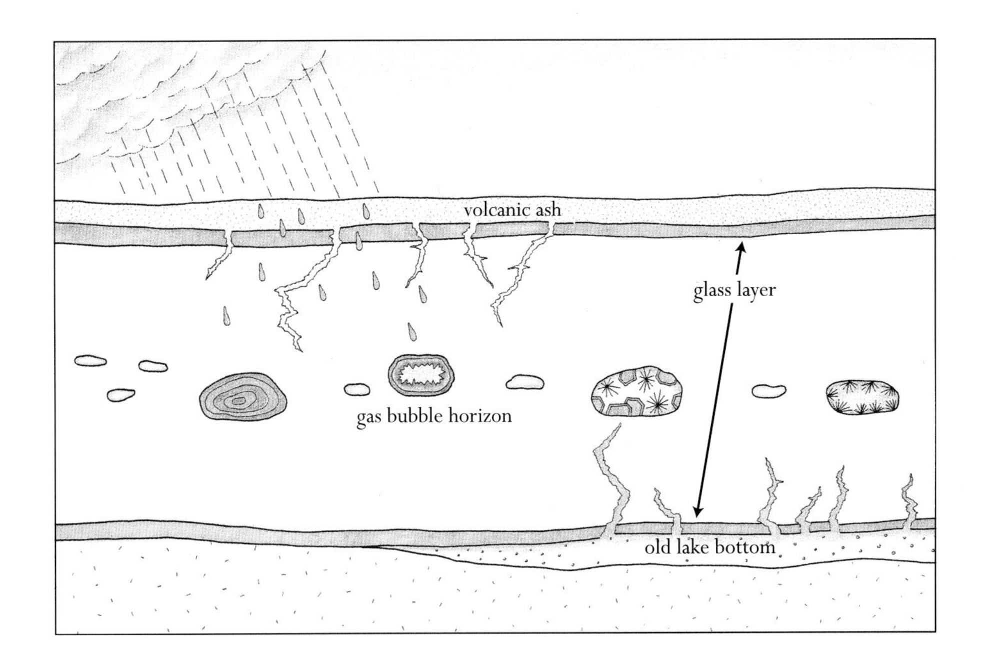

with the glass faster than with the rock (since the atoms in the glass are not bonded together tightly in crystal structures) and leaches iron, magnesium, aluminum, silicon, calcium, potassium, and sodium ions from it. Similar reactions occur when water filters through volcanic ash. After the entire body has cooled sufficiently, various minerals begin to crystallize from the hydrothermal solution that has filled the open spaces [FIGURE G]. Often much of the iron and magnesium is removed as a thin layer of blue-green clay precipitates. If the solution is unusually high in dissolved silica, minerals such as amethyst, jasper, bloodstone, or agate may fill the opening. When the temperature cools to about 200°C, zeolite minerals will begin to crystallize as their levels of saturation are exceeded.

Zeolites are a group of about fifty minerals, the more common of which include analcime, chabazite, heulandite, natrolite, mesolite, mordenite, and stilbite [see PLATES 6, 53, and 54]. Most zeolites are aluminosilicates of calcium, sodium, or potassium, as are the minerals prehnite, datolite, pectolite, and apophyllite [PLATE 6], which frequently accompany them. Since water is an integral part of their structure, zeolites are stable only at relatively low temperatures. That's why they generally form at temperatures below 200°C. Many zeolites can exchange cations (atoms that have lost electrons) with surrounding solutions because they have large "holes" in their crystal structures. This feature has led to their commercial application as molecular sieves in devices ranging from water softeners to medical dialysis units●

PLATE 58 *(left)*

Fluorapatite on Quartz.
Panasqueira, Portugal. 6 x 8 cm.

Fluorapatite is a common accessory mineral in high-temperature tin-tungsten deposits known as greisens. The hot, aqueous solutions that carry these metals, as well as the fluorine and phosphorus required to form fluorapatite, are derived from granitic magma.

PLATE 59 *(above)*

Cassiterite.
Cornwall, England. 5 x 6 cm.

Because high-temperature minerals such as cassiterite (tin oxide) usually precipitate early from mineralizing solutions, they occur relatively near their place of origin, as exemplified by the granite-hosted tin deposits of Cornwall, England.

chapter fourteen

Beyond Pegmatites

IN Part II we learned that as a magma cools, minerals crystallize from the melt, changing the composition of the remaining liquid by removing specific elements from it, and that near the end of the differentiation period, fluids capable of generating pegmatites with exotic minerals may form. Sometimes, however, pegmatites are not formed, or if they are, the differentiation process may continue beyond the pegmatite stage. The last remaining liquids become enriched in water and volatile components such as fluorine, chlorine, and sulfur, along with metals such as tin, tungsten, gold, silver, uranium, and other elements that could not be incorporated into the structures of the earlier-formed minerals because of their unusual sizes or ionic charges. Such remaining liquids represent one more source of hydrothermal solutions and are the type most frequently cited as "classic" examples of hydrothermal solutions. ◖ These solutions emanate from the magma and travel away from it, following permeable rocks or openings along faults and fracture systems and forming hydrothermal vein deposits. Accompanying steam may develop sufficient pressure locally to blast its way through to an area of lower pressure, creating a **breccia pipe** (breccia is a rock composed of sharp, angular fragments cemented together). The resulting rapid decrease in pressure causes the solution to cool, which in turn may cause various minerals to precipitate. Breccias also may be created by grinding and fracturing of rocks due to movements along faults in the Earth's crust or by collapse when supporting rock is dissolved away. The open spaces between the blocks of broken rock in breccias provide an ideal environment for solutions to migrate and crystals to form. The chrysocolla specimen in PLATE 79 graphically illustrates this process.

PLATE 60

Tetrahedrite (black) **with Bornite** (purplish), **Chalcopyrite** (golden), **and Quartz** (white).
Herodsfoot mine, Liskeard, Cornwall, England. 5 x 6 cm.

Except for quartz, all these minerals are important ores of copper and are commonly mined from vein deposits formed by precipitation from hot, metal-bearing, aqueous solutions.

PLATE 61

Nickel-Skutterudite.
Schneeberg, Saxony, Germany. 3 x 4 cm.

As hot, metal-bearing solutions that emanate from magmas migrate farther from their sources, silver and nickel cobalt arsenide minerals such as this one may precipitate.

PLATE 62

Rhodochrosite (pink) **and Fluorite** (green).
Silverton, Colorado. 3 x 4 cm.

Studying the distribution of certain isotopes of oxygen, hydrogen, and other elements reveals that meteoric as well as magmatic solutions may have played a role in the formation of these and other minerals in the mines of the San Juan Mountains, Colorado.

The most obvious proof that solutions forming a hydrothermal mineral deposit evolved from a magma is a direct association of the deposit with igneous rocks. Often such an association is clear, but if the solutions had to travel some distance from their parental magmas, the connection is less apparent. A clue to the heritage of such deposits may be gleaned from the minerals they contain. Because they evolve by differentiation, the minerals in such deposits are those last formed by that process, such as quartz, as well as minerals containing the elements that were the "leftovers" isolated by the "law of constant rejection." As the solutions move farther from their sources and begin to lose heat to their surroundings, minerals are formed as dissolved components reach their respective levels of saturation and precipitate. Solutions derived from granitic magmas frequently precipitate fluorine-bearing species such as fluorite, topaz, and fluorapatite [PLATE 58]; oxides of uranium, tungsten, and tin [PLATE 59]; sulfides of iron, copper, lead, zinc, silver, antimony, arsenic, and other metals [PLATES 60, 61, and 64]; and even some native metals, such as gold, silver, copper, arsenic, and bismuth [see PLATES 63 and 74].

Similar patterns of hydrothermal mineralization occur in widely separated locations. Mines in the counties of Devon and Cornwall in southwestern England have been exploited for tin since the Bronze Age. There the major ore of tin is cassiterite [see PLATE 59], which occurs in veins in granite, along with a wealth of other metallic minerals: major ores of tungsten, such as wolframite and scheelite; copper-bearing sulfide minerals, such as chalcopyrite, chalcocite, bournonite, bornite, and tetrahedrite [see PLATE 60]; the lead and zinc sulfides galena and sphalerite; and dozens of others.

PLATE 64

Stibnite.
Baia Sprie, Transylvania, Romania.
6 x 9 cm.

Antimony minerals such as stibnite usually precipitate from hydrothermal solutions at relatively low temperatures, which helps explain their occurrence in veins distant from their parental magmas.

PLATE 65

Pyrite and Quartz.
Butte, Montana. 2.5 x 4 cm.

The simple iron sulfide pyrite is perhaps the most common metallic mineral in hydrothermal vein deposits, proving that not all such deposits contain valuable metals.

PLATE 63

Gold.
Botes, Transylvania, Romania.
6 x 6 cm.

Many gold-bearing quartz veins such as those in Transylvania are believed to form from hot, aqueous solutions derived from magmas.

An equally important and historic mining center is in the Erzgebirge highlands near the border of Germany and the Czech Republic, where tin, copper, silver, and other economically important metals have been mined since 1168. These deposits also consist of veins in or emanating from granitic rocks. As in England, in the Erzgebirge highlands species such as cassiterite, wolframite, scheelite, apatite, and topaz occur in the earlier-formed tin- and tungsten-rich veins, whereas sphalerite, chalcopyrite, and galena, along with a number of silver-, nickel-, uranium-, cobalt-, and arsenic-bearing minerals [see PLATE 61] are found in later-formed veins, at greater distances from their source.

Similar mineralogical trends are visible in the tungsten mines of Panasqueira, Portugal, among other places. In each of these regions, the granites intruded tectonically active zones along folded mountain chains or plate boundaries. High-temperature minerals like wolframite and cassiterite precipitate from the solution before minerals such as the sulfides, whose elements remain in solution by forming complex ions that are stable at lower temperatures. Acting like a time-released drug, this process is what gives rise to the elemental zoning observed in these deposits.

Other genetically similar deposits are found in volcanic and tectonically active zones throughout the world. Some, such as those in the San Juan Mountains of Colorado or in Transylvania, Romania, have also been significant producers of gold, silver, and other minerals [PLATES 62 to 64]. One such group, known as **porphyry copper** deposits because they occur in or adjacent to quartz-monzonite porphyries or related granitelike rocks, can be found all along the Pacific Rim, from the Canadian Cordillera, through the Rockies, into Arizona and New Mexico, and extending south as far as Peru and Chile [PLATES 8, 62, and 65].

Because of their obvious close association with volcanos and igneous rocks, these deposits have traditionally been considered examples of hydrothermal solutions derived from magmatic sources. Today, however, we know that this is not always completely true. The fluid inclusions contained in the minerals from some of these deposits and their oxygen and hydrogen isotope compositions point to the same conclusion: meteoric water has played an important role in their genesis. At some localities the data suggest that meteoric and magmatic hydrothermal waters mixed; at others the magma probably only provided the heat required to raise the temperature of meteoric water sufficiently to enable it to leach metals from the surrounding rock and redistribute them in cooler, fractured rocks it encountered as it migrated away from the heat source. The occurrence of a mineral in a vein of an igneous rock does not alone prove that it was formed by an igneous process●

chapter fifteen

The Effect of Tectonism

WHEN plates collide, tremendous compressional forces cause some to break and others to be folded like an accordion, depending on their composition and degree of brittleness. When the accompanying heat and pressure are great enough, existing minerals recrystallize, producing metamorphic rocks. Among other effects, this same heat and pressure can liberate water from the rocks, which flows into areas of lower pressure such as fractures or other open spaces. As it flows through the rocks, the heated water dissolves some of the minerals it encounters, forming a hydrothermal solution. The minerals that precipitate from such hydrothermal solutions depend on several factors: the initial composition of the fluid released, the composition of the rocks through which they flow, how long they are in contact with these rocks, the pressure and temperature, and the rate at which the pressure and temperature change at the site of crystallization. ◖ Because they are derived from the rocks in which they are found, the minerals in this type of hydrothermal deposit must have a chemical makeup consisting of combinations of the elements in those rocks. (No matter how hard you squeeze a tea bag, it will never give you coffee.) This point is well illustrated by two famous and very different localities, one in central Arkansas, the other just 35 kilometers from the Arctic Ocean, in Canada's Yukon Territory. At the Arkansas location, about 300 million years ago the collision of the African and North American plates caused the birth of the Ouachita Mountains, leading eventually to the creation of the entire Appalachian range. Today, we find hundreds of widespread veins of quartz crystals [PLATE 66] in the Ouachitas, especially in the area between Hot Springs and Mount Ida, Arkansas. Quartz crystals are abundant here because

PLATE 66

Quartz.
Hot Springs, Arkansas. 5 x 9 cm.

Because quartz is the major constituent of the rocks from which they were derived, the hydrothermal solutions that once filled fractures in the rocks near Hot Springs, Arkansas, could produce few other minerals.

PLATE 67

Vivianite.
Big Fish River, Yukon Territory. 4 x 5 cm.

The hydrothermal solution from which these crystals of vivianite precipitated derived iron and phosphorus from the rocks around it.

the rock that was tilted and fractured by the ancient tectonic forces that formed the Ouachitas is relatively pure sandstone, composed almost entirely of one mineral: quartz. The hydrothermal water that was activated as a result of the tectonic activity had little else available to dissolve and precipitate.

Rocks of the Rapid Creek-Big Fish River area in the Yukon Territory host more complex examples of minerals formed from hydrothermal solutions activated by tectonic processes. These rocks consist mainly of iron-rich sandstones, mudstones, and shales that contain an unusually high amount of phosphorus. Deformation and fracturing of these rocks occurred during uplift of the northern Rocky Mountains about sixty million years ago. Hydrothermal solutions activated by the tectonism flowed through the rocks, leaching silica, iron, phosphorus, and other elements from them. When the solutions reached the open fractures, these elements were redeposited as quartz, siderite, and a large suite of unusual phosphate minerals, among them some of the world's finest lazulite and wardite crystals. Calcium-bearing phosphates like collinsite formed in fractures in the calcium-rich mudstone, iron-rich phosphates such as vivianite [PLATE 67] formed in veins in the iron-rich rocks, and complex assemblages formed in veins cutting across several types of rock.

The quartz veins in Arkansas and the phosphate minerals in the Yukon both formed at relatively low temperatures and pressures. When the prevailing temperature and pressure confining the hydrothermal solution are higher, other minerals form instead. If those parameters later change, the changes are reflected in the minerals that follow. Nowhere is this better illustrated than in the famous **alpine cleft** deposits found throughout the Alps in Switzerland, Austria, France, and Italy.

High in the Swiss Alps near the Grimsel and Furka passes, some of the world's finest crystals of quartz and pink fluorite have been collected from clefts in granitic rocks. One of the most famous finds was made over a century ago by a team of **strahlers** near the Tiefen Glacier. Strahlers are highly skilled mountain guides who specialize in finding and collecting minerals from alpine clefts, a Swiss tradition that has prevailed to the present. Suspicious about a quartz vein high above them in the cliff face, the Tiefen Glacier team made their way up the wall, where they excavated the vein for several days. Eventually they encountered a small, dark hole, which when opened up, proved to be one of the largest alpine clefts ever discovered. Measuring 6 x 4 x 2 meters, the cleft was full of soft, dark green chlorite embedded with huge, perfectly formed crystals of quartz weighing more than 100 kilograms and nearly a meter long. For the next eight days almost the entire population of the nearby village of Gutannen pitched in to assist in removing the crystals. Some of these wondrous specimens can be seen today in the Natural History Museum in Bern.

Although they seldom find crystals of such proportions, present-day strahlers continue to collect equally fine but smaller crystals of quartz, pink fluorite, hematite [PLATE 68], and other minerals from similar occurrences throughout the Alps. Clefts containing minerals with more varied compositions, such as anatase, brookite, titanite [PLATE 69], or epidote [PLATE 70], occur in

PLATE 69

Titanite (yellow) **and Orthoclase** (white).
Stubachtal, Salzburg, Austria.
3.5 x 5 cm.

The tiny, green flakes of chlorite adhering to the other minerals on this specimen are characteristic of alpine cleft deposits.

PLATE 70

Epidote.
Untersulzbachtal, Salzburg, Austria.
4 x 4.5 cm.

Overlooking the Grossvenediger Glacier high in the Austrian Alps, this famous alpine cleft deposit has produced what most experts agree are the world's finest crystals of epidote.

PLATE 68

Hematite.
Cavradi, Grisons, Switzerland.
3.5 x 5 cm.

Crystals of hematite such as this one, known as eisenrosen, or "iron roses," are classic examples of alpine cleft minerals. The tiny red crystals aligned on their faces are the mineral rutile.

chapter sixteen

Summary of Water's Role in Mineral Formation

BECAUSE of its unique structure, water causes many ionic compounds, including some minerals, to dissociate into their component ions, creating an aqueous solution. Increasing the temperature may enable more ions to be dissolved, but eventually saturation occurs, preventing further dissolution. When their limit of solubility is exceeded, minerals precipitate out of solution as individual cations and anions bond together to form crystals. The slower the precipitation, the greater the likelihood that large, well-formed crystals will develop. Decreasing temperature or confining pressure, evaporation, and mixing with other solutions are common natural processes that induce precipitation. ◖ As meteoric water flows over and through rocks, it dissolves some of their minerals, causing the rocks to weather and be eroded. The dissolved minerals are carried by rivers and streams to the sea or to an inland basin, where a playa lake may form. Tectonic uplift also may isolate part of an ocean, forming an inland sea. If the rate of evaporation exceeds that of rainfall, the lake or sea dries up and forms an evaporite deposit. The first minerals to precipitate are the least soluble; the most-soluble minerals precipitate last. Important deposits of salt, gypsum, and borax form in this manner. ◖ Erosion transports sediments to topographically low areas known as basins, where they accumulate. When burial is sufficiently deep, heat mobilizes the water within the sediments, creating a hydrothermal solution. Evaporites in the sediment may dissolve to form a brine that enables metals such as lead, zinc, or copper present in the sediments to form complex ions with either chlorides or organic molecules in the brine. Such metal-bearing solutions can be driven considerable distances by changes in temperature or pressure, dissolving even more metals until they cool or encounter anions that cause the metals to precipitate. Lead-zinc MVT deposits, as well as the native copper deposits in the Keweenaw Peninsula of Michigan, probably formed in this way.

Hydrothermal solutions also may evolve as an end product of magmatic differentiation, particularly in granitic magmas. These solutions are typically enriched in volatile components such as fluorine, sulfur, and water and carry metals whose ions are of unusual size or charge, which precluded their incorporation by earlier-formed minerals. Hydrothermal solutions may also be produced by tectonic forces as water is driven from their enclosing rocks into fractures created by the forces. Minerals that form in this type of deposit always contain the same elements that are present in the host rocks because the solutions from which they precipitate are derived from those rocks ●

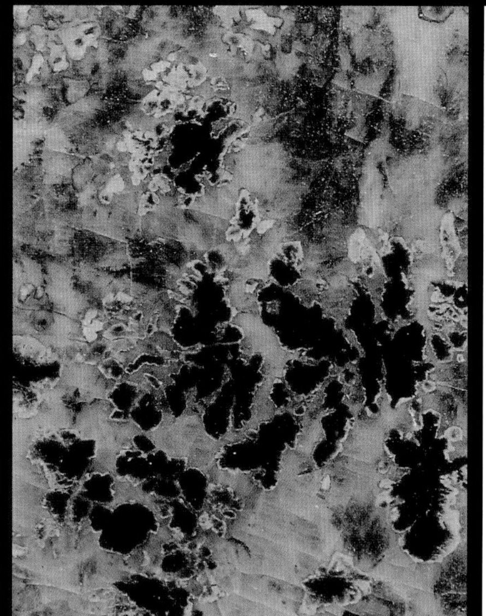

PART IV

Chemical Alteration

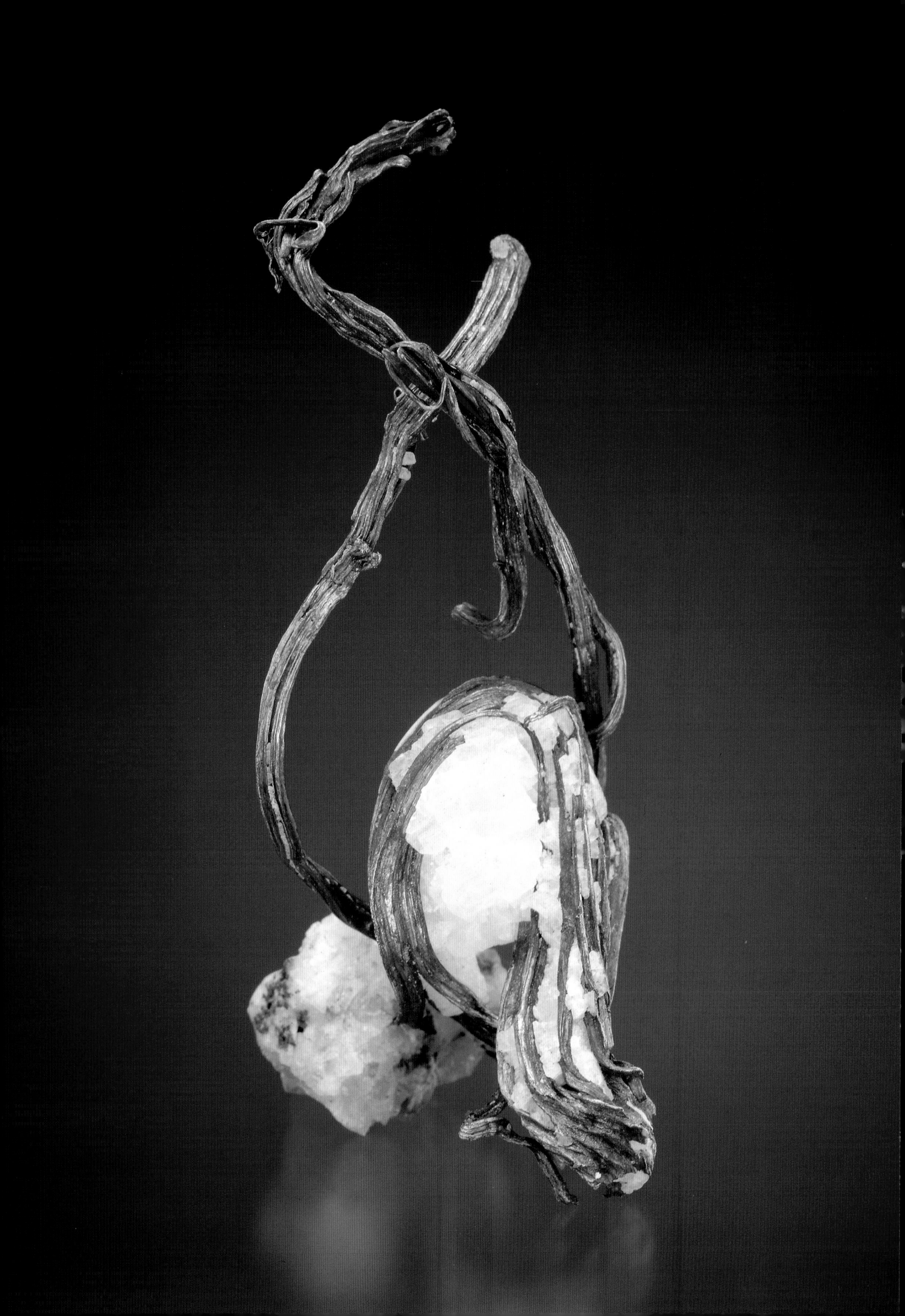

chapter seventeen

Equilibrium and Chemical Reactions

MOST of us are already familiar with the effects of chemical alteration. It's the reason we have to take our cars to body shops to repair the damage from rust, and why we periodically have to polish our tarnished silverware. Iron in the steel used to make cars reacts with oxygen in the air to form iron oxide, or rust; silver reacts with sulfur in foods, in water, or in the atmosphere to make silver sulfide, which appears as a dark tarnish. Why do these reactions occur? In the conditions to which they are exposed, neither the iron nor the silver is stable. They "want" to be in a more stable, "comfortable" state for the imposed conditions, a state where there is no tendency to change spontaneously. Chemists know this state as **equilibrium.** ℂ In nature, native iron is comparatively rare at the Earth's surface, but where it does occur it rusts, just as the iron in our automobiles does. Likewise, iron meteorites nearly always have a well-developed outer coating of iron oxide [see PLATE 1], and native silver, which is relatively common in nature, almost always has a dark, tarnished surface of silver sulfide (the mineral acanthite) if it has been exposed to groundwater or to the atmosphere [PLATES 74 and 140]. ℂ Minerals are merely some of nature's solid chemicals, and they react with various chemical elements and compounds in their environment, including each other. The laws governing these reactions are the same as those that govern all other chemical reactions. Water promotes chemical reactions between minerals by aiding in their dissolution and providing a medium in which their constituent ions are free to move about and react. Dissolving salt in water results in a solution containing sodium cations and chloride anions. What happens, though, if the water already has other cations or anions in it? What determines whether a given

PLATE 74

Native Silver and Calcite.
Silver Islet, Lake Superior, Ontario.
2 x 5 cm.

The tarnish on these natural wires of silver formed by reaction with sulfur, just as it does on the silverware in your kitchen drawer.

cation will react with a particular anion to form a new compound?

Chemicals or minerals react because the **products** of their reaction are more stable in the existing environment than are they, the **reactants.** Like a ball rolling down the stairs, a chemical reaction does not stop until the final "step" is reached. The final step in this case is the point at which one of the reactants has been completely consumed or there is a change in the existing conditions that effectively establishes a new environment. Sometimes a reaction that should take place fails to come about. For example, a ball can remain precariously balanced on the top step for a very long time. We know the ball inevitably will roll down the stairs because it is unstable in its present position, but exactly when this will happen is uncertain. All that is needed is a little boost to get things going. To chemists this boost is known as **activation energy.**

In the mineral world, crystals of sulfur [see PLATE 135] may be in contact with oxygen in the air for eons without changing because without activation energy sulfur reacts with oxygen so slowly that nothing seems to be happening. Heat accelerates the process, producing sulfur dioxide by the chemical reaction commonly known as burning. When a ball rolls to the bottom of the stairs, some of its stored **(potential)** energy is converted to mechanical **(kinetic)** energy. In giving up stored energy, the ball (or in our case, the sulfur and oxygen molecules) reaches a lower, more stable energy state for the existing conditions.

A ball will roll down the stairs just as well in your house as in mine. Sulfur burns equally well in a test tube in the laboratory or on the edge of a vent from an active volcano. The chemical alteration of minerals may take place in any geological setting or in any kind of rock—igneous, sedimentary, or metamorphic. The only requirements are a condition of disequilibrium and a source of activation energy. Heat is one of the all-time great activators of chemical reactions, but it is not the only form of energy. Light, too, can initiate chemical changes in certain sensitive minerals such as proustite or vivianite. Even mechanical energy can initiate some chemical reactions; that's why dropping a bottle of nitroglycerin is not recommended!

At equilibrium there is no tendency for spontaneous change. Reactants and products are in a state of balance. It's as if we had ten equally energetic volleyball players, five on either side of the net, each throwing a ball over the net. Players on the left represent reactants; those on the right represent products. Under these conditions the number of balls thrown from left to right by the "reactants" is exactly counterbalanced by those coming in the opposite direction from the "products." The "game" is in equilibrium. How can we make more balls travel to the right? One way is to call out another player or two from the reactants' bench. Or we could let those players rest and achieve the same effect by removing one or two members of the products team from the court. The same is true for chemical reactions in equilibrium.

A chemical reaction may be driven in one direction or the other by increasing or decreasing the concentrations of reactants or products. As we saw in Chapter 11, this is the principle that governs the formation of evaporite minerals. At saturation the precipitating crystals of solid halite are in a state of equilibrium with their dissolved component ions, Na^+ (sodium) and Cl^- (chloride). Adding water, a reactant, causes the reaction to shift toward its products (i.e., dissolved ions), so more halite dissolves. Introducing more dissolved ions, the products, causes the opposite reaction to occur: halite precipitates. The removal of a reactant (water) by evaporation can also reverse the reaction and cause halite to precipitate, as it does in a playa lake ●

PLATE 75

Tincalconite pseudomorph after Borax.
Boron, California. 6 x 9 cm.

This specimen of chalky-white tincalconite was formed by dehydration of transparent, lustrous borax crystals.

PLATE 76

Metatorbernite pseudomorph after Torbernite.
Musonoi mine, Kolwezi, Shaba, Zaire. 2 x4.5 cm.

Like the tincalconite in PLATE 75, these metatorbernite crystals formed by dehydration of torbernite and have retained the outward crystal form of that mineral.

chapter eighteen

Drying Out

MOST chemical reactions responsible for the creation of minerals involve water. Among the simplest are **hydration-dehydration** reactions. You are probably already familiar with one example from the mineral world,but you may not realize it. Calcium sulfate exists naturally in two common forms: gypsum ($CaSO_4 \cdot 2H_2O$), which contains two water molecules per formula unit, and anhydrite ($CaSO_4$), which has no water. Gypsum is quarried and roasted to drive off its water, then powdered and sold as plaster of Paris. What do we do with it? Add water to make gypsum again! Because the roasting drives off water, it is a **dehydration** reaction. The second reaction, adding water, is **hydration.** ◐ Calcium sulfate is not the only mineral with variable states of hydration. Common in mineral collections are large, well-formed groups of opaque, chalky-white crystals labeled as borax from Boron, California, or other localities. But such labels are wrong. Borax ($Na_2B_4O_7 \cdot 10H_2O$) contains ten molecules of water per formula unit, and its crystals are transparent and glassy. Although the specimens in question may have been borax when they were collected, with time they have dehydrated into the mineral tincalconite ($Na_2B_4O_7 \cdot 5H_2O$), which is more stable in the dry conditions of storage [PLATE 75]. Similarly, the heavily hydrated uranium minerals autunite and torbernite often dehydrate into meta-autunite and metatorbernite [PLATE 76]. ◐ Specimens such as these, that although different from the original mineral, have retained its crystal shape, are called **pseudomorphs** (*pseudo* = false, *morph* = shape). They are direct, tangible evidence of chemical alteration. Because they are composed of one mineral formed *after* the other, pseudomorphs are commonly described as minerals "after" the original mineral—for example, tincalconite after borax or metatorbernite after torbernite ●

chapter nineteen

Neutralization Reactions

LIKE water, acids and bases may influence the solubility of some compounds, and hence their tendency to participate in a chemical reaction. In water hydrogen ions (H^+) are bonded to hydroxyl (OH^-) ions to make HOH, more commonly written as H_2O. In **acids** hydrogen ions are bonded to different anions, such as Cl^- (chloride) in hydrochloric acid (HCl), or SO_4^{2-} (sulfate) in sulfuric acid (H_2SO_4). **Bases** contain OH^- bonded to a cation other than H^+. A familiar example is household lye, which is sodium hydroxide (NaOH). Aqueous solutions of acids therefore contain higher concentrations of H^+ than do aqueous solutions of bases, which contain more OH^-. The concentration of H^+ or OH^- in the solution, which determines how strongly acidic or basic it is, is measured on the **pH scale.** ◖ The pH scale ranges from 0 to 14. The lower the pH value, the more acidic the solution (i.e., the greater the concentration of H^+); the higher the pH value, the more basic the solution (i.e., the greater the concentration of OH^-). At pH=7, at ordinary room temperature, the concentrations of H^+ and OH^- are equal, and the solution is **neutral,** neither acidic nor basic. Pure water (HOH) is neutral because it has equal concentrations of H^+ and OH^-. ◖ Acids and bases react with many minerals, as well as with each other. The reaction of a base with an acid is called **neutralization.** In water lye dissociates into Na^+ and OH^-, and hydrochloric acid dissociates into H^+ and Cl^-. When a solution of lye is added to a hydrochloric acid solution to neutralize it, the H^+ and OH^- react to make H_2O, leaving the Na^+ and Cl^-dissolved in the resulting solution. ◖ Neutralization reactions occur in nature as well as in the laboratory. As rain droplets coalesce and fall through the atmosphere, for example, small amounts of oxygen, carbon dioxide, sulfur dioxide, and other gases may dissolve in them, producing weak acids. By the time the rain reaches the Earth's surface it is ready to begin chemically altering some minerals. As the acidic rainwater seeps into the ground, it dissolves salts and organic acids, increasing its acidity. The resulting solution, called **groundwater,** can chemically weather minerals in the rocks that it contacts. As a result, feldspars are decomposed into clays, sulfide minerals into sulfates, and silicate minerals like olivine or pyroxene into serpentine.

We benefit from these reactions in many ways. Not only do they help make soil and provide the mineral nutrients essential for plant growth, but when conditions are favorable, they also produce deposits of bauxite or garnierite, important ores of aluminum and nickel. The large bauxite deposits of Arkansas and Suriname and garnierite deposits of New Caledonia probably formed in this manner. On the down side, as rain falls through polluted air, it dissolves more of some gases (such as sulfur dioxide) than we would like, resulting in **acid rain.** Fortunately, nature has a way of controlling that, too—at least sometimes.

The pH of a solution greatly affects the solubility of some minerals. In areas where the underlying rock, known as **bedrock,** is limestone, lakes appear to be less affected by acid rain than those where the bedrock consists of igneous rocks such as granite. Why? Limestone is made of calcite, which dissolves more readily in acid solutions than in neutral water because of the increased number of H^+ ions available to help dissociate it. The abundant H^+ ions in an acidic lake combine with the HCO_3^- supplied by the dissolved calcite to form H_2CO_3, which breaks down into H_2O and CO_2. In the process the pH level increases to a more desirable, neutral value. The leftover culprits of acidity in acid rain, SO_4^{2-} ions, combine with Ca^{2+} to make calcium sulfate, which is relatively harmless to the lake.

This natural process has taught us that dumping crushed limestone into lakes suffering from the effects of acid rain can help restore their pH to neutral. The same neutralization reaction occurs on the outsides of some buildings, where acid rain has reacted with the $CaCO_3$ in their concrete or mortar, producing a white, powdery, surface coating of calcium sulfate, or gypsum. You seldom have to travel far to find minerals!●

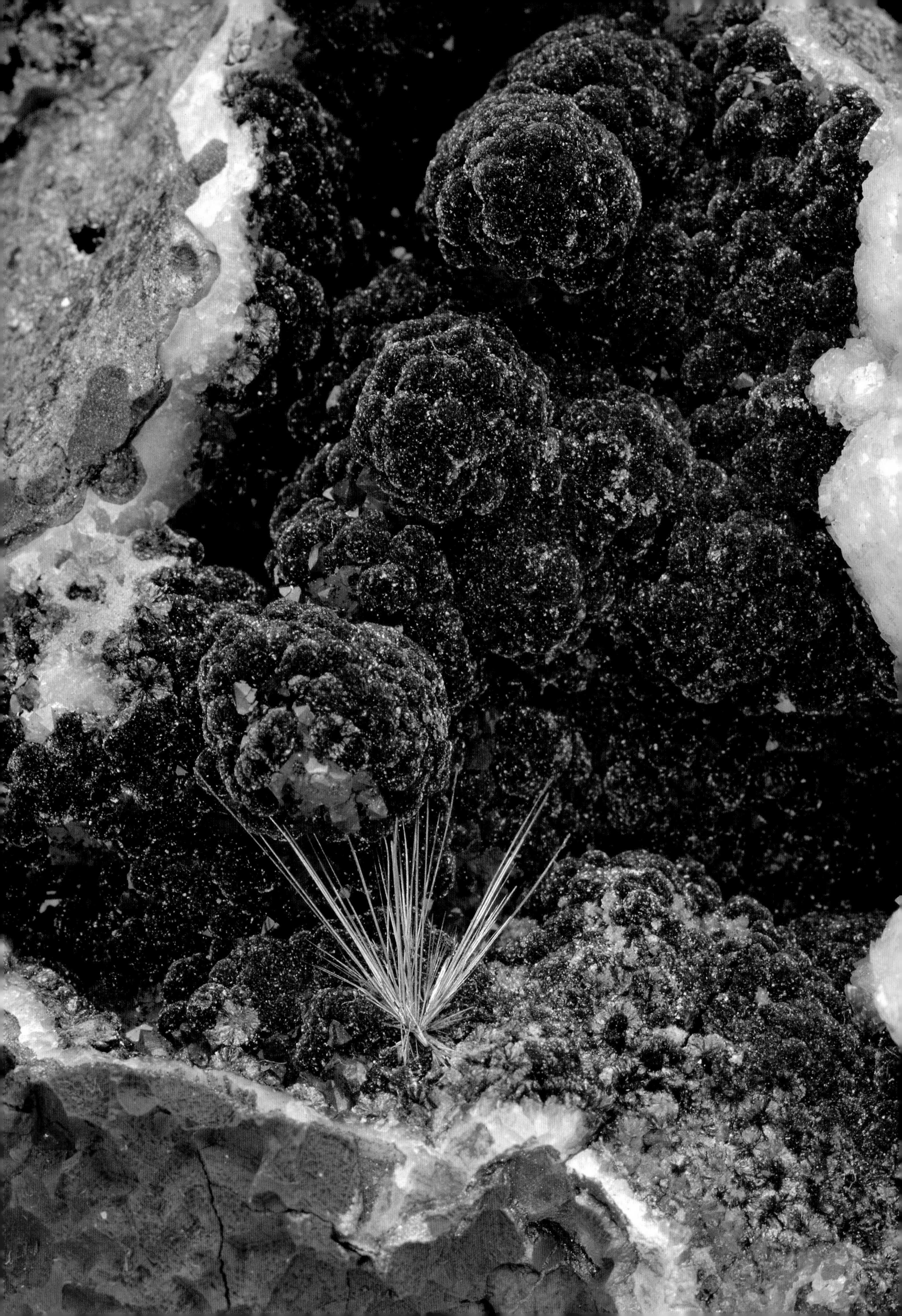

chapter twenty

Oxidation-Reduction Reactions

WHEN iron rusts or sulfur burns, the iron and sulfur combine with oxygen and are therefore said to be **oxidized.** The terminology seems logical, since they combine with oxygen. What really happens, however, is that the oxygen robs the iron and sulfur atoms of some of their electrons. In the broad, chemical sense, **oxidation** is the loss of electrons from an atom. In rusting, atoms of neutral, metallic iron (Fe^0) are oxidized to either ferrous (Fe^{2+}) or ferric (Fe^{3+}) cations by giving up either two or three of their electrons to oxygen atoms. Similarly, when sulfur burns, each neutral sulfur atom shares four of its electrons with two oxygen atoms to make the compound sulfur dioxide (SO_2). Because both iron and sulfur *lose* electrons, they are considered oxidized. But what about the oxygen? It experiences an opposite effect; it *gains* electrons. The gain of electrons is known as **reduction.** ◖ Whenever something is oxidized, something else must be reduced. This is what happens in **oxidation-reduction** reactions ("redox" reactions for short). In the rusting of iron, iron is oxidized and oxygen is reduced. When sulfur burns, sulfur is oxidized and oxygen is reduced. The total number of electrons lost by oxidation equals those gained by reduction, since formation of a stable compound requires that the total positive and negative charges of its constituent atoms are equal. It's like buying and selling: without seller and buyer agreeing on an equitable price, there can be no transaction. ◖ Other, more-complicated redox reactions can occur between iron minerals, as a colleague and I learned while studying the mineralogy of the Sterling iron mine near Antwerp, New York. This old, water-filled hematite mine was one of our favorite collecting spots. Although we never found any of the spectacular specimens of millerite

PLATE 77

Stilpnomelane (velvety green-brown) and **Millerite** (brassy needles).
Sterling mine, Antwerp, New York.
4 x 4 cm.

Millerite from the Sterling mine is world famous, but its associated minerals are more important in deciphering the origin of the deposit. The oxidation of some of the iron in the stilpnomelane was accompanied by the reduction of part of the iron in the hematite (red) to form the mineral magnetite (black).

[see PLATE 77] for which the locality is famous, we did collect other interesting minerals that provided clues about the role of redox reactions in their formation. Since these minerals were all iron-rich, the first question to address was the source of the iron that formed the hematite ore. The answer was in the surrounding rock.

The Sterling mine is situated in a body of granitic gneiss (a metamorphic rock with the composition of a granite) that contains local concentrations of the iron-sulfide minerals pyrite and pyrrhotite. Groundwater containing dissolved oxygen readily attacks these minerals, oxidizing them to soluble iron sulfates and liberating sulfuric acid in the process. Immediately adjacent to the pyrite-bearing gneiss is a large body of marble. If such an acidic solution were to encounter marble, the calcite in the marble would neutralize the solution, thereby increasing its pH. A higher pH (i.e., an increase in OH^- concentration) would cause precipitation of iron hydroxide, which is relatively unstable and, with time, dehydrates into hematite or goethite. At the Sterling mine stalactitic and botryoidal hematite appears to be one of the earliest-formed minerals, suggesting that such precipitation from solution did occur. Perfectly formed crystals of hematite coat some of the stalactites or line crystal pockets in the ore, suggesting that they formed after the initial precipitation.

However, there is something strange about these hematite crystals: they are magnetic, but hematite should not be attracted by a magnet. Why are these hematite crystals magnetic? Rather than hematite, these crystals are magnetite pseudomorphs after hematite. In hematite (Fe_2O_3) all the iron is ferric (Fe^{3+}), but in magnetite (Fe_3O_4) one of the iron atoms is ferrous (Fe^{2+}). The only way to change hematite into magnetite is by reduction, which implies that oxidation is occurring somewhere else, but where? The answer this time was among the "interesting" minerals we had collected: stilpnomelane [PLATE 77].

Stilpnomelane may contain both ferrous and ferric iron. The Sterling mine contains two kinds of stilpnomelane—one green, the other golden brown—but both must have formed after the hematite crystals, since stilpnomelane is always found coating the crystals. When we analyzed the two stilpnomelanes, we found that the green variety, which tended to be associated with the magnetite-poor ore, contained significantly less ferric iron than the golden-brown variety, which was associated with the magnetite-rich ore. Thus, the electrons made available from the oxidation of Fe^{2+} to Fe^{3+} in the stilpnomelane probably reduced some of the Fe^{3+} in the hematite to Fe^{2+}, resulting in the formation of magnetite.

With many minerals oxidation proceeds in a much less complicated, more direct manner. When groundwater containing dissolved oxygen percolates downward following fractures in a hydrothermal vein or rock containing native metals or metal-sulfide minerals, those previously formed, or **primary** minerals, often are oxidized. The minerals that form from them are referred to as **secondary,** and the portion of the vein affected by the oxidation is called an **oxide zone.** Oxide zones normally develop in the upper portion of a vein,

PLATE 78

Pyrite in Shale.
Goldfield, Western Australia.
8 x 10 cm.

The occurrence of pyrite in black, organic-rich shales is common, indicating the prevalence of oxygen-poor, reducing conditions.

FIGURE H

The Development of an Oxide Zone.

As groundwater containing dissolved oxygen trickles through fractures in veins bearing metallic-sulfide minerals, many of the metallic minerals are oxidized. Oxide zones usually develop above the water table, where more air is available. In the more-reducing conditions below the water table metal-sulfide minerals may precipitate again, forming a minable zone of enriched ore. (After Mason and Berry, 1968, Elements of Mineralogy, page 141.)

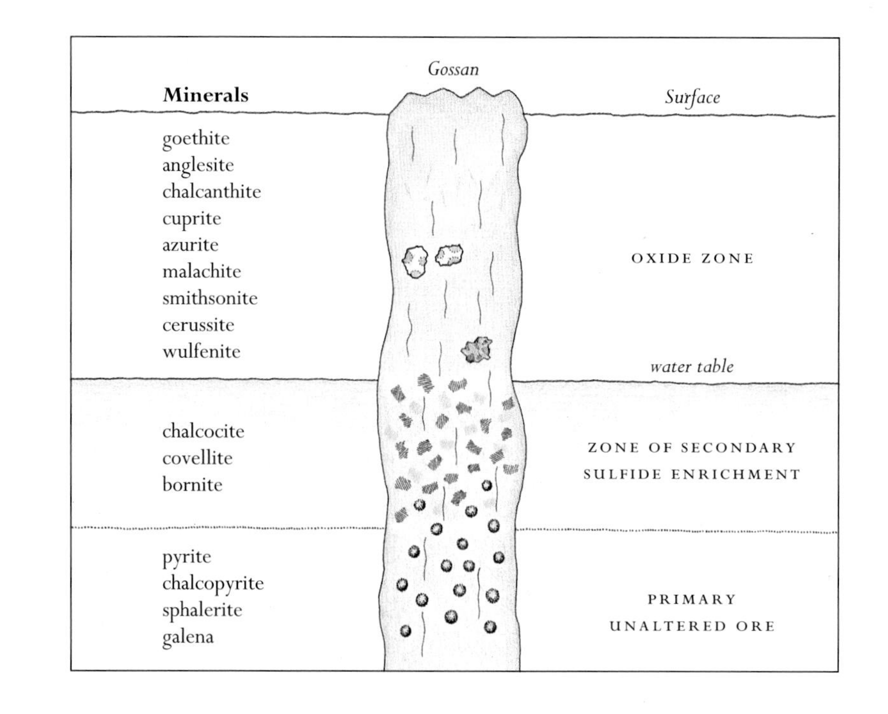

above the water table, where oxidizing conditions tend to prevail because air is usually present [FIGURE H]. Below the water table conditions are more reducing, since the available oxygen is limited to only the small quantity that remains dissolved in the water. Native metals and sulfide minerals are particularly susceptible to oxidation and commonly form oxides or sulfates, some of which may dissolve in the water, causing the metals to be leached away or to migrate below the water table, where under reducing conditions they may precipitate again as different minerals. Zones of **secondary enrichment** formed in this manner may concentrate metals into economically important ore deposits. The rocks in and around the vein from which the soluble minerals have been leached acquire a porous, often rusty appearance, forming a **gossan,** or "iron hat." A gossan often denotes the presence of an ore deposit beneath it. Such structures are often staked by prospectors in search of future mines.

Somewhat analogous are the sediments at the bottom of a lake or sea, where the chief source of oxygen is that dissolved in the water above the sediment. The accumulation and subsequent decay of organic matter depletes the water of available oxygen and forms black, carbon-rich sediments. Therefore, as one goes deeper into the sediment, conditions become increasingly reducing. That's why minerals such as pyrite, that favor reducing conditions, form deeper in the sediment than do minerals such as goethite, hematite, or siderite, that require a more oxidizing environment. The universal presence of pyrite nodules in black, carbonaceous shale [PLATE 78] illustrates the influence of reduction on iron mineralization in such geologic environments●

chapter twenty-one

Lights, Camera, Action!

THE cast is ready and knows the script, but has the stage been properly set? Just because all the right ingredients are present does not mean they will react. The physical and chemical parameters must not be at or near equilibrium, and reaction must lead to a more stable assemblage. What are some of the physical and chemical factors that determine which minerals form? The compositions of the primary minerals determine the elements available to be released as cations or anions by oxidation. The concentration of dissolved oxygen and the presence or absence of acids (i.e., the pH) may affect the degree of oxidation possible. Dissolved cations or anions already present in the groundwater may cause the precipitation of some minerals if their solubility limits are exceeded. The presence or absence of a particular anion in the solution often determines whether an available cation is transported by the solution because some anions may render the cation insoluble, thus removing it from the cast of players. ◖ Temperature also has an effect on the solubility and hence on the concentrations of cations and anions, as well as on the rates of their reactions. Usually an increase in temperature increases the rate of a chemical reaction. The grain size of the reactant minerals also influences the rate of reaction. Small grains collectively provide more surface area and thus react more quickly than larger grains do. The availability of fractures and the porosity of the rock determine where and how far solutions can penetrate. Locally, the permeability and composition of a rock can be highly variable and can cause the deposition of quite different minerals only centimeters apart. ◖ Finally, there is our old geological friend, time. The longer the reactants are exposed to each other, the longer they can react.

PLATE 79 *(above)*

Chrysocolla (blue) **and Tenorite** (black) **in Breccia.**
Silver City, New Mexico. 7.5 x 9 cm.

The fractures in this rock provided convenient avenues for copper-bearing solutions to enter and deposit these minerals.

PLATE 80 *(left)*

Cuprite (red) **on Native Copper.**
Ray, Arizona. 2.5 x 4.5 cm.

Oxygen dissolved in groundwater is probably responsible for the oxidation of this copper into cuprite.

Canada has thousands of metal-sulfide mineral deposits, but almost no oxide zones are associated with them, in spite of abundant fresh air and water. Why? The hard, dense Precambrian rock that hosts most of these deposits is relatively impermeable to oxygen-laden groundwater, the relatively low average temperatures impede oxidation reactions, and erosion by the glaciers from the last great ice age probably scoured away any shallow oxide zones that might have developed.

Many factors may be involved in even the simplest geochemical system. For example, if water containing dissolved oxygen comes in contact with native copper, the most likely minerals to form are copper oxides such as tenorite or cuprite [PLATES 79 to 81]. Whether the outcome is tenorite or cuprite depends largely on how much oxygen is dissolved in the water; cuprite requires less oxygen to form than does tenorite. Because groundwater also frequently contains dissolved carbon dioxide, two other minerals, the carbonates azurite and malachite [PLATES 7 and 82], commonly form. Together these two species probably account for most of the blue and green stains associated with copper-bearing mineral deposits. Which of the two forms usually depends on the concentrations of oxygen, carbon dioxide, and water. Although azurite has a broader range of stability than does malachite among these chemical parameters, it favors drier, less-oxidizing conditions, which are not always readily provided or maintained in oxide zones (because drier areas require more air than water and therefore generally have readily available oxygen). An increase in pH, on the other hand, as may occur when groundwater encounters limestone, can sometimes favor malachite stability. Malachite thus

PLATE 81

Cuprite (red) **and Chrysocolla** (blue).
Mashamba West, Shaba, Zaire.
2.5 x 4 cm.

Similar geochemical trends occur in widely separated oxide zones throughout the world. The precipitation of the copper silicate chrysocolla following copper oxides is demonstrated by both the tenorite in PLATE *79 and this cuprite.*

PLATE 82

Azurite and Malachite.
Apex mine, St. George, Utah.
8 x 10 cm.

Hollow molds of former gypsum crystals coated by blue azurite, partially replaced by green malachite, record a history of chemical change.

PLATE 83

Smithsonite.
Kelly mine, Magdalena, New Mexico.
4 x 7 cm.

When groundwater with dissolved oxygen filters through a near-surface ore deposit containing sphalerite or other zinc-rich sulfide minerals, it may oxidize some of them, putting zinc into solution. If the solution also contains dissolved carbon dioxide, or if it encounters limestone, the zinc-carbonate mineral smithsonite may form.

PLATE 84 *(right)*

Hemimorphite.
Iglesias, Sardinia, Italy. 6 x 8 cm.

Although pure hemimorphite and smithsonite [see PLATE *83] are colorless or white, both minerals occur in a wide range of colors resulting from trace amounts of elements such as cobalt or copper.*

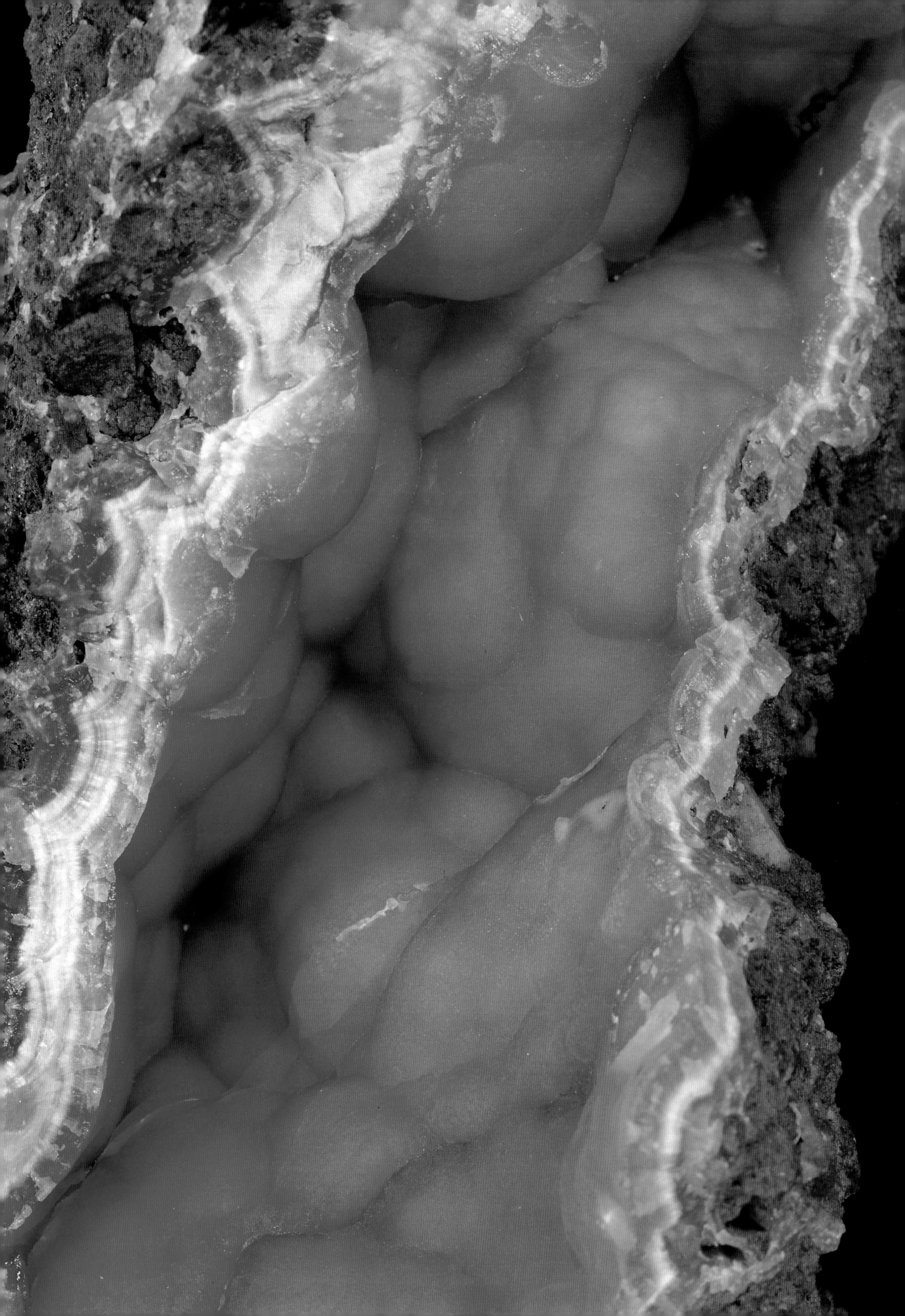

PLATE 85

Anglesite and Galena.
Blanchard claims, Bingham,
New Mexico. 5 x 6 cm.

Oxidation and replacement have formed concentric bands of anglesite surrounding cores of unaltered galena in this specimen. Under ideal conditions, crystals such as the one in plate 86 may form.

PLATE 86

Anglesite.
Touissit, Morocco. 6 x 8 cm.

When dissolved lead and sulfate ions are present in neutral to acidic solutions in the oxygen-rich environment above the water table, crystals of anglesite such as this one may form.

PLATE 87

Cerussite.
Tsumeb, Namibia. 4.5 x 7 cm.

Resembling delicate snowflakes, these twinned crystals of cerussite probably formed when lead, leached from galena, reacted with groundwater containing dissolved carbon dioxide.

is often more abundant than azurite and replaces it as pseudomorphs [see PLATE 82]. These examples show that the composition of the groundwater and of the rock it encounters can be just as important in determining what secondary minerals form in an oxide zone as the composition of the primary minerals themselves.

Now let's turn the camera to some other actors. So far, we have considered what minerals might form when a single element (e.g., copper) is exposed to a relatively "pure" groundwater containing only dissolved oxygen and carbon dioxide. Most hydrothermal vein deposits, however, contain several metal-bearing minerals, and groundwater nearly always contains a considerable variety of dissolved cations and anions. The ubiquitous presence of pyrite, chalcopyrite, galena, and sphalerite in hydrothermal vein deposits provides iron, copper, lead, and zinc ions to most oxide zones. More-complex sulfide minerals, such as tetrahedrite or tennantite, provide antimony and arsenic.

With a greater diversity of cations and anions available to react, a greater number of secondary minerals are able to form. Among the simpler and more common of these are the zinc carbonate and zinc silicate minerals smithsonite and hemimorphite [PLATES 83 and 84] and the lead sulfate and lead carbonate minerals anglesite and cerussite [PLATES 85 to 87]. Lead is rather immobile because it forms relatively few compounds that are soluble in water. Unless it can form complex ions with sulfur or chlorine, which keep it in solution, lead normally does not travel far. Instead it forms relatively simple secondary minerals close to its point of origin in oxide zones. Gold, too, is a very insoluble and inert element that must form complex ions in order to travel in solution. As a result, it forms no know secondary mineral species.

PLATE 88 *(left)*

Olivenite on Quartz.
Wheal Unity, Gwennap, Cornwall, England. 4 x 5 cm.

Originally deposited by hot, aqueous solutions, earlier-formed sulfide minerals containing copper and arsenic were oxidized at lower temperatures to make this copper arsenate mineral.

PLATE 89 *(right)*

Aurichalcite.
79 mine, Hayden, Arizona.
5.5 x 6.5 cm.

The presence of limestone at this mine was probably instrumental to the formation of aurichalcite [$(Zn,Cu)_5(CO_3)_2(OH)_6$], since it provided a source of carbonate (CO_3) and the means to neutralize acidic solutions in which aurichalcite would dissolve.

Dozens of other secondary copper, lead and zinc minerals are found in oxide zones, each with limits of stability with respect to oxidation and pH. For example, secondary minerals of zinc—such as the arsenate hydroxides, adamite [$Zn_2(AsO_4)(OH)$] and legrandite [$Zn_2(AsO_4)(OH){\cdot}H_2O$] [PLATE 90], and the carbonate hydroxides, aurichalcite [$(Zn,Cu)_5(CO_3)_2(OH)_6$] [PLATE 89] and rosasite [$(Cu,Zn)_2(CO_3)(OH)_2$]—favor basic conditions and require a highly oxidizing environment for their formation. All these minerals contain OH^- as an essential ingredient, and both their arsenic and their copper exist in states of high oxidation. It is not surprising that all four of these minerals occur both at the Ojeula mine, near Mapimi, Durango, Mexico, and at Tsumeb, Namibia, since at each locality primary zinc, copper and arsenic sulfide minerals (e.g., sphalerite, tennantite or arsenopyrite) were exposed to aerated groundwater which oxidized them, providing the cations and anions essential to form these minerals. Limestone ($CaCO_3$) at each locality probably provided the dissolved carbonate required for the aurichalcite and rosasite, and neutralized any acids, thereby increasing the pH of the solution, and so its $(OH)^-$ concentration. The chemically similar species, olivenite [$Cu^{2+}{}_2(AsO_4)(OH)$] [PLATE 88], is well known from Tsumeb, and has also been reported from Mapimi.

Virtually all metallic minerals are subject to chemical alteration. Given the proper pH range and oxidizing environment, dissolved aluminum, silicate or phosphate ions in the groundwater may react with copper, for example, to form minerals such as chrysocolla, liroconite, or turquoise [PLATES 79, 91, and 92]. Lead may combine with less-common anions to form equally colorful

PLATE 90

Legrandite.
Ojuela mine, Mapimi, Durango, Mexico. 4 x 6 cm.

Oxidation of primary sulfide minerals such as sphalerite and arsenopyrite provided the essential ingredients to form this rare secondary zinc arsenate mineral.

PLATE 91

Liroconite.
Wheal Gorland, Gwennap, Cornwall, England. 2.5 x 3 cm.

The world's finest examples of this rare copper aluminum arsenate mineral are from Wheal Gorland.

PLATE 92

Turquoise.
Gabbs, Nevada. 5.5 x 6 cm.

Nodules of turquoise such as this one form when acidic, copper-bearing groundwater reacts with aluminum- and phosphorus-rich minerals in porous volcanic rocks.

PLATE 93

Crocoite.
Dundas, Tasmania, Australia.
8 x 8 cm.

The occurrence of these crystals of crocoite on a rusty gossan matrix is typical of many minerals found in oxide zones.

PLATE 94

Wulfenite.
Mammoth-St. Anthony mine,
Tiger, Arizona. 4 x 6 cm.

This colorful lead molybdate mineral occurs in the oxide zones of many of Arizona's old copper, lead, and silver mines.

PLATE 95

Pyromorphite.
Cornwall, England. 3.5 x 4.5 cm.

The presence of pyromorphite, a lead phosphate chloride, on a porous, spongy quartz matrix suggests that the small cavities now in the quartz were probably once occupied by lead-bearing sulfide minerals that were leached away by groundwater containing dissolved oxygen, chloride, and phosphate.

PLATE 96

Mimetite.
San Pedro Corralitos, Chihuahua, Mexico. 7 x 10 cm.

The outward structure of this mimetite suggests that it precipitated from an aqueous solution, but its composition indicates that chemical alteration played an equally important role in its formation.

PLATE 97

Vanadinite.
Mibladen, Morocco. 3 x 5 cm.

Because pyromorphite and mimetite [see PLATES 95 and 96] have atomic structures and color ranges similar to vanadinite, identifying these three minerals without chemical analysis can be extremely difficult.

PLATE 98

Cuprosklodowskite.
Musonoi mine, Kolwezi, Shaba, Zaire. 6 x 6 cm.

Under the right conditions the oxidation of earlier-formed copper- and uranium-bearing minerals can result in the formation of this rare copper uranyl silicate mineral.

PLATE 99 *(left)*

Erythrite on Quartz.
Schneeberg, Saxony, Germany.
2-cm crystals.

Known to miners as "cobalt bloom," erythrite forms by the oxidation of cobalt- and arsenic-bearing minerals such as cobaltite and skutterudite. The purplish mineral in the matrix of the nickel-skutterudite specimen in PLATE 61 is probably erythrite.

PLATE 100 *(right)*

Chalcanthite.
Helvetia, Arizona. 2.5 x 3.5 cm.

This water soluble sulfate of copper forms by the oxidation of copper sulfide minerals such as chalcopyrite [PLATE 8], but survives only in relatively dry climates.

minerals, such as wulfenite, crocoite, pyromorphite, mimetite, or vanadinite [PLATES 93 to 97]. Uranium, cobalt, nickel, and antimony, as well as many other metals, also form secondary mineral species [PLATES 98, 99, and 102]. Some of these minerals are locally abundant, but globally they are uncommon because relatively few locations simultaneously have all the elements and the restricted chemical parameters essential for their formation.

Although chemically complex, these minerals are the natural and predictable consequence of the oxidation of a hydrothermal vein deposit. Theoretical modeling through laboratory simulations of such complex systems is possible but complicated. It is extremely difficult to anticipate and account for all factors in even the simplest natural environments. Even the climate may control the presence or absence of some species, such as chalcanthite [PLATE 100], which is readily soluble in water and therefore preserved only at localities in arid regions, such as Chuquicamata, Chile, or the southwestern United States. All too frequently more variables are operative in natural geochemical systems than are suspected. Perhaps that's why the theoretically impossible frequently seems to occur in nature ●

chapter twenty-two

Which Comes First ?

STUDYING the order in which secondary minerals form in oxide zones provides important clues to the chemical parameters that prevailed during their formation. This order, or succession, of mineral species, known as the **paragenetic sequence,** is discerned by observing what minerals crystallize on the surfaces of or replace other minerals. Pseudomorphs are perfect examples. The minerals constituting the pseudomorph formed after the original species, whose shape is retained. Goethite after pyrite [PLATE 101], for example, is proof that the iron sulfide pyrite formed first and was subsequently oxidized and replaced by goethite, an iron oxide hydroxide. ◖ Deciphering the paragenetic sequence usually requires only a good microscope, a notebook, and the patience to make and record careful observations on as many specimens as possible from the occurrence. Since the idea is to look for repeated, characteristic patterns, large numbers of **representative** samples must be examined. Otherwise the true picture may be distorted. For a geologist to determine the paragenetic sequence of a locality based on only a few samples from one small area would be equivalent to a visitor from outer space who, after arriving on this planet in the middle of the Sahara Desert, reports that the Earth is a hot, dry, planet devoid of life. When done carefully, simple paragenetic observations can reveal much about the genesis of a mineral deposit. One of my favorite mineral localities—a small, abandoned antimony mine in Ham Sud Township, Wolfe County, Quebec—illustrates this point well. ◖ At this locality veins of quartz containing native antimony (Sb) bear cavities containing small crystals of stibnite (Sb_2S_3), some of which appear to be replaced by kermesite (Sb_2S_2O). Crystals of valentinite and

PLATE 101

Goethite pseudomorph after Pyrite.
Windover, Nevada. 3.5 x 7 cm.

The perfect preservation of the distinctive, striated crystal faces on these former pyrite crystals by goethite leaves little doubt concerning their identity.

PLATE 102

Valentinite (white), **Kermesite** (red), **Stibnite** (black), **and Native Antimony** (silver).
Ham Sud Township, Wolfe County, Quebec. 4 x 5 cm.

Deciphering which minerals form before or after others may provide clues to changing chemical conditions during their formation. In this specimen native antimony was first oxidized by sulfur-bearing solutions to make stibnite, an antimony sulfide. As oxidation proceeded, solutions progressively richer in oxygen formed kermesite, an antimony sulfur oxide, followed by valentinite, an antimony oxide.

senarmontite (both Sb_2O_3) grow on the kermesite [PLATE 102], and the whole assemblage is frequently coated by stibiconite [$Sb^{3+}Sb^{5+}{}_2O_6(OH)$]. Take a minute to examine the formulas for these minerals. Initially only the native element, antimony, was present. Early in the paragenetic sequence sulfur reacted with antimony to make stibnite, but the sulfur was gradually replaced by oxygen as the prominent anion in the system, and kermesite and other oxide minerals formed.

That conditions became progressively more oxidizing is demonstrated by the sequential increase in the number of electrons lost by antimony, from zero in native antimony to three in stibnite, kermesite, valentinite, and senarmontite, and finally to five in the stibiconite. The presence of OH^- in stibiconite further suggests an increase in pH and/or a drop in temperature (minerals formed at high temperatures seldom contain OH^-) occurred. These observations are in accord with the geological setting. The veins themselves are locally sheared and fractured, indicating movement during or after they formed. The fracturing created openings for oxygenated water to enter and form the observed sequence of secondary minerals as erosion of overlying rock exposed the veins ●

chapter twenty-three

Chemical Alteration in Igneous and Metamorphic Rocks

CHEMICAL alteration is not confined to oxide zones and near-surface environments. In granitic pegmatites minerals like tourmaline, beryl, microcline, spodumene, or uraninite that form early in the crystallization sequence frequently may appear corroded or replaced by other minerals. Such minerals are no longer stable in the fluids present near the end of the crystallization of the pegmatite. Tourmaline may be replaced by micas or cookeite, beryl by bertrandite, spodumene by eucryptite, and uraninite by gummite, a bright orange mixture of secondary uranium oxides [PLATE 103]. The dissolution of these minerals replenishes the fluid with lithium, boron, beryllium, aluminum, uranium, and other elements to form even more exotic, rare minerals that are stable at lower temperatures. ℭ Probably the most famous minerals to form in this manner are the secondary phosphate species that arise from the alteration of primary phosphate minerals like amblygonite or triphylite [see PLATE 31]. The secondary phosphates that form depend largely on what cations are available in the fluid, on the temperature, and on whether conditions are oxidizing or reducing. Other physical and chemical constraints may be important for individual species. The available cations are a function of what primary minerals were attacked by the solution. If the environment is not very oxidizing, triphylite may be altered into ferrous iron phosphates such as ludlamite or vivianite, but under oxidizing conditions ferric iron phosphates like heterosite, rockbridgeite, or strengite may form. None of these minerals requires the addition of elements other than those present in the original primary phosphates from which they are derived. In most pegmatites, however, many other cations are available, so

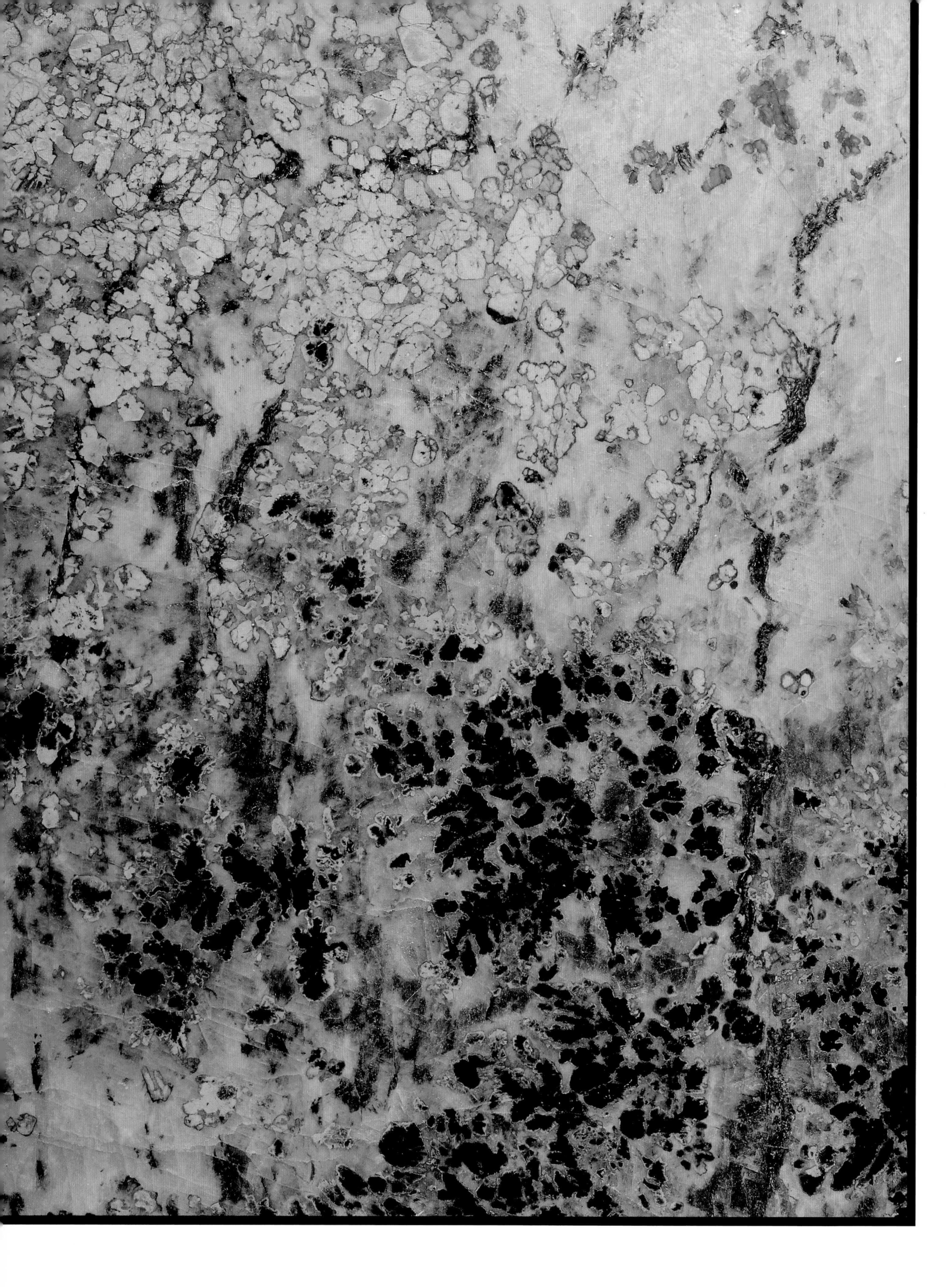

PLATE 104

Eosphorite on Rose Quartz.
Taquaral, Minas Gerais, Brazil.
3.5 x 5 cm.

Eosphorite is a common secondary phosphate mineral that forms by the chemical alteration of earlier-formed phosphate minerals, such as lithiophilite and triphylite [see PLATE *31], in some granitic pegmatites.*

PLATE 105

Hydroxylherderite.
Virgem da Lapa, Minas Gerais, Brazil.
1.5-cm crystal.

The corroded white feldspar under this crystal of hydroxylherderite provides evidence of chemical attack and dissolution of earlier-formed minerals in this granitic pegmatite. Hydroxylherderite is a secondary phosphate mineral formed by the dissolution of earlier-formed calcium-, beryllium- and phosphorus-bearing minerals.

PLATE 103 *(left)*

Gummite (orange) **and Uraninite** (black).
Ruggles mine, Grafton, New Hampshire. 6 x 7 cm.

Gummite forms by the chemical alteration of uraninite. All of the orange mineral in this polished slab of pegmatite was once black, as proven by the remnants of unaltered uraninite that form the cores of some of the grains of gummite.

PLATE 106

Tremolite replacing Diopside.
DeKalb, New York. 2 x 3 cm.

Water and carbon dioxide are probably responsible for the chemical alteration of part of this green, glassy diopside crystal first into tremolite (white), and later the tremolite into talc (gray).

many chemically more-complex minerals can form. For example, adding aluminum to triphylite that is in the process of being altered may result in the formation of childrenite or eosphorite [PLATE 104], whereas beryllium, from the dissolution of beryl, might form a secondary beryllium phosphate such as hydroxylherderite [PLATE 105].

Evidence of chemical alteration is visible in many metamorphic rocks, too. Certain minerals, which formed originally under conditions of much higher temperature and pressure, are subsequently exposed to more-aqueous conditions at lower temperatures and pressures, in which they are not as stable. As expected, they react to form species that are more stable in the new conditions. Because these reactions occur in response to a declining temperature and pressure, the process is sometimes referred to as **retrograde metamorphism.**

One of the best examples of retrograde metamorphism I have encountered is at a small outcrop of metamorphosed sedimentary rocks in a pasture near DeKalb, New York. This locality is famous for its transparent, green, gem-quality crystals of diopside [see PLATE 115], which formed under conditions of high temperature and pressure when calcium-, magnesium-, and silica-rich sediments were metamorphosed, over a billion years ago. The diopside occurs in small crystal pockets filled with a mixture of tremolite, calcite, talc, and quartz. Unfortunately, fresh, gem-quality crystals are uncommon here because most of the crystal pockets have been exposed to solutions that have altered much of the diopside into the mineral tremolite. Many crystals show further breakdown of the tremolite into talc [PLATE 106], quartz, and calcite, probably due to continued reaction with carbonated water. Other common examples of retrograde metamorphic reactions include the alteration of micas into chlorites or vermiculite, cordierite into chlorite or mica, and olivine into serpentine ●

chapter twenty-four

Replacement Deposits

REGARDLESS of their origins or how they are mobilized, solutions containing dissolved minerals nearly always bring about some sort of chemical reaction as they travel through permeable rocks. Frequently the reaction results in the replacement of one or more minerals by others. The specific reactions and how much rock gets replaced depend not only on physical and chemical parameters such as temperature, pressure, pH, and oxidation potential, but also on the compositions of the solution and of the rock being invaded, the permeability of the rock, and how long the solution and rock are in contact. The longer the ions can penetrate the rock by diffusion, the more extensive the replacement will be. Given sufficient time to equilibrate with their surroundings, such solutions can replace enormous volumes of rock, completely changing their composition and appearance. The wall-rock alteration that accompanies the emplacement of a hydrothermal mineral deposit may provide an important exploration guide, since a zone of alteration exposed on the surface may be easier to detect than an ore body buried beneath it.

❡ One of the classic examples of a rock formed by chemical replacement is dolomitic limestone, or dolostone. Dolomite [PLATE 107], the essential mineral constituent of dolostone, is calcium magnesium carbonate [$CaMg(CO_3)_2$]. Although several theories regarding the formation of dolomite have been proposed, none is universally accepted. All explanations agree, however, that it involves the replacement of some calcium in the calcite ($CaCO_3$) of the limestone by magnesium traveling in solution. Seawater is generally considered the primary source of this magnesium. Evidence of dolomite formation is often present in limestone (or dolostone) quarries in

PLATE 107

Calcite (golden) **on Dolomite** (pink).
Sainte-Clotilde-de-Châteauguay, Quebec. 20 x 20 cm.

The occurrence of large golden calcite crystals on pink dolomite in the Beekmantown dolostone of Quebec and in MVT deposits in the midwestern United States suggests that similar environments prevailed at each location.

PLATE 108

Quartz.
Middleville, Herkimer County, New York. 4 x 5 cm.

It's easy to see how these quartz crystals got the name "Herkimer diamonds"!

the form of a "pocket layer," a particular layer of sediment that contains crystal-lined cavities called pockets or **vugs.** Quarries in the Beekmantown, Lockport, and Little Falls dolostones that cover much of central New York and southeastern Ontario and Quebec frequently encounter such layers in their operations. In addition to dolomite, these pockets often contain crystals of quartz [PLATE 108] and minor amounts of the same kinds of minerals found in MVT deposits, since they form in a similar manner. Where magnesium replacement is more extensive, magnesite ($MgCO_3$) forms. Such magnesite deposits have been found in Brumado, Bahia, Brazil, and Eugui, Navarra, Spain.

Carbonate-rich rocks such as limestone, dolostone, or marble are frequently victims of other chemical replacement reactions, involving acidic, metal-bearing solutions. These solutions sometimes cause the formation of ore deposits of lead, zinc, or iron by gradually dissolving the carbonate minerals. The addition of the carbonates to the solution changes its pH and composition, causing the metals to precipitate in situ, thereby replacing the original rock or specific minerals in it. Some important examples of deposits formed in this way include the Magma mine at Superior, Arizona; various mines in the mining districts of Leadville, Colorado, and Santa Eulalia, Chihuahua, Mexico; and the Nanisivik mine on the western end of Baffin Island, Northwest Territories, Canada. The Nanisivik mine is famous for

PLATE 109

Pyrite.
Nanisivik, Northwest Territories.
3.5 x 5 cm.

At the Nanisivik mine beds of dolostone were replaced by sphalerite and pyrite, some of which contains pockets filled with crystals such as these. The unusual shape of these crystals was inherited from earlier-formed, twinned crystals of marcasite, over which they grew.

its unusual pyrite crystals [PLATE 109], which occur in ice-filled pockets. The pockets are filled with ice because the mine itself is situated in permafrost (permanently frozen ground) that extends to a depth of nearly 500 meters!

The unusual shape of these pyrite crystals is related to the way they formed. At Nanisivik, metal-bearing, sulfide-rich solutions permeated a bed of dolostone and replaced it with pyrite, sphalerite, and galena, creating the ore that is mined there. Originally acidic, the solutions became more basic by reaction with the dolostone. Marcasite and pyrite (both FeS_2) have different crystal structures. Because marcasite exists in more acidic conditions than those favored by pyrite, at Nanisivik marcasite preceded pyrite as the stable form of FeS_2. Flat, star-shaped, twinned crystals of marcasite are common at Nanisivik and often form a substrate upon which the pyrite crystals grow. The orientation and shape of the pyrite crystals is therefore inherited from the marcasite crystals, which act as a template. Most pyrite crystals, which usually do not overgrow and replace marcasite, more closely resemble those from Butte, Montana [see PLATE 65], than those from Nanisivik.

Hundreds of other examples of chemical alteration are known in the mineral world. Whether pyrite-replaced beds of limestone or tincalconite-replaced borax crystals, pseudomorphs provide irrefutable evidence that a change has taken place. As long as the physical and chemical parameters are favorable and the products of the reaction are more stable for the given conditions than are

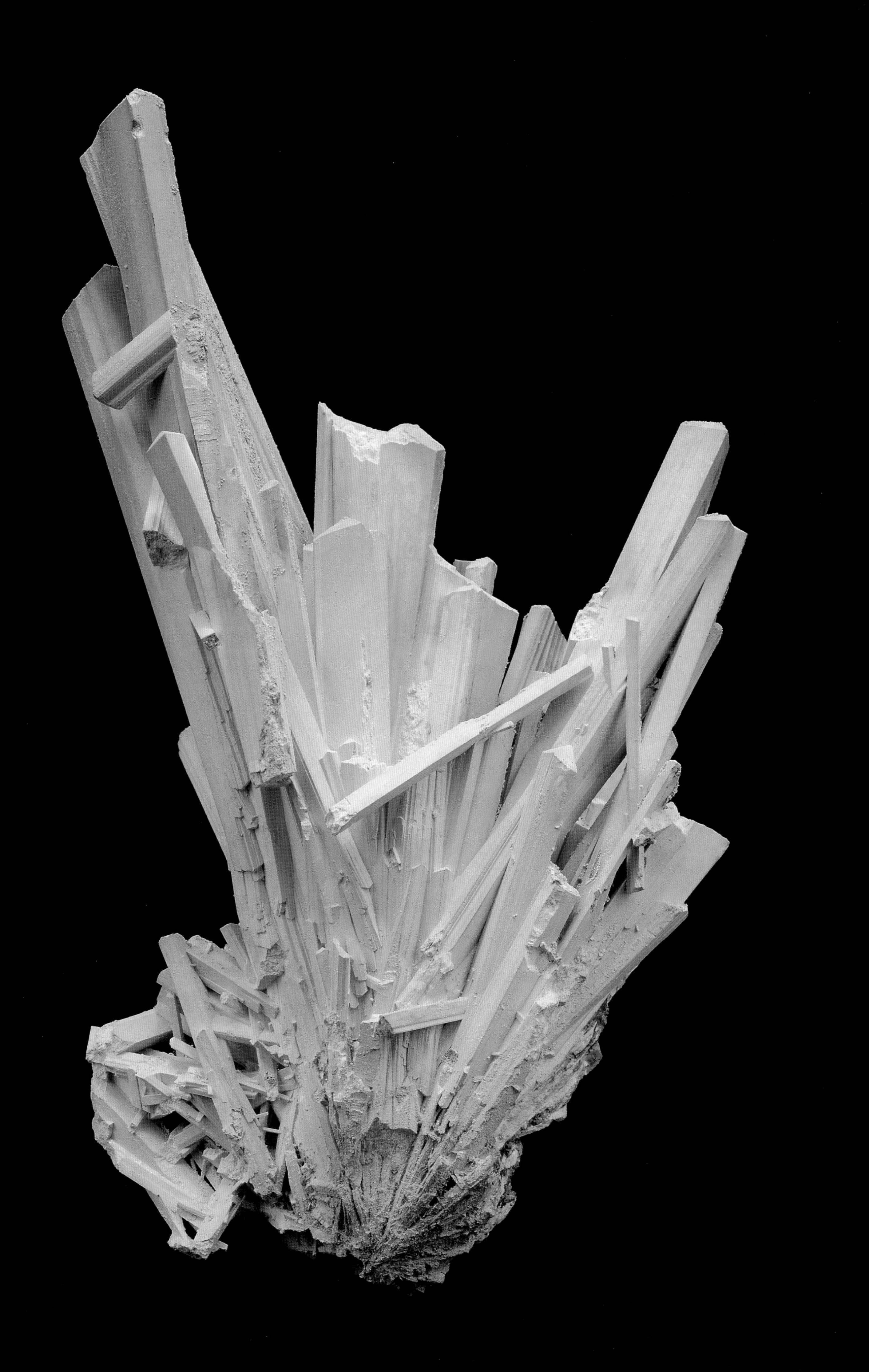

PLATE 111

Quartz replacing Riebeckite.
Girqualand West, Namibia. 5 x 8 cm.

The texture of the original fibrous riebeckite has been perfectly preserved by quartz to form this popular gem material, tigereye.

the reactants, almost any type of mineral can replace any other. Pseudomorphs may or may not make immediate "chemical sense," however. For example, it seems logical that stibiconite replaces stibnite [PLATE 110], since both minerals are antimony compounds, but why should quartz (SiO_2) replace chemically unrelated minerals such as fluorite (CaF_2) or barite ($BaSO_4$), as it does at Ouray, Colorado, or riebeckite (a Na-Fe-Mg silicate) to form the popular gemstone tigereye [PLATE 111], as it does at Griqualand West, Cape Province, South Africa? The formation of such pseudomorphs often involves a series of reactions induced by compositional changes in the solutions responsible for their creation. The minerals that make up the pseudomorphs must be more stable under the changed conditions, but the number of intermediate reactions required to arrive at these pseudomorphs may never be known. Nor can we be sure that the pseudomorphs we see represent a final, stable equilibrium for the conditions, since our collecting may have intercepted them en route to final equilibrium ●

PLATE 110

Stibiconite pseudomorph after Stibnite.
Catorce, San Luis Potosi, Mexico.
13 x 22 cm.

Probably once shiny and metallic like the stibnite in PLATE 64, these crystals were chemically altered into white, powdery stibiconite by oxygenated groundwater.

chapter twenty-five

Summary of the Chemical Alteration of Minerals

IN reviewing why minerals are subject to chemical alteration, remember that minerals are themselves nothing more than chemicals and that they are stable only under certain sets of conditions. When the conditions change, so do the minerals. The chemical factors that affect these conditions are varied. They include parameters such as pH, the oxidizing or reducing potential of the system, and the presence or absence of specific chemical constituents. Sometimes reactions require activation energy to progress. Another essential requirement is water, which is capable of dissociating ionic compounds into their constituent cations and anions, thus making them more reactive. Physical parameters—most importantly pressure and temperature, as we will see in Part V—also affect overall conditions. Natural geological environments provide a great range of temperatures, pressures, and chemical parameters, which result in a corresponding variety of possible reactions among minerals. ¶ Some reactions, such as hydration or dehydration of a species due to changes in temperature or relative humidity, or the simple oxidation of a metal into its oxide, are relatively simple; others are much more complicated. From the Earth's surface to the water table, conditions are generally oxidizing; below the water table they are more reducing. The oxidation of metallic minerals in hydrothermal veins often results in an oxide zone in the uppermost parts of the deposit. Sulfides are oxidized to sulfates, many of which are water-soluble and therefore can be carried in aqueous solutions through permeable rocks and can react with other elements to make more complex mineral species. When the oxidized metal ions encounter reducing conditions, as they might below the water table, they may precipitate as sulfides again, sometimes in concentrations of economic importance.

Chemical alteration also occurs in magmatic and metamorphic environments, where early-formed minerals are exposed to changing conditions in which they may no longer be stable. Different minerals react in different ways when exposed to changing chemical parameters. The availability of activation energy, the presence of water, the permeability of the rock, the size of the individual grains of minerals in the rock, variations in acidity, oxidizing or reducing nature of the environment, the presence or absence of specific cations or anions, and the amount of time available for reactions to occur all influence chemical reactions. Each mineral has limits of tolerance to these variables, and each will react when those limits are exceeded in an attempt to reach a lower-energy, more stable state of equilibrium with its surroundings. No mineral is immune to chemical alteration given the right physical and chemical conditions ●

PART V

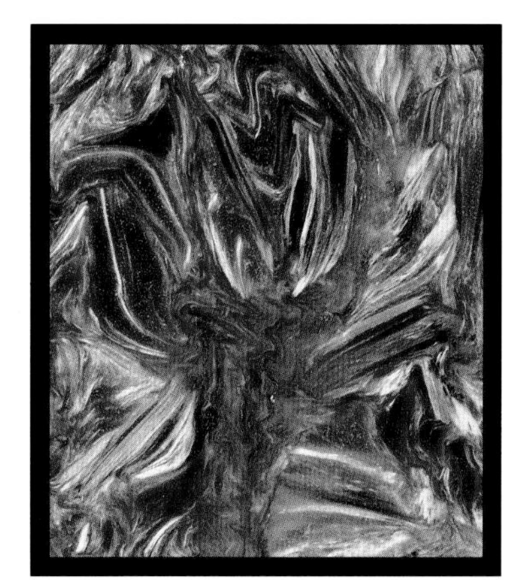

Recrystal-lization

chapter twenty-six

Heat and Pressure

IT'S three days after Thanksgiving, and we have a problem: how do we serve the leftover turkey this time? Should we heat it up again and serve it plain, or should we slice it and mix it with other ingredients to make turkey salad or a hot casserole? Whatever we decide, one fact remains the same: the turkey has already been cooked once and will require either reheating or the addition of new ingredients to make a new dish. Like our leftover turkey, previously formed minerals may be changed by reheating and/or combining them with other ingredients to make new minerals. Since the minerals already crystallized once before, they are said to **recrystallize** when heat and pressure involves them in the process a second time. ◖ Where does recrystallization take place? In all probability at this very moment minerals are recrystallizing about seven kilometers below our feet or anywhere else with sufficient heat and pressure to cause individual minerals to react chemically with one another or reorganize the atoms in their structures to make new minerals. Beyond the first one or two kilometers from the surface, the temperature in the Earth's crust increases with depth at an average rate of about 30°C per kilometer. We all know that rocks are heavy. The weight of several kilometers of rock is enormous and is the primary reason that pressure increases with depth. Depending on the density of the rock, a typical increase of pressure with depth in the Earth's crust ranges from 250 to 300 atmospheres per kilometer (1 atmosphere = 1,013 millibars, or 14.7 pounds per square inch at sea level). On the average, then, the minerals in the rocks seven kilometers below us are being subjected to temperatures and pressures of about 200°C and 2,000 atmospheres. Exposing minerals to this much heat and pressure is the equivalent of placing your hand on a hot coil on your kitchen stove and then parking 540 cars on top of it. It's enough heat and pressure to make you want to change the situation; it's also enough heat and pressure to initiate recrystallization in most rocks.

There are, of course, other sources of heat and pressure. When a magma intrudes a rock, for example, much heat is generated, and the pressure may increase locally because of the physical force applied by the magma. On a grand scale we know from plate tectonics that the continents are on the move. The pressure from weight and the stored thermal energy in the Earth's crustal plates is tremendous, and changes in pressure and temperature will be inevitable when these plates come in contact with one another. Furthermore, the changed conditions may persist for tens of millions of years. Because the plates themselves are recycled, so are the rocks in them. Today's marble may have been yesterday's limestone, today's garnet yesterday's beach. The recrystallization of minerals is an ongoing process.

In Parts I, II, and III, I emphasized that minerals are stable only within specific ranges of chemical and physical conditions but concentrated on the chemical conditions. In this section I will show how the two important physical parameters, heat and pressure, influence the recrystallization of minerals. Heat and pressure are two different things. **Heat** is a form of energy; **pressure** is the amount of force acting on a given unit of area. Like changes in chemical parameters, changes in temperature and/or pressure may cause an existing assemblage of minerals to become unstable and to recrystallize into one that is stable under the new conditions. Because heat and pressure are two different variables, they can affect the stability of an assemblage of minerals by acting independently or together ●

chapter twenty-seven

Metamorphism

WHEN subjected to heat and pressure, a rock may change both physically and chemically. The directed compression imposed by the collision of the Earth's plates causes crystals of platy minerals like mica or elongate crystals like amphiboles to align in parallel planes, changing the texture of the rock. (You can witness the effect by squeezing randomly stacked cards or pencils between two books.) If high enough, the pressure will crush and smear the mineral grains, increasing their total surface area and releasing trapped fluids. These fluids are critical for recrystallization. Because we usually focus our attention on the minerals, we too easily forget about the fluids. ◖ Without fluids there would be very little recrystallization because the reactions would proceed too slowly, even in terms of geological time. Increased surface area of mineral particles and the presence of fluids facilitate chemical reactions, which may be activated by heat. Some minerals respond to heat and pressure by reorganizing the atoms in their structures into a more stable arrangement; others react chemically with one another to make a new assemblage of minerals. The net result is that the whole character of the rock may be changed. Recrystallization by heat and pressure can induce both textural (physical) and mineralogical (chemical) changes in rocks. ◖ As in all other geological processes, time is also a critical factor. The longer a rock is exposed to heat and pressure, the more pronounced their effects will be. Like crystals growing in a magma or in an aqueous solution, crystals forming by recrystallization grow larger with time. The changes induced by recrystallization, however, must be accomplished in the solid state. The rock may behave like plastic, preserving swirls and folded layers created by imbalances in pressure, but it

PLATE 112

Kyanite pseudomorphs after Andalusite.
Lisens Alp, Tyrol, Austria. 6 x 8 cm.

An increase in pressure or a decrease in temperature was needed to transform these former andalusite crystals into kyanite.

must not melt, for if it does, a magma forms and crystallization proceeds by the igneous process differentiation. It is the solid-state transformation of one rock into another in response to changes in the physical and chemical conditions around it that geologists call **metamorphism** (meaning "change in form"). The process that brings about the change is recrystallization. Both the geographic extent and relative intensity of metamorphism can vary widely. The effects of metamorphism may be seen over very large areas **(regional** metamorphism), or they may be confined to within a few meters or kilometers of an igneous intrusion that brings about only a local increase in heat and pressure **(contact** metamorphism).

It's easy to see how melting sets limits for recrystallization at high temperatures, but what about at low temperatures? Drawing the line between high and low temperatures is not easy. Both metamorphic and sedimentary rocks form under heat and pressure. How then do we determine which temperature-pressure ranges result in sedimentary rocks and which induce metamorphism? The only objective criterion is the presence or absence of specific minerals. We recognize the onset of metamorphism, for example, by the appearance of one or more metamorphic **indicator** minerals.

The presence of indicator minerals and the changed textures in the rock enable a geologist to recognize where recrystallization has taken place. Some minerals, such as chlorite or quartz, may be stable in both sedimentary and metamorphic rocks, so taken alone, their presence tells us little about the temperature or pressure conditions under which the rock formed. Others, such as andalusite, kyanite [see PLATE 5], or sillimanite, are ideal indicators because their presence proves that a recrystallization reaction indicative of a specific range of pressure and temperature has occurred.

Andalusite, kyanite, and sillimanite have the same formula: Al_2SiO_5. The arrangement of the aluminum, silicon, and oxygen atoms within them, which is dependent upon pressure and temperature, is what makes them different. In general, andalusite forms under conditions of lower pressure than the others, kyanite forms at higher pressure, and sillimanite at higher temperature and higher pressure. These minerals demonstrate how temperature and pressure may act independently or together to recrystallize Al_2SiO_5. Andalusite may recrystallize into kyanite by a decrease in temperature or an increase in pressure, or into sillimanite by an increase in either or both of these variables. Similarly, kyanite may recrystallize into sillimanite by a decrease in pressure or an increase in temperature.

The famous andalusite crystals from Lisens Alp, Tyrol, Austria, are proof that such changes occur [PLATE 112]. Although for many years they were thought to be among the best examples of large, sharp andalusite crystals in the world, when X-rayed, many of these specimens exhibit patterns indicative of kyanite! Rather than pure andalusite, these crystals are kyanite pseudomorphs after andalusite, and could have formed only by an increase in pressure or a decrease in temperature after crystallization of the original andalusite.

Andalusite, kyanite, and sillimanite are good examples of indicator minerals because each forms within specific ranges of temperature and pressure. The mutual coexistence of two of these minerals in equilibrium greatly narrows the range, and the rare occurrence of all three in equilibrium defines a unique point at about 600°C and 6 kilobars of pressure. This type of theoretical information provides a powerful tool for solving geological problems, but it is not without its limitations. Experimental conditions only ***approximate*** natural ones, and the values obtained from them must be used with caution. The presence of trace elements and other chemical components in the natural environment, as well as unanticipated physical complications can change things dramatically!

Nevertheless, if we are mindful of such limitations and account for them, a sound, theoretical understanding of what *should* happen to certain minerals when exposed to specific temperatures and pressures provides us with a means to test geological hypotheses. At present the study of metamorphic rocks is one of the most important branches of the geological sciences because it is crucial to our understanding and assessment of the plate tectonic model of Earth dynamics. If we can predict how specific minerals should recrystallize as they accompany a plate on its way down to the mantle, we can look for the predicted results among the rocks in plates that have made the journey •

chapter twenty-eight

Composition

INDEPENDENT of how much or how long heat and pressure are applied to a rock is one other important and by now obvious factor that determines what minerals will form: the chemical composition of the rock being recrystallized. (After all, we can't make a ham sandwich from leftover turkey!) The fewer the initial ingredients, the fewer the potential products of recrystallization. In the simplest case, of course, only one mineral is present. Pure sandstones or limestones, which are composed almost entirely of the minerals quartz or calcite, respectively, fall into this category. When sandstone and limestone recrystallize into their metamorphic equivalents, quartzite and marble, their textures may change and coarsen, but their constituent minerals remain the same. Recrystallization of chemically more-diverse rocks, such as shale or impure limestone, has a greater variety of potential outcomes. ◖ Shale is composed largely of clay minerals and quartz (SiO_2). Clay minerals have various compositions and contain aluminum, sodium, potassium, calcium, magnesium, iron, and other elements. The quartz and clay minerals in shale are stable over a broad range of pressure up to about 300°C, so until that temperature is reached, few new minerals form. The physical appearance of the shale may change as fluids are driven off and it becomes more compact and brittle, but as the temperature increases, chemical changes produce various metamorphic indicator minerals. One of the first is pyrophyllite, which forms by the reaction of quartz with the clay mineral kaolinite. At around 400°C pyrophyllite breaks down into Al_2SiO_5 (in the form of either andalusite or kyanite, depending on the pressure),.quartz, and water. ◖ By this time the shale has probably become much harder and more compact and acquired a

PLATE 113

Almandine in Mica Schist.
Wrangell, Alaska. 2.5-cm crystals.

When aluminum-rich sediments are subjected to moderately high temperature and pressure, crystals of almandine garnet may form as the sediments are transformed into metamorphic rocks.

PLATE 114

Cordierite.
India. 13 x 14 cm.

Known as iolite or water sapphire, this attractive blue gem mineral forms when aluminum- and magnesium-rich rocks are subjected to high temperature and pressure.

shiny luster, the result of alignment of the platy clay minerals. It has become the metamorphic rock, slate. With increasing temperature and pressure, the slate will become even shinier, and its layers will probably deform as it recrystallizes into yet another rock, phyllite. Between 400 and 500°C chlorites and micas begin to appear, and the rock is transformed from phyllite into schist. Crystals of almandine garnet [PLATE 113] also may form at this time, especially with an increase in pressure. From 500 to 700°C additional minerals, such as staurolite, cordierite, or sillimanite [PLATES 5 and 114] may form, along with clots or distinct bands of feldspars and quartz, changing the rock from a schist into a gneiss.

In the late 1800s a Scottish geologist, George Barrow, found this precise sequence of minerals in northern Scotland, where shales have been metamorphosed by high heat and pressure. Starting where the effects of metamorphism were the least and progressing to where they were the most intense, Barrow found that the rocks changed from shale to slate, to phyllite, chlorite schist, mica schist, and finally gneiss. He also observed that the minerals in these rocks appeared in the following order: chlorite, biotite mica, almandine garnet, staurolite, kyanite, sillimanite. A very similar sequence exists between eastern Vermont and the White Mountains of New Hampshire, and many others have been documented elsewhere.

The composition of an impure limestone is quite different from that of shale and consists predominantly of calcite with variable amounts of dolomite, quartz, clays, and other minerals. Since these minerals provide calcium, magnesium, silicon, and aluminum as "ingredients," these same elements must constitute the new minerals formed by recrystallization. Exposing an impure limestone to increasing temperature and pressure usually results in the silicate minerals talc, tremolite, diopside, and forsterite. Because the carbonate content of limestone normally exceeds its silicate content, "excess" calcite left over from the reactions that form these minerals is common, which explains why they are often enclosed in coarsely recrystallized calcite [see PLATES 123 and 128].

Because the carbonates calcite and dolomite release carbon dioxide when they break down, the concentration of CO_2, as well as that of H_2O, becomes important in regulating the possible reactions. At very high temperatures, calcite and quartz may react to form the calcium silicate wollastonite and CO_2. However, since CO_2 is a product of the reaction, it must be removed for the reaction to proceed. (Remember the volleyball players from Chapter 17; the game could be won by adding players to the reactants team or removing them from the products team.) An environment of low pressure (likely provided by contact metamorphism), into which the CO_2 can escape, or the presence of water to dilute the CO_2 concentration accomplishes this task.

When limestones rich in silica and magnesium recrystallize, rocks such as tremolite-diopside schists (often with interbedded quartzite) may result. Such rocks are found at a number of localities in St. Lawrence County, New York, where well-formed crystals of diopside and tremolite [PLATES 106 and 115]

PLATE 115

Diopside.
DeKalb, New York. 4 x 4 cm.

These gem-quality crystals of diopside probably formed by the recrystallization of a sandy dolostone under conditions of high heat and pressure.

PLATE 116

Lazurite.
Sar e Sang, Nuristan, Afghanistan. 2-cm crystal.

Known to the gem trade as lapis lazuli, lazurite has been mined in northeastern Afghanistan since biblical times.

PLATE 117

Corundum, variety ruby.
Mysore, India. 3 x 4 cm.

These rubies probably formed by the recrystallization of aluminum-rich rocks at very high temperatures and pressures.

PLATE 118

Rhodonite in Galena.
Broken Hill, New South Wales, Australia. 8 x 10 cm.

Manganese and silica are required to form this colorful metamorphic mineral.

PLATE 119

Franklinite (black) **and Zincite** (red) **in Calcite.**
Franklin, New Jersey. 3.5-cm crystal.

Home to over 300 different mineral species, the unique ore body at Franklin, New Jersey, was formed when sediments rich in zinc and manganese were recrystallized by heat and pressure nearly a billion years ago.

occur in small pockets or seams. The famous diopside locality dear DeKalb and uvite locality near Richville (Gouverneur township) are classic examples. By contrast, if little silica is available, the presence of aluminum may lead to the formation of the oxide minerals spinel or corundum, since it is the silicate minerals that normally incorporate the aluminum. The ruby deposits of Mogok, Burma, which have yielded fabulous gem-quality ruby corundum and spinel crystals for centuries, are probably the most famous example. The presence of evaporite minerals in carbonate-rich sediments may lead to the formation of another important gem, lapis lazuli, which is composed of the ultramarine blue mineral lazurite [PLATE 116].

Any rock—igneous, sedimentary, or metamorphic—can be recrystallized. The minerals that form depend on the composition of that rock, the amount of heat and pressure to which it is subjected, the availability of fluids to promote recrystallization, and the allowance of sufficient time for the process to occur. The diversity of minerals that can result is enormous. Rocks such as peridotite, that are rich in magnesium silicate minerals, may recrystallize to form deposits of talc, serpentine, or chrysotile asbestos. Given the right conditions, the recrystallization of aluminum-rich rocks yields rubies or sapphires [PLATE 117]. Seafloor sediments containing local concentrations of manganese may recrystallize into rhodonite [PLATE 118]. At Franklin, New Jersey, metamorphism of manganese- and zinc-rich sediments in contact with limestone has produced a unique zinc ore deposit consisting of coarsely crystallized franklinite, willemite, and zincite [PLATE 119], along with numerous more-exotic species. Hundreds of other examples of regional metamorphic minerals could be cited, from actinolite to zussmanite, but the basis of their formation is the same: each is more stable in the imposed conditions of temperature and pressure than were the initial minerals from which it formed ●

chapter twenty-nine

Complex Recrystallization

SO FAR, I have talked only about minerals that form during the large-scale, regional recrystallization of rocks, driven by heat and pressure supplied by burial or tectonic forces. All the essential ingredients of recrystallization have been contained in the rocks themselves. Essentially, I have been only reheating the turkey. Now it's time to make a casserole by adding ingredients from outside sources. This is exactly what happens during **contact metamorphism** within a few meters of a small igneous intrusion or a few kilometers of a larger one [FIGURE 1]. ℭ The magma provides not only heat but also new ingredients to the rock that it intrudes by the injection and diffusion of fluids. While the heat induces recrystallization, the fluids bearing chemical components may react with one or more of the minerals in the rock, chemically replacing them. Both processes are commonly grouped together under the umbrella of contact metamorphism because both are confined to an area in contact with the intrusion. This contact zone of recrystallized and chemically replaced rock physically manifests itself as a band or **aureole** around the intrusion. Because any type of magma can intrude any type of rock, the possibilities for many different and exotic mineral species to form are great. Let's look at a few recipes for mineral casseroles. ℭ As we learned in Part II, the magmatic fluids that form granitic pegmatites become enriched in volatiles and light elements such as lithium, boron, and beryllium by the "law of constant rejection." When such a magmatic fluid intrudes a rock, these elements tend to diffuse into the rock because their small atomic size and mass usually permits their access to even the smallest permeable spaces. These elements form highly reactive ions that chemically alter the rock and form

PLATE 120

Beryl, variety emerald.
Malisheva mine, Yekaterinburg, Sverdlovsk Oblast, Russia.
3-cm crystal.

Emeralds in mica-schist deposits have a complex genesis, involving both magmatic and metamorphic processes.

FIGURE I

Contact Metamorphism.

When a magma intrudes a rock, chemical components are often exchanged, and the local increase in heat and pressure recrystallizes the rock in contact with the magma. A common example is the intrusion of a dolomitic limestone by a silicate magma to form a rock known as skarn. Silicon and aluminum from the magma react with calcium and magnesium in the limestone to make minerals such as diopside and grossular in a halo, or aureole, around the magma.

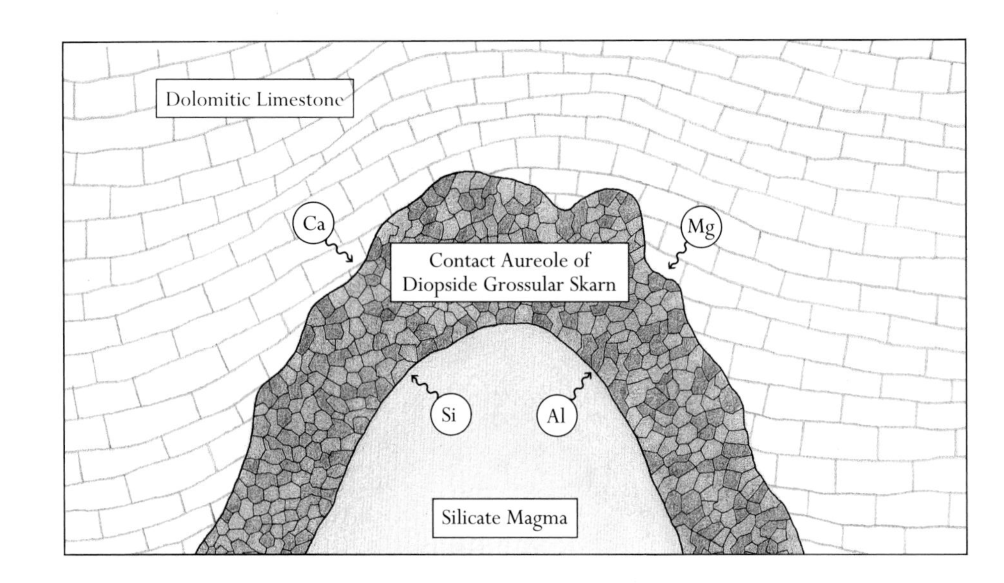

new minerals. Lithium-bearing amphiboles such as holmquistite, and the magnesium tourmaline, dravite, which are often found in rocks adjacent to intrusions of granitic pegmatites, are two examples. The lithium or boron required to make these minerals is supplied by the invading magmatic fluid; most of the other elements are provided by minerals already present in the rock.

When a beryllium-bearing fluid from a granitic magma invades a silica-poor rock such as amphibolite or serpentinite, a mica schist-type beryl deposit may result. The significance of such deposits is that the rocks that are intruded often contain minor concentrations of chromium or vanadium, two elements that may be incorporated by the beryl to produce emeralds [PLATE 120]. The legendary emerald mines of Cleopatra, near the Red Sea in Egypt, are of this type and are among the oldest mines known, having produced emeralds for Cleopatra, and possibly even King Tutankhamen, over 3,300 years ago. Similar deposits in Russia, Brazil, and other countries are the major producers today. In some of these deposits the rare variety of chrysoberyl, alexandrite, is also mined. Named for Russian emperor Alexander II, this unique gemstone appears green in daylight but red in artificial incandescent light●

chapter thirty

Skarns: Trash or Treasure?

IN Part IV we learned that at low temperatures limestone is susceptible to chemical replacement by metal-bearing solutions. At higher temperatures, near intruding magma, limestone typically recrystallizes into coarse-grained marble, while iron, aluminum, silicon, sodium, potassium, and water supplied by the magma react with the limestone to make silicate minerals containing those elements. The resulting recrystallized rock, called **skarn** [see FIGURE 1], may host a wide variety of mineral species. The word skarn is derived from the Swedish for "waste" or "trash." It originally referred to the silicate-rich band of waste rock (i.e., the contact metamorphic aureole) surrounding more-valuable ore. ◖ One person's trash is another's treasure. Some skarns may contain economically recoverable ore minerals such as magnetite, cassiterite, scheelite, sphalerite, or chalcopyrite, or provide local concentrations of manganese, phosphorus, uranium, and other metals. Many skarns contain large, well-formed crystals enclosed in recrystallized calcite, or marble. The crystals of ilmenite from Arendal, Norway, or Bancroft, Ontario, on the other hand, may not occur in sufficient quantities to be valued for their titanium content, but they are valued by scientists and collectors as some of the largest well-formed crystals of ilmenite known. Likewise, the spinel crystals from Amity, New York, may be too dark and opaque for use as gemstones, but they are among the finest large crystals of spinel in the world. ◖ The minerals that form in a skarn depend on temperature and pressure, as well as on the compositions of both the magmatic fluid and the original limestone with which it reacts. The heat transferred to the limestone from the magma may increase the temperature in the contact zone to nearly that of

PLATE 121

Uvite.
Powers farm, Pierrepont, New York.
7 x 12 cm.

The origin of the skarnlike assemblage of minerals at this famous locality is not well understood because much of its geology is concealed by overlying sand that was deposited by glaciers approximately ten thousand years ago.

PLATE 122

Uvarovite.
Outokumpu, Finland. 2-cm crystals.

Most garnet is red, but when rocks containing small amounts of chromium are recrystallized under heat and pressure, the rare green garnet, uvarovite, may form.

PLATE 123

Diopside (green) **and Phlogopite** (brown) **in Calcite.**
Bird's Creek, Ontario. 5 x 9 cm.

These calcium- and magnesium-rich silicate minerals are typical of skarns that form in dolomitic limestones.

PLATE 124

Fluor Silicic Edenite.
Wilberforce, Ontario. 4 x 4 cm.

The fluorine in this complex species of amphibole is a reminder of the importance of volatile components such as fluorine and water in promoting recrystallization reactions.

the magma itself. Therefore, it is not unusual to find feldspars, garnets, pyroxenes, or olivines in a contact aureole, since these minerals also are stable in high-temperature magmatic environments. The concentrations of water, carbon dioxide, fluorine, and other volatile components have a great influence on the total pressure and regulation of possible reactions between the various minerals present. The fewer the chemical constituents available, the fewer and simpler the minerals that can form. The simplest of these are anhydrous minerals, such as magnetite or wollastonite, that contain only one or two metal cations. Vesuvianite and uvite tourmaline [PLATE 121] are examples of chemically more-complex species found in skarns.

Given the elements most often involved in the formation of a skarn, and recalling some of the minerals formed by magmatic processes, it is not too difficult to predict which minerals might be found in skarns. For example, garnets form at relatively high temperatures and include species rich in calcium, aluminum, and iron, elements common in skarns. The calcium-rich garnets grossular, andradite, and uvarovite [PLATE 122] thus all occur in skarns. Like garnets, the pyroxenes, amphiboles, and micas most often encountered in skarns are rich in calcium, iron, and magnesium. Examples include the pyroxenes hedenbergite and diopside [PLATE 123]; the amphiboles tremolite, actinolite, and edenite [PLATE 124]; and the micas biotite and phlogopite [see PLATE 123]. Composed of similar elements, the minerals epidote and vesuvianite also are common in some skarns.

Potassium and plagioclase feldspars both occur in skarns, as does scapolite, because the combination of an impure limestone and a magmatic fluid rich in silica provides all the ingredients necessary to make these minerals. The calcium-dominant scapolite, meionite [PLATE 125], is especially prevalent in skarns because of the universal abundance of calcium in limestones. Large, well-formed crystals of scapolite have been found in hundreds of localities. One very different and interesting occurrence, however, stands out from all the rest: Mars. Some geologists believe that scapolite is present on Mars because infrared spectra from certain areas of the planet's surface closely resemble those of scapolite. Assuming that scapolite does exist on Mars, the requirements for its formation on the Earth, particularly the fluid phase necessary to promote crystallization, suggest that at one time in the geological past Mars had a very different atmosphere and perhaps even lakes or oceans, the remains of which may now be stored in its polar ice caps.

Because they result from recrystallization, chemical alteration, and replacement occurring simultaneously, skarns are among the most complex deposits known. In some of the billion-year-old metamorphosed sediments of northern New York State and southeastern Ontario and Quebec are a number of classic mineral localities that contain typical skarn minerals. Specimens of uvite, titanite, zircon, uraninite, fluorapatite, and other species from these occurrences can be viewed in nearly every major natural history museum in the world [PLATES 121 and 123 to 128]. Marble is usually present at or nearby most of these localities, but in some cases the skarns form curious, calcite-cored veins that cut

PLATE 125

Meionite.
Leslie Lake, Pontiac County, Quebec. 2.5 x 4 cm crystal.

This calcium-rich scapolite may have formed by the reaction of plagioclase feldspar with calcite under heat and pressure.

PLATE 126

Zircon.
Silver Queen mine, near Perth, Ontario. 4-cm crystal.

In skarns, the occasional occurrence of minerals such as zircon, that are typical of magmatic deposits, is a reminder of their complex origin.

PLATE 127

Uraninite.
Cardiff mine, Wilberforce, Ontario. 2-cm crystal.

Some of the largest well-formed crystals of uraninite in North America come from metamorphic calcite-fluorite veins in the billion-year-old rocks of the Wilberforce area in southeastern Ontario.

PLATE 128

Fluorapatite in Calcite.
Yates mine, Otter Lake, Quebec. 2 x 6 cm crystal.

Did the phosphorus required to form this fluorapatite crystal come from nearby sediments or from the igneous rocks with which its calcite matrix was in contact?

PLATE 129 *(left)*

Diopside (green) **and Grossular** (orange).
Jeffrey quarry, Asbestos, Quebec.
3 x 3 cm.

Diopside and grossular garnet commonly form when calcium-bearing hydrothermal solutions acting under heat and pressure convert granitic dikes in magnesium-rich rocks into rodingites.

PLATE 130

Vesuvianite.
Jeffrey quarry, Asbestos, Quebec.
3.5 x 11 cm.

Large, gemmy vesuvianite crystals such as these from Quebec are rare, but nearly identical smaller ones have been found in rodingites in Vermont, Italy, Pakistan, and Russia.

Asbestos, Quebec; Eden Mills, Vermont; and Asbest, Russia, are indistinguishable from each other, suggesting that similar geological conditions once existed at each of these places. Grossular from each of these localities ranges in color from pale pink to bright orange; some specimens are even emerald green. Green grossular owes its color to chromium, originally present in the magnesium-rich rocks as chromite and made available to the grossular by the fluids moving through and chemically altering the rocks. The cores of many of the green grossular crystals from Asbestos, Quebec, contain a tiny grain of chromite. Andradite garnet also occurs in some rodingites. The most famous locality is probably Val Malenco, Lombardy, Italy, where the green, gem variety known as demantoid has been found. Excellent crystals of epidote, clinozoisite, prehnite, pectolite, and many other minerals also have been found in rodingites.

Skarns and rodingites are not the only rocks formed by the action of chemically reactive fluids during metamorphism. The movement of hydrothermal solutions along contact zones between rocks of different com-

PLATE 131

Diopside (pale green) **with Clinochlore** (micaceous green) **and Grossular** (orange).
Mussa Alp, Piedmont, Italy.
3.5 x 4.5 cm.

The same tectonic forces that created the Alps also mobilized hydrothermal solutions to make these minerals and others in a belt of serpentinite that stretches across northern Italy into Austria.

PLATE 132

Charoite.
Chary River, northwest Aldan, Yakut, Russia. 8 x 10 cm.

Complex reactions between alkalic rocks and marble resulted in this attractive, unique gem mineral.

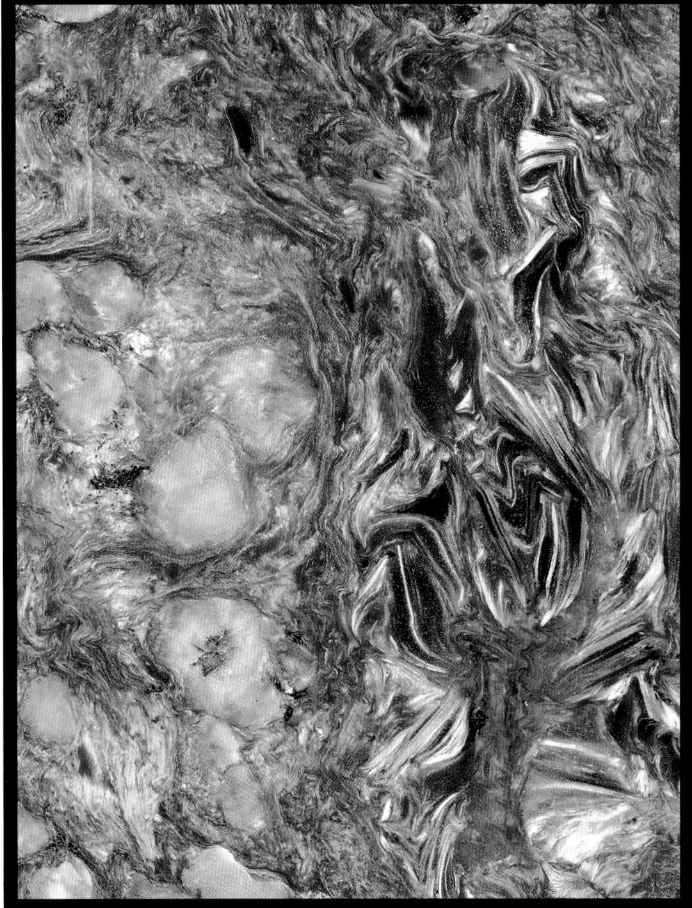

FIGURE J

The Formation of Rodingites.

a. When plates collide, pieces of oceanic crust are sometimes thrust over continental crust. b. With time the oceanic crust may become incorporated into a mountain chain, as the plates continue to move. c. Granitic magma formed by melting sediments dragged down a subduction zone rises and intrudes the old slab of oceanic crust as dikes. d. Heat and pressure cause hydrothermal solutions to be driven from the rock. The solutions chemically alter the oceanic crustal rock (e.g., peridotite) into serpentinite and the granitic dikes into rodingites.

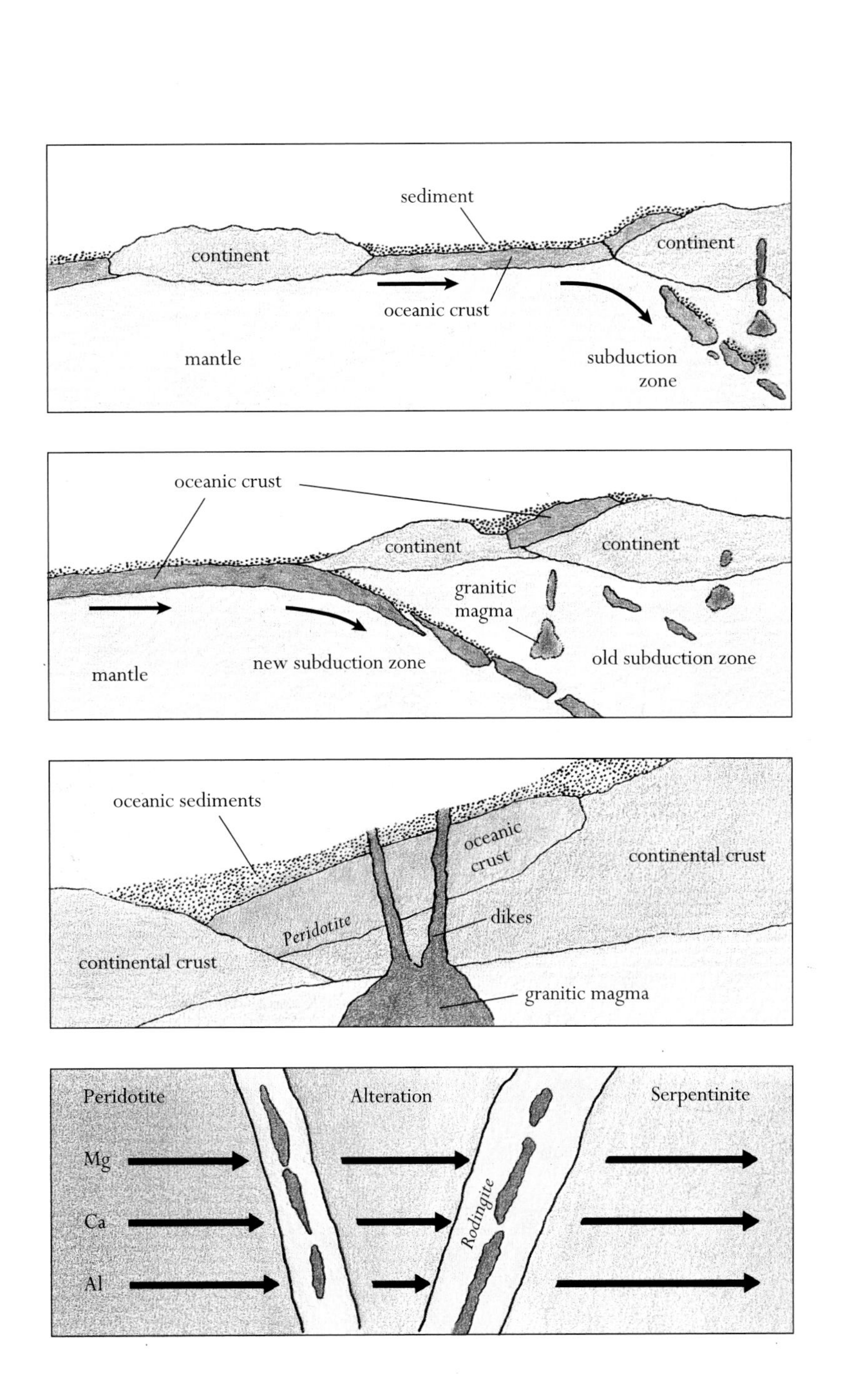

positions frequently results in the chemical alteration of one or both rocks. This type of alteration is especially evident in tectonically active areas such as folded mountain chains, where changes in temperature and pressure are experienced. The commercial deposits of talc in the Appalachians of Vermont and in southern Quebec probably formed in such a manner along contacts between serpentinite and various indigenous rocks, as did deposits of nephrite jade in similar geological settings. The huge almandine garnet crystals exposed at the Barton mine on Gore Mountain, near North Creek, New York, also are believed to be the result of chemical replacement along a contact zone, in this case between an olivine metagabbro and metasyenite. Charoite [PLATE 132] is found in chemically altered rocks at the contact of a nepheline aegirine syenite with limestone in the Chary River area, northwest Aldan, Yakut, Russia. Occurring as large masses of interlocking lavender crystals associated with several other rare minerals, charoite is used as both a gemstone and an ornamental rock. To date, the deposit remains unique ●

chapter thirty-two

Summary of Recrystallization

MANY minerals are not stable when exposed to high temperature and pressure and respond by reorganizing their own atoms into new, more-stable structures or by reacting chemically to form new minerals that are more stable under the prevailing conditions. This natural process of recycling old minerals into new ones is known as recrystallization. Recrystallization is facilitated by the presence of (usually water-rich) fluids in the rock, although the process itself occurs in the solid state. The metamorphic rocks that result are changed both chemically and physically. The parallel alignment of platy and rodlike minerals in response to directed pressure usually imparts a layered structure to these rocks. ❡ The three most important factors that determine which minerals form during recrystallization are temperature, pressure, and the chemical components available from the minerals and fluids in the rock. After metamorphism, rocks often show progressive zones of changing mineral assemblages that correlate with gradients in temperature and pressure. The presence of indicator minerals denotes specific conditions of temperature and pressure. The greater the number of chemical components involved, the greater the variety of minerals that may form. ❡ In more-complex systems, additional chemical components may be introduced by the intrusion of a magma or its accompanying diffusion. Since the area affected is confined to that in contact with the intrusion, the resulting process is called contact metamorphism. Within the contact zone, or aureole, there is usually evidence of chemical replacement of one mineral by another. Contact metamorphism is characterized by high temperature but relatively low pressure. Skarns are one of the most familiar examples of contact metamorphism, and result when a magma intrudes a carbonate-rich rock, such as limestone or dolostone. The carbonate rock supplies calcium, magnesium, and carbon dioxide to the system, while the magma supplies silicon, aluminum, iron, sodium, potassium, and various volatile components. The resulting skarn is thus rich in silicates of those elements. The presence of other minor chemical constituents enables numerous more exotic minerals to form. The conditions of contact metamorphism may be mimicked by other geological processes, such as the incorporation of pieces of limestone by a magma and their subsequent recrystallization under conditions of high temperature and low pressure.

Various geological settings may provide the conditions necessary for recrystallization. Both temperature and pressure increase as one progresses into the Earth's crust from its surface. Thus, simple burial of sediments in deep basins may be enough to initiate recrystallization. As sediments and oceanic crustal rocks are carried along the Earth's tectonic plates, they are also subjected to changes in heat and pressure. These changes are especially evident at plate boundaries, where plates collide. Hydrothermal fluids may be released by such tectonic activity and chemically alter the rocks they contact, making new minerals. Magnesium-rich rocks such as peridotites are often transformed into serpentinite during the process, while associated igneous rocks such as granite may be converted into rodingites. Rocks of dissimilar compositions also may be placed in contact with each other by activity along faults. Fluids migrating along their contact under the elevated temperatures and pressures of metamorphism may promote recrystallization. Numerous geological environments are conducive to recrystallization, each providing a mechanism for change in its governing parameters: temperature, pressure, and available chemical components ●

Summary

PART VI

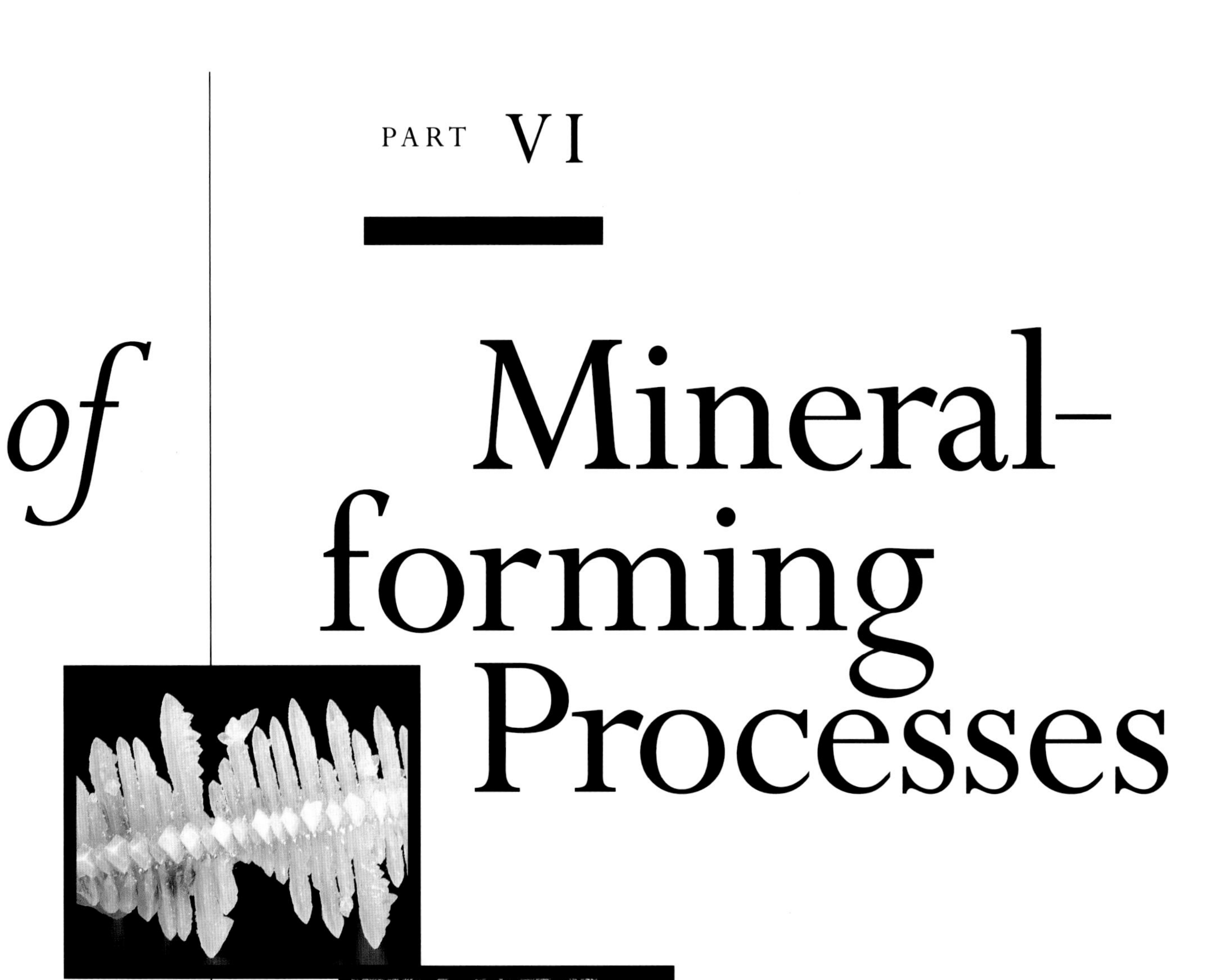

of

Mineral-forming Processes

PLATE 133

Beryl, variety emerald.
Muzo, Colombia. 4-cm crystal.

Several mineral-forming processes were required to make this crystal of emerald.

PLATE 134

Cubanite.
Henderson mine, Chibougamau, Quebec. 3-cm crystal.

It is hard to believe that the iron, copper, and sulfur constituting these lustrous, rare crystals of cubanite were probably once a black cloud of sulfide "smoke" spewed from a hydrothermal vent kilometers beneath the ocean.

chapter thirty-three

Interaction of Mineral-forming Processes

MINERALS are chemical compounds having specific ranges of stability with respect to physical and chemical factors such as heat, pressure, pH, and oxidation. These ranges are broad for some minerals, such as quartz and calcite, but narrow for others, such as diamond and gypsum. Often more than one geological environment provides conditions conducive to the formation of a particular mineral. That's why vivianite and other phosphate minerals may occur in such radically different geological settings as granitic pegmatites in Brazil or quartz veins in shale in northern Canada. More than one geological or mineral-forming process may be required to make some minerals. Nowhere is this better illustrated than in the famous emerald deposits of Colombia, or the volcanogenic massive sulfide deposits of the Canadian shield. ℭ The emerald mines of the Muzo and Chivor districts of Colombia have been actively worked since the mid-1500s, but only now are we beginning to understand how the deposits probably formed. A good part of the reason for our longstanding ignorance is that these deposits are totally unlike any other known occurrences of beryl, including other emeralds. Instead of occurring in granitic pegmatites or mica schists, emeralds in Colombia [PLATE 133] occur in veins of calcite and albite in black, organic-rich shales and limestones—sedimentary rocks normally indicative of relatively low temperatures and pressures. Even a few rare specimens of emerald replacing fossil snails have been found! Because all other occurrences of beryl are related to magmatic fluids, for years a similar origin was surmised for the solutions that deposited the Colombian emeralds. However, recent research offers a very different and plausible explanation. ℭ By studying the compositions and distribution of specific isotopes in fluid inclusions in these emeralds, we now know that the hydrothermal solution from which the emeralds crystallized does not have a magmatic origin at all. Instead it is more akin to a localized, higher-temperature version of a Mississippi Valley Type deposit. In MVT deposits metals are leached from sediments in geological basins by brines, which carry them in solution to their site of deposition. There chemical reactions (perhaps involving organic molecules) cause the metal-bearing minerals to precipitate.

In Colombia, tectonic forces deformed and fractured the organic-rich shale and limestone that hosts the emerald deposits, creating openings and releasing brines trapped in the sediments. The beryllium and other essential elements needed to form the emeralds were probably present in the sediments themselves and entered the brine complexed with organic molecules. As the somewhat acidic brine cooled in the fractures, it reacted with the limestone, which neutralized the brine and caused various minerals, including emeralds, to precipitate. The fortuitously present sulfur (probably in the form of hydrogen sulfide, H_2S) scavenged the available iron and precipitated it as pyrite, thereby preventing iron from being incorporated by the crystallizing emeralds. If that had happened, the color of the resulting beryl would probably have been a much more ordinary blue-green.

Volcanogenic massive sulfide deposits have an even more complex history. To see how some of these deposits form, we must journey to the bottom of the ocean where two of the Earth's plates are actively spreading apart, such as along the Mid-Atlantic or Juan de Fuca Ridges. Scientists traveling in deep-sea submersible vessels have visited both locations, collecting samples and recording their findings on videotape. In addition to witnessing underwater volcanic eruptions and discovering some new, interesting life forms, these scientists discovered and filmed a geological wonder that has revolutionized how geologists interpret the origins of many ore deposits: **black smokers.** Like smokestacks on a factory roof, these strange pipelike structures billow out clouds of what appears to be dense, black smoke, which, when analyzed, is found to be a very fine precipitate of sulfide minerals.

As seawater seeps down through cracks in the basaltic oceanic crust on the ocean floor, it is warmed by the magma below [FIGURE K]. The resulting hot, salty solution dissolves some of the metal-bearing minerals in the basalt. As if in a giant percolator, the solution is heated and forced back up through the cracks in the rock, dissolving more minerals, until ultimately it encounters the ice-cold seawater. Unable to remain in solution because of the abrupt change in temperature, the metals precipitate out as metal-sulfide minerals such as pyrite, chalcopyrite, galena, and sphalerite. The larger the chimneys grow, the more unstable they become, and eventually they topple over in a mound, but the finely dispersed sulfide "smoke" may spread out over a considerable area and settle as finely banded layers on the seafloor.

The finely dispersed sulfide minerals produced by black smokers are gradually buried by other marine sediments and carried away from their birthplace as the plates slowly drift apart. Eventually the sulfide-rich sediments reach a continent, where they are either carried downward along with the subsiding plate or are scraped off onto the continent. In the first case they descend toward the mantle and are heated until they melt, forming a sulfide-rich magma. Under pressure the

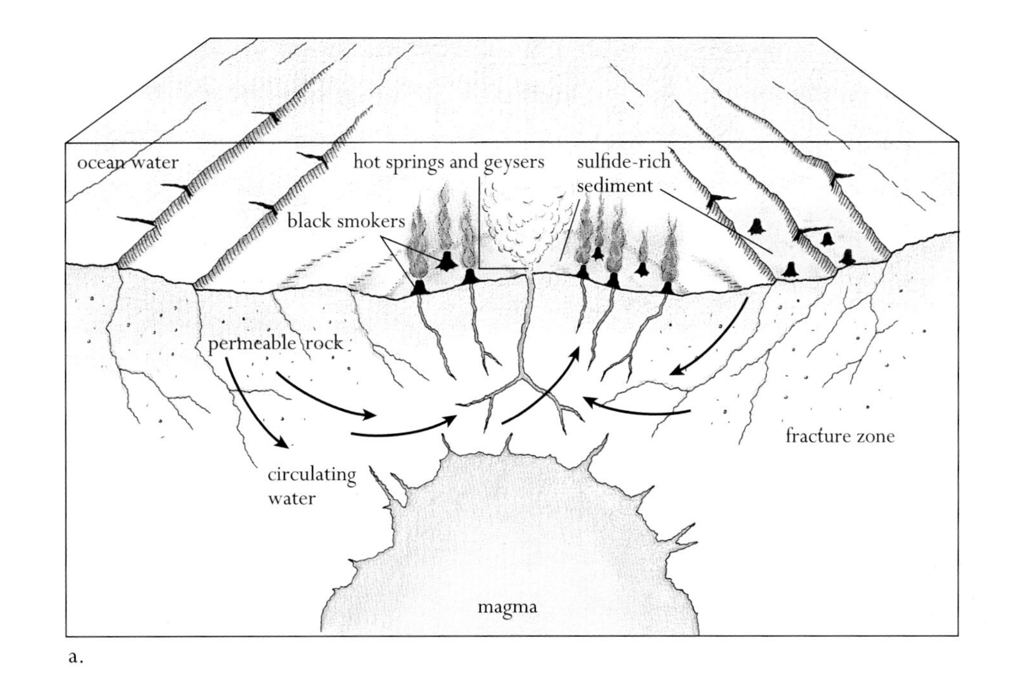

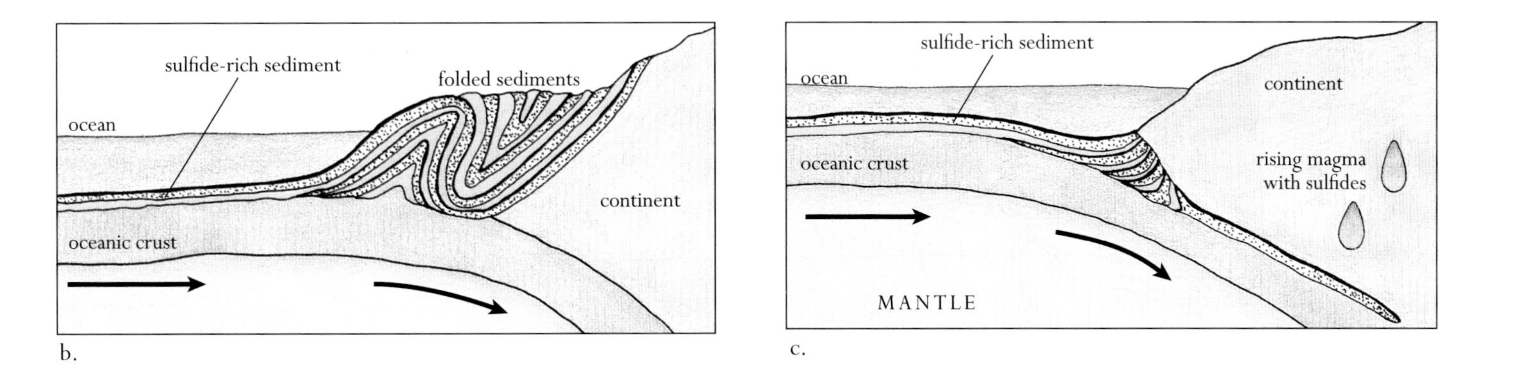

FIGURE K

Volcanogenic Massive Sulfide Deposits.

a. Near mid-oceanic ridges where two plates spread apart, magma heats seawater, which infiltrates fractures in the oceanic crust and dissolves metals from the rock. The heated water is forced upward under pressure and erupts from the seafloor as what is known as a black smoker. The "smoke" is really a fine precipitate of sulfide minerals that settles to the bottom as sediment. b. The sulfide-rich sediment is carried by the moving plate to a continent, where it may be scraped off and recrystallized into a massive sulfide deposit in a forming folded mountain chain.
c. Alternatively, the sediment may be carried down a subduction zone and melted to form a sulfide-rich magma, which rises and forms a massive sulfide deposit by crystallization and segregation.

magma rises along faults and fractures into the crust above it, where it eventually may form an ore deposit. In the second scenario, the sulfides are subjected to substantial heat and pressure from the tremendous force exerted by the converging plates and are recrystallized. Continued tectonic activity causes the sulfides to be folded and fractured as they are uplifted and incorporated into a forming mountain chain.

The fracturing also opens channels for tectonically activated hydrothermal solutions to enter and precipitate a second generation of sulfides, locally enriching the grade of the ore. The cubanite crystals in plate 134 probably formed in such a manner. As uplift continues, erosion wears down the rocks covering the deposit, exposing it to the surface and providing some lucky prospector with an opportunity to become a millionaire. Some of the world's richest mines are thought to have formed in this way. The repeated association of such massive sulfide deposits with volcanic rocks and marine sediments along both modern and ancient continental margins is more than just coincidence; it's a consequence of the recycling of the Earth's minerals by plate tectonics●

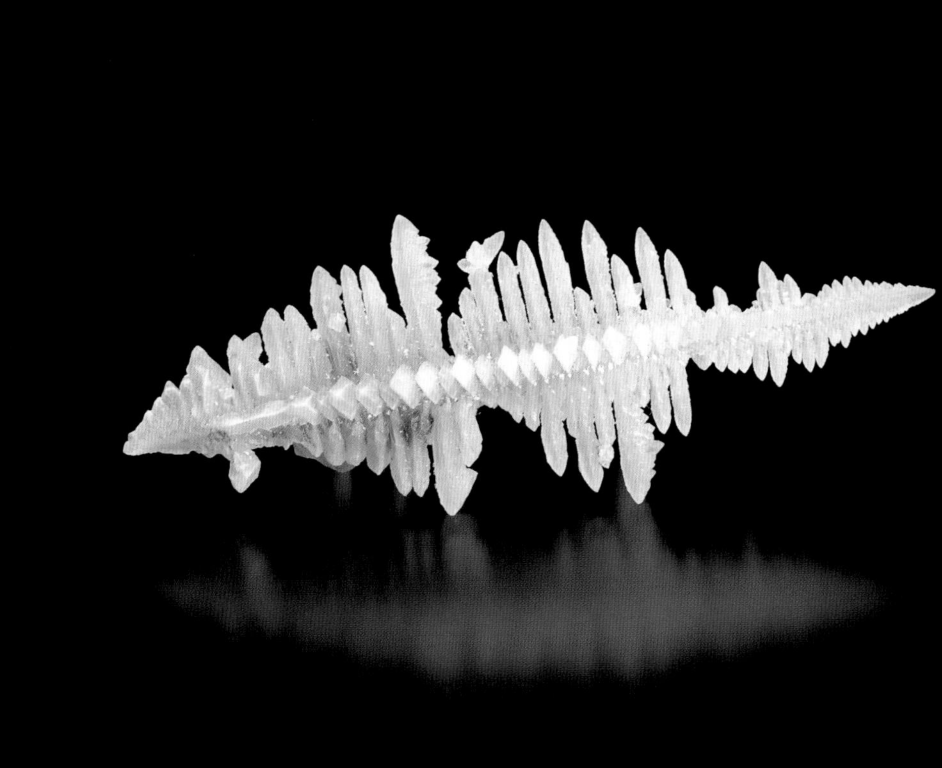

PLATE 135

Sulfur.
Perticara, Sicily, Italy. 3-cm crystals.

Once thought to be of volcanic origin, we now know that these crystals of sulfur formed by a biological process, as bacteria chemically converted the sulfate in minerals like anhydrite into elemental sulfur.

PLATE 136

Sal Ammoniac.
Ravat, Tadshikistan. 1.5 x 5 cm.

This type of skeletal crystal commonly results from rapid growth, which is characteristic of sublimation, the transition of a substance directly from a vapor to the solid state. In this case the vapor came from burning beds of coal.

chapter thirty-four

Biogenic Minerals

IN addition to the tectonic forces that recycle rocks and make minerals, we must reckon with another "force," the rather incredible "geological force" of life itself. In Part IV I gave examples of how oxidation-reduction reactions govern the genesis of certain minerals. Another very interesting, though less obvious, variant of such reactions—that provided by certain microorganisms—also leads to the formation of minerals. Plants and animals have respiratory systems that oxidize food aerobically (i.e., using oxygen) to produce energy. Anaerobic bacteria, however, do not have this capability. They derive their energy by the process of **chemosynthesis.** ◖ Anaerobic bacteria can take advantage of elements with multiple oxidation states, such as iron and sulfur, by effecting a change in the oxidation state. The energy released in the process sustains the life of the bacteria. For example, some bacteria can oxidize native sulfur or sulfide-bearing minerals into sulfates. The reverse process is accomplished by other bacteria, which reduce sulfates to sulfides. The white staining frequently seen on old limestone or concrete buildings is a result of the first of these reactions. Sulfate produced by bacteria oxidizing sulfides in the limestone reacts with calcium in the limestone or concrete to make gypsum (calcium sulfate), which forms the powdery white coating. The products of sulfur-reducing bacteria commonly occur as pyrite or other sulfide minerals in dark, organic-rich shales [see PLATE 78]. Because the genesis of such minerals is biologically driven, they are said to have a **biogenic** origin. ◖ Another, and economically more important, example of a biogenic mineral is native sulfur itself [PLATE 135]. Once thought to be of volcanic origin, most of the world's major sulfur deposits, such as those in Poland, in Sicily, or

along the Gulf coast of the southeastern United States, are now known to have been created by sulfate reduction by microorganisms. The biogenic origin of native sulfur was determined using isotope analysis. The most abundant isotope of sulfur is ^{32}S, but ^{34}S is also relatively common. Bacteria preferentially act on the lighter ^{32}S isotope, which causes a relative increase in the concentration of the heavier isotope, ^{34}S, in the "unused" sulfate. By measuring the concentrations of these two isotopes in the sulfur and sulfate, we can establish whether or not the reaction has been driven by bacteria.

In the Sicilian sulfur deposits, it is believed that beds of gypsum and anhydrite (both calcium sulfates) provided the sulfate that was reduced to sulfide (probably in the form of H_2S) while organic material was oxidized, liberating carbon dioxide. The carbon dioxide dissolved in water, forming carbonic acid (H_2CO_3), which reacted with the calcium ions derived from the gypsum and anhydrite to form aragonite or calcite (both $CaCO_3$). The frequent presence of hydrocarbons in close association with all these minerals lends credence to the theory that they are biogenic.

The crystals of sulfur, sal ammoniac [PLATE 136], and other minerals from Ravat, Tadshikistan, have different biological roots. These minerals form from vapors given off by burning beds of coal below the Earth's surface. Following fractures to the surface, the vapors cool rapidly, and the minerals form by **sublimation** (i.e., crystallization from a vapor directly to a solid). Certain calcium phosphate minerals (e.g., hydroxlapatite, monetite or brushite) are thought to derive their phosphorus from accumulations of bird droppings (called guano) from the large colonies of birds that inhabit some coral islands. The droppings are a source of soluble phosphorus that reacts with the calcium carbonate in the underlying coral to produce these minerals.

Similar reactions are known to occur in limestone caves that contain large accumulations of bat excrement (also called guano). Iron and aluminum phosphates such as strengite, phosphosiderite, and variscite have been found in guano deposits where iron- or aluminum-rich minerals are present beneath the droppings. Near Kerch, on the Crimean peninsula of Ukraine, well-formed crystals of vivianite and anapaite occur in casts of fossil shells [PLATE 137]. Vivianite occurs with other phosphate species in fossil shells at Mullica Hill, New Jersey, and the large crystals of vivianite found in clay in Richmond, Virginia, in the 1960s were reported to have been associated with fossilized whale bones.

Microorganisms are not the only life forms responsible for biogenic minerals. Interference in geological systems by fungi, plants, and animals—including humans—also causes minerals to form. The ancient mines of Laurium, Greece, located on the Aegean Sea approximately 40 kilometers southeast of Athens, have been worked for lead, silver, and zinc since 600 B.C. Slag produced from smelting the ore was dumped into the sea, where unrecovered lead reacted with seawater to form laurionite, penfieldite, and other rare minerals in cavities created by gas bubbles in the slag. All sorts of secondary sulfate minerals form on the walls of mines, where mining operations

PLATE 137

Anapaite.
Kerch Peninsula, Crimea, Ukraine.
4 x 5 cm.

Did the dissolution of the fossil shells lead to the formation of this specimen of anapaite by providing the necessary calcium or by locally changing the pH to a range more favorable for anapaite to form?

PLATE 138

Cuprite on Copper.
Britannia mine, Britannia Beach, British Columbia. 9 x 17 cm.

Are these minerals? The copper precipitated from water flowing over an iron pipe unintentionally left behind in the process of mining. The cuprite formed naturally by reacting with oxygen from the air, just as it probably did to form the crystals in PLATE 80.

have exposed primary sulfide minerals to moisture and air. Without human intervention, none of these minerals would have formed where they did.

As we drive down the highway in our cars and trucks, exhaust contaminated with lead from gasoline additives settles onto the roadside, where it reacts with phosphates in the soil to make plumbogummite. And how many kinds of minerals do you suppose might eventually form in our landfills and garbage dumps? Will any of them become an ore body? Even within our own bodies mineral processes are at work. Hydroxylapatite forms an essential part of our bones and teeth, and the less fortunate among us may be manufacturing struvite or weddellite in the form of urinary stones.

The formation of minerals by biogenic processes presents mineralogists with a philosophical problem: when is a mineral not a mineral? There are probably as many definitions of minerals as there are textbooks about them, but all the definitions have some common criteria: minerals must occur naturally, be a solid phase, and have a definite chemical composition and crystal structure. Some definitions also include caveats that minerals must be inorganic or be formed by inorganic processes. Hence purists may question the validity of biogenic or post-mining products [PLATE 138] as mineral species. Even most purists, however, have a hard time accepting that imposing such a rigid definition excludes such classic examples as the sulfur crystals from Sicily, the calcite forming the chalk in the white cliffs of Dover, England, and perhaps all marine dolomite from minerals.

Where human activities result in the formation of minerals, we must consider intent. If the intent of the activity is to produce the species (e.g., a furnace-grown diamond or ruby), then most mineralogists would probably consider it a synthetic compound, not a mineral. By this definition, the biogenic aragonite constituting a natural pearl would be considered a mineral, but the same compound in a cultured pearl would not because, without deliberate human intervention, the oyster would not have produced the pearl ●

chapter thirty-five

What Have We Learned?

YOU now know that some minerals form by a single process, others by several. You have also seen that considerable overlap and interaction may exist between these processes and the forces (including life!) that drive them. My decision to include or exclude any given mineral from the discussion of the main geological processes has been arbitrary and, of course, is limited by how much scientists know about its genesis. Often I can attribute the genesis of a mineral to a particular mineral-forming process with confidence based on its physical structure, associated species, or analogy to similar specimens of known origins. These conclusions usually are correct, but they may be wrong, especially if extended too far. ¶ For example I am confident that the famous rhodochrosite stalactites from Catamarca Province, Argentina [PLATE 139], form in open voids by precipitation from aqueous solutions because all other stalactites form in that way and because rhodochrosite is common in deposits of hydrothermal origin. I am not comfortable, though, speculating on whether the source of the aqueous solution that formed these specimens is meteoric, magmatic, or tectonic because I have never visited this locality, performed any fluid-inclusion or isotope studies on the rhodochrosite, or read an account of such studies. Similarly, I believe that the famous large stibnite crystals from the Ichinokawa mine on the island of Shikoku, Japan, as well as the classic examples of wire silver from Kongsberg, Norway [PLATE 140], also formed by precipitation from aqueous solutions, but I can only guess about the source of the antimony or silver required to make these minerals, or about the derivation of the solutions. ¶ The twentieth century has seen incredible advances in science and technology, ranging from humankind's first powered

PLATE 139

Rhodochrosite.
Catamarca, Argentina. 6 x 8.5 cm.

Its formation by precipitation from aqueous solution may be inferred from the structure of this rhodochrosite stalactite, but the sources for the solution or the manganese and carbonate it must have contained cannot.

PLATE 140

Wire Silver.
Kongsberg, Norway. 8 x 10 cm.

Although this silver specimen is from one of the world's most famous mineral localities and silver mining districts, many of the specifics concerning its origin remain a mystery.

flight to its first footsteps on the moon. By understanding how minerals form in nature, we are now able to synthesize many of them and can even improve some of their desired properties by doing so. We have made great strides in understanding how nature operates, but as each new discovery answers a question, it nearly always raises numerous others. We have gained considerable knowledge of how matter behaves in relatively simple, closed systems, and from the laws of thermodynamics we can successfully predict which minerals should be stable at a given set of parameters for that system. Unfortunately, nature seldom provides such simple, closed systems. A quick look at the silicon-oxygen system proves the point.

Silica (SiO_2) may appear as one of several different minerals, depending on temperature and pressure. At atmospheric pressure, low quartz, high quartz, tridymite, and cristobalite crystallize at progressively higher temperatures. At significantly higher pressures, coesite or stishovite may form. (The existence of the latter two species was predicted before their actual discovery in nature.) At the Earth's surface, low quartz is the stable form of SiO_2. Tridymite should exist only at temperatures from 870 to 1470°C, yet tridymite still persists in volcanic rocks formed in that temperature range at the Earth's surface tens of thousands of years ago, because it lacks the activation energy required to recrystallize into low quartz.

Opal, another silica-rich mineral, forms under near-atmospheric conditions of low pressure and temperature and should therefore form as low quartz. X-ray studies of opal indicate, however, that the much-higher-temperature mineral cristobalite, is the species usually present. How can this be? The answer is that although they are SiO_2, the opals formed in a complex system containing more than just silicon and oxygen. The system includes water, aluminum, and alkali metals, which, when incorporated in only minor amounts into the structure of the crystallizing SiO_2, favor a cristobalite-like structure.

The world of mineralogy is full of surprises and unanswered questions. Knowing positively how a given mineral has formed and why is the exception rather than the rule. Because minerals and the elements they contain have become essential to our everyday lives, research will undoubtedly continue to reveal new information about their properties, potential uses, and formation in nature. Knowledge of how minerals form is fundamental to locating future ore deposits and to synthesizing substitutes for crucial minerals whose natural occurrences are being depleted. *Homo sapiens* is in a unique position: no other organism has such an ability to understand and control its environment. Understanding the Earth's basic geological processes and their relationship with the biosphere and with human activity is essential to our own survival ●

Recommended Reading

American Geological Institute. 1962. *Dictionary of Geological Terms.* Garden City, NY: Dolphin Books, Doubleday. A concise, handy reference for all audiences.

Bancroft, Peter. 1973. *The World's Finest Minerals and Crystals.* New York: Viking Press. Intended primarily for mineral collectors, this is an easily read text.

Bancroft, Peter. 1984. *Gem and Crystal Treasures.* Fallbrook, CA: Western Enterprises. A unique blend of travel, history, and descriptive mineralogy focusing on most of the world's most important mineral localities; for both general and informed audiences.

Brownlow, Arthur H. 1979. *Geochemistry.* Englewood Cliffs, NJ: Prentice-Hall, 1979. A modern, broad-based approach to geochemistry for an advanced audience.

Carmichael, Ian S. E., Francis J. Turner, and John Verhoogen. 1974. *Igneous Petrology.* New York: McGraw-Hill. A thorough discussion of magmatic processes and the genesis of igneous rocks; advanced level.

Cattermole, Peter, and Patrick Moore. 1985. *The Story of the Earth.* Cambridge: Cambridge University Press. A modern physical geology text written for a general or informed audience.

Cocks, L. R. M., and P. H. Greenwood, eds. 1981. *Chance, Change and Challenge: The Evolving Earth.* Cambridge: Cambridge University Press. A compendium of papers on the evolution of the Earth and its geological systems; for an informed audience.

Degens, Egon T. 1989. *Perspectives on Biogeochemistry.* New York: Springer-Verlag. A comprehensive review of the interaction between the living and nonliving world; advanced level.

Desautels, Paul E. 1968. *The Mineral Kingdom.* New York: Madison Square Press, Grosset and Dunlap. A nontechnical approach to the fascinating world of minerals; written for a general audience.

Ehrlich, Henry Lutz. 1981. *Geomicrobiology.* New York: Marcel Dekker. A technical discussion of some of the more important biogenic minerals; advanced level.

Fleischer, Michael, and Joseph A. Mandarino. 1991. *Glossary of Mineral Species 1991.* Tucson: The Mineralogical Record. A handy, alphabetical listing giving the chemical formula and a published reference for all known mineral species.

Frye, Keith, ed. 1981. *The Encyclopedia of Mineralogy.* Stroudsburg, PA: Hutchinson Ross. A comprehensive, advanced reference manual.

Guilbert, John M., and Charles F. Park, Jr. 1986. *The Geology of Ore Deposits.* New York: W. H. Freeman. A comprehensive review explaining the genesis of nearly every kind of ore deposit; intended for an informed to advanced audience.

Jackson, Kern C. 1970. *Textbook of Lithology.* New York: McGraw-Hill. A thorough treatment of the fundamentals of physical geology for a general but informed audience.

Kearey, Philip, ed. 1993. *The Encyclopedia of the Solid Earth Sciences.* Oxford: Blackwell Scientific Publications. A modern, comprehensive reference manual for all audiences.

Keller, Peter C. 1990. *Gemstones and Their Origins.* New York: Van Nostrand Reinhold. An informative, well-illustrated text explaining the genesis of some of the world's most important gem deposits; for both general and informed audiences.

Lapedes, Daniel N., ed. 1978. *McGraw-Hill Encyclopedia of the Geological Sciences.* New York: McGraw-Hill. A comprehensive reference manual for all audiences.

Mason, Brian. 1958. *Principles of Geochemistry.* New York: John Wiley and Sons. One of the classic texts describing the chemistry of the Earth and its geological processes; for an advanced audience.

Mason, Brian, and L. G. Berry. 1968. *Elements of Mineralogy.* San Francisco: W. H. Freeman. A traditional but popular mineralogy textbook; for an informed audience.

Mottana, Annibale, Rodolfo Crespi, and Giuseppe Liborio. 1978. *Guide to Rocks and Minerals.* New York: Simon and Schuster. A handy pictorial guide for a general or informed audience.

Nickel, Ernest H., and Monte C. Nichols. 1991. *Mineral Reference Manual.* New York: Van Nostrand Reinhold. A handy, alphabetical listing giving chemical and physical data with reference citations for all known mineral species.

Pough, Frederick H. 1960. *A Field Guide to Rocks and Minerals.* Boston: Houghton Mifflin. One of the most popular texts among amateur mineralogists; for general and informed audiences.

Redfern, Ron. 1983. *The Making of a Continent.* New York: Times Books, New York Times Book Company. A well-illustrated text explaining the principles of plate tectonics; for general to informed audiences.

Roberts, Willard Lincoln, Thomas J. Campbell, and George Robert Rapp, Jr. 1990. *Encyclopedia of Minerals.* 2d ed. New York: Van Nostrand Reinhold. A comprehensive reference manual for all audiences.

Sinkankas, John. 1964. *Mineralogy for Amateurs.* New York: Van Nostrand Reinhold. A comprehensive, easily read, popular text for general and informed audiences.

Sofianides, Anna S., and George E. Harlow. 1990. *Gems and Crystals from the American Museum of Natural History.* New York: Simon and Schuster. A well-illustrated, easily read text focusing on the occurrence, properties, history, and lore of common and uncommon gemstones; for general or informed audiences.

Strahler, Arthur N. 1981. *Physical Geology.* New York: Harper and Row. A thorough treatment of the fundamentals of physical geology for a general but informed audience.

Tindall, James R., and Roger Thornhill. 1975. *The Collector's Guide to Rocks and Minerals.* New York: Van Nostrand Reinhold. An easily read text intended for a general audience.

Turner, Francis J. 1968. *Metamorphic Petrology: Mineralogical and Field Aspects.* New York: McGraw-Hill. A thorough, advanced discussion of metamorphism and metamorphic rocks.

Verhoogen, John, Francis J. Turner, Lionel E. Weiss, and Clyde Wahrhaftig. 1970. *The Earth: An Introduction to Physical Geology.* New York: Holt, Rinehart and Winston. A comprehensive, advanced physical geology text.

Wilk, Harry, and Olaf Medenbach. 1986. *The Magic of Minerals.* New York: Springer-Verlag. Superb photographs highlight this introduction to the world of minerals; for general and informed audiences.

Winkler, Helmut G. F. 1979. *Petrogenesis of Metamorphic Rocks.* 5th ed. New York: Springer-Verlag. A thorough, advanced discussion of metamorphism and metamorphic rocks.

Appendix: Some Additional Minerals and How They Form

There are currently about 3,800 known mineral species. In selecting those discussed in the preceding chapters, I have tried to strike a balance between minerals and localities that are familiar, and those that best illustrate a particular geological process. As a result, some important minerals had to be omitted, along with many additional examples of ways that given species might form. The following table lists some of these minerals, their localities, and modes of genesis. Although some, such as carletonite or cumengite, are globally rare species, they have been included because they are often seen in collections. I have used general mineral names to imply a series of species, especially when the species have common modes of origin (e.g., "allanite" implies allanite-(Ce) and allanite-(Y); "axinite" implies ferro-, magnesio-, or manganaxinite). Similarly, polymorphs with similar modes of origin, such as atacamite and paratacamite, have been listed simply as "atacamite." The process/type code used in the table is given below, and the numbered references follow the table. Where possible, I have tried to provide nontechnical references that describe the occurrence of the species and its associated minerals. References giving more complete and scientific descriptions for each species may be found in the *Encyclopedia of Minerals, Mineral Reference Manual,* and *Glossary of Mineral Species,* which are cited in the Recommended Reading list. A visit to nearly any university or museum library will provide many more.

Processes and Types

M · crystallization from magma

- ag · agpaitic pegmatite
- cb · carbonatite
- df · differentiation
- gr · granitic pegmatite
- vx · volcanic extrusive

C · chemical alteration

- bi · biogenic
- ox · oxidation-reduction
- re · chemical replacement
- hp · hydrothermal alteration in pegmatite

P · precipitation from aqueous solution

- al · alpine cleft
- ev · evaporite
- mvt · Mississippi Valley type
- mt · meteoric
- hy · hydrothermal (mg · magmatic, me · meteoric, tc · tectonic, bs · basalt flow mineralization)

R · recrystallization

- rd · rodingite
- rm · regional metamorphism
- sk · skarn
- cm · contact metamorphism
- vs · recrystallized volcanogenic massive sulfide deposit

Species	Locality	Process/type	References
Acanthite	Cobalt, Ontario	P/hy	14,78,148
	Freiberg, Saxony, Germany	P/hy-mg	5,6,121,180 ,202
	Guanajuato, Mexico	P/hy	6,151,202
Andorite	Oruro, Bolivia	P/hy-mg	5,32,202
Ajoite	Ajo, Arizona	C/ox	194
	Messina District, South Africa	C/ox	21
Allanite	Madawaska, Ontario	M/gr	171
	Fahlun, Sweden	M/gr	42,212
	Olden Twp., Frontenac County, Ontario	R/cm	171
	Otter Lake, Quebec	R/cm	115
	Trimouns, France	P/al	173
	Eifel District, Germany	M/vx	83
Alstonite	Alston, Cumbria, England	P/mvt + C?	149
	Hardin County, Illinois	P/mvt + C?	122,177
Anhydrite	Naica, Chihuahua, Mexico	P/hy	151
	Simplon Tunnel, Switzerland	P/al	5,149,203
	Stassfurt, Saxony, Germany	P/ev	19,149
Annabergite	Laurium, Greece	C/ox	105
Anthophyllite-gedrite	Fiskenaesset, Greenland	R/rm	159
	Nuuk, Greenland	R/rm	2
Antlerite	Chiquicamata, Chile	C/ox	33,149
Artinite	Staten Island, New York	P/hy-tc?	91,167
	San Benito County, California	P/hy-tc	31
Astrophyllite	St. Peters Dome, Colorado	M/ag	140,152
	Mont Saint-Hilaire, Quebec	M/ag	85,125
	Narssârssuk, Greenland	M/ag	159
Atacamite	Wallaroo, South Australia	C-ox	162
	Atacama, Chile	C-ox	33
Augelite	North Groton, New Hampshire	C/hp	135,184
	Rapid Creek, Yukon Territory	P/hy-tc	176
	Oruro, Bolivia	P/hy-mg	32
	Mono County, California	P/hy	223
Augite	Salzburg, Austria	P/al	6,183
	Ariccia, Roma, Italy	M/df	42
	Franklin, New Jersey	R/sk	58,59,146
	Eifel District, Germany	M/vx	83
Austinite	Gold Hill, Utah	C/ox	106
	Mapimi, Durango, Mexico	C/ox	151
Autunite	Spokane, Washington	C/ox	166
	Autun, Saône-et-Loire, France	C/ox	149
Axinite	Bourg d'Oisans, Isère, France	P/al	49
	New Melones Lake, California	P/al	160
	Polar Urals, Russia	P/hy-tc	185
	Luning, Nevada	R/sk?	49
	Bungo Province, Kyushu, Japan	R/cm	181
	Vitória da Conquista, Bahia, Brazil	P/hy-tc	25
Babingtonite	Westfield, Massachusetts	P/hy-bs	147
	Bombay, Maharashtra, India	P/hy-bs	40
	Paterson, New Jersey	P/hy-bs	156,157
Beryllonite	Stoneham, Maine	C/hp?	135,155
	Paprok, Nuristan, Afghanistan	C/hp?	175
Betafite	Betafo area, Madagascar	M/gr	148
	Bancroft, Ontario	R/sk (M/cb?)	102,171
Bismuthinite	Jefferson County, Colorado	M/gr	148
	Cornwall, England	P/hy-mg	52,148
	Kingsgate, New South Wales Australia	P/hy-mg	53
Blödite	Soda Lake, California	P/ev	154
Boleite	Boleo, Baja California, Mexico	C/ox	6,151
Boracite	Lüneburg, Hannover, Germany	P/ev	19,149
Boulangerite	Noche Buena, Zacatecas, Mexico	P/hy	151
	Trepca, Serbia	P/hy	120
	Madoc, Ontario	P/hy-me?	16,130
Brazilianite	Minas Gerais, Brazil	C/hp	5,6,135
	North Groton, New Hampshire	C/hp	41,135,184
	Rapid Creek, Yukon Territory	P/hy-tc	176
Brewsterite	Strontian, Scotland	P/hy	196
	Yellow Lake, British Columbia	P/hy-bs	196
	Harrisville, New York	P/hy-tc	172

Species	Locality	Process/type	References
Brochantite	Bisbee, Arizona	C/ox	68
	Bingham, New Mexico	C/ox	192
Brucite	Brewster, New York	P/hy-tc	88,91
	Texas, Pennsylvania	P/hy-tc	63,148
	Asbest, Russia	R/rd	174
	Asbestos, Quebec	R/rd	77
Buergerite	San Luis Potosi, Mexico	M/vx	47,151
Cacoxenite	Hagendorf, Bavaria, Germany	C/hp	135
	Indian Mountain, Alabama	P/mt	12,67
	Coon Creek, Arkansas	P/mt	103
Calaverite	Cripple Creek, Colorado	P/hy-mg	6,187,217
	Calaveras County, California	P/hy	154,217
Caledonite	Tiger, Arizona	C/ox	6,17
	San Bernardino County, California	C/ox	39
	Leadhills, Lanarkshire, Scotland	C/ox	64
Cancrinite	Bancroft area, Ontario	C/re	84
	Mont Saint-Hilaire, Quebec	M/ag	85,125
	Eifel District, Germany	M/vx	83
	Point of Rocks, New Mexico	M/vx	43
Carletonite	Mont Saint-Hilaire, Quebec	R/cm	85,125
Carnotite	Colorado Plateau area, Utah, Colorado, New Mexico, Arizona	C/ox	78,149
Cavansite	Owyhee Dam, Malheur County, Oregon	P/hy-bs	208
	Poona, Maharashtra, India	P/hy-bs	111
Chondrodite	Amity, New York	R/rm	99
	Brewster, New York	R/sk	88,91
Chlorargyrite	Broken Hill, New South Wales, Australia	C/ox	127,224
Cinnabar	Almaden, Ciudad Real, Spain	P/hy	37
	Hunan & Kweichow Prov., China	P/hy	5,6,168,181
	Coast Range, California	P/hy	154
	Idria, Slovenia	P/hy-mg	7
Clinoclase	Tintic District, Utah	C/ox	126,149
	Liskeard, Cornwall, England	C/ox	52
	Majuba Hill, Nevada	C/ox	92,93
Cobaltite	Cobalt, Ontario	P/hy	14,78
	Tunaberg & Håkansboda, Sweden	R/sk	1,6
Conichalcite	Mapimi, Durango, Mexico	C/ox	151
	Gold Hill, Utah	C/ox	106
Covellite	Butte, Montana	P/hy-mg	6,148
	Summitville, Colorado	P/hy-mg	109
	Alghero, Sardinia, Italy	P/hy	148
Creedite	Santa Eulalia, Chihuahua, Mexico	P/hy	6,151
Cryolite	Ivigtut, Greenland	M/gr	6,159
	Mont Saint-Hilaire, Quebec	M/ag	85,125
	Francon quarry, Montreal, Quebec	M/cb	29
Cumengite	Boleo, Baja California, Mexico	C/ox	6,151
Cyanotrichite	Grand Canyon, Arizona	C/ox	117
	Moldawa, Banat, Romania	C/ox	149
Danburite	San Luis Potosi, Mexico	P/hy	151
	Russell, New York	R/sk	91
	Anjanabonoina, Madagascar	M/gr	221
Dawsonite	Francon quarry, Montreal, Quebec	M/cb	29
	Mte. Amiata, Tuscany, Italy	P/hy	149
	Mont Saint-Hilaire, Quebec	M/ag	85,125
Descloizite	Grootfontein, Namibia	C/ox	215
	Santa Eulalia, Chihuahua, Mexico	C/ox	151
Diaspore	Chester, Massachusetts	P/hy-tc	123
	Mugla Province, Turkey	P/hy-tc	136
Dresserite	Francon quarry, Montreal, Quebec	M/cb	29
Dumortierite	Alpine, California	R/rm	154
Dyscrasite	St. Andreasberg, Harz, Germany	P/hy-mg	6,121,148
	Příbram, Czech Republic	P/hy-mg	6,107
Edingtonite	Bathurst, New Brunswick	P/hy-tc	76
	Dumbartonshire, Scotland	P/hy-bs	196
	Ice River, British Columbia	M/ag	74
Enargite	Butte, Montana	P/hy-mg	6,148
	San Juan County, Colorado	P/hy-mg?	109,141
	Quiruvilca, La Libertad, Peru	P/hy-mg	118
Enstatite (Hypersthene)	Jackson County, No. Carolina	M/df	222
	North Creek, New York	R/rm	66,91
	Eifel District, Germany	M/vx	83
	Summit Rock, Oregon	M/vx	82
Epistilbite	Berufjord, Iceland	P/hy-bs	196
	Bombay, Maharashtra, India	P/hy-bs	40,196
Euxenite	Quadeville, Ontario	M/gr	102,171
	Madawaska, Ontario	M/gr	171
	Ampangabe, Madagascar	M/gr	148
	Iveland, Aust-Agder, Norway	M/gr	148,212
Fergusonite	Madawaska, Ontario	M/gr	102,171
	Iveland, Aust-Agder, Norway	M/gr	212
	Baringer Hill, Llano County, Texas	M/gr	186
Forsterite	Zabargad Island, Egypt	P/hy-me + C	6,100,214
	Notre-Dame-du-Laus, Quebec	R/cm	198
Francevillite	Mounana, Gabon	C/ox	28
Gadolinite	Falun, Sweden	M/gr	42,212
	Hiterö, Vest-Agder, Norway	M/gr	42,212
	Brunet & Llano Counties, Texas	M/gr	38,186
	Graubünden, Switzerland	P/al	70
Gahnite	Rowe, Massachusetts	R/rm	42
	Topsham, Maine	M/gr	57
	Franklin, New Jersey	R/sk	58,59,146
Glauberite	Camp Verde, Arizona	P/ev	195
	Searles Lake, California	P/ev	153,154
Glaucophane	Coast Range, California	R/rm	154
Gmelinite	Two Islands, Nova Scotia	P/hy-bs	196
	Flinders, Victoria, Australia	P/hy-bs	18,196
	Paterson, New Jersey	P/hy-bs	156,157,196
Gormanite - souzalite	Rapid Creek, Yukon Territory	P/hy-tc	176
Goyazite	Minas Gerais, Brazil	C/hp	135
	North Groton, New Hampshire	C/hp	135,184
	Rapid Creek, Yukon Territory	P/hy-tc	176
	Binnental, Valais, Switzerland	P/hy-tc	6,69,188
Graftonite	Grafton, New Hampshire	M/gr	41,135
	Black Hills, South Dakota	M/gr	135,170
Graphite	Ticonderoga, New York	R/cm	91,114
	Sri Lanka	R/rm	148
	Crestmore, California	R/sk	89

Species	Locality	Process/type	References
Greenockite	Bishopton, Renfrew, Scotland	P/hy	148
	Granby, Missouri	P/mvt	148
	Paterson, New Jersey	P/hy-bs	156,157
	Llallagua, Potosi, Bolivia	P/hy-mg	6,8
Gyrolite	Poona, Maharashtra, India	P/hy-bs	40
Halotrichite	Mojave, California	C/ox	154
Hambergite	San Diego County, California	M/gr	54,112
	Anjanabonoina, Madagascar	M/gr	219
	Gilgit, Pakistan	M/gr	98
Hanksite	Searles Lake, California	P/ev	153,154
Harmotome	Strontian, Scotland	P/hy	196
	Andreasberg, Harz, Germany	P/hy	196
	Vaasa, Finland	P/hy	196
Hauerite	Raddusa, Sicily, Italy	P/mt + C/bi?	5,148
Hauyne	Eifel District, Germany	M/df	83
	Mte. Somma, Vesuvius, Italy	M/df	42
Hessite	Botes, Transylvania, Romania	P/hy-mg	6,148
Hornblende	Aussig, Bohemia, Czech Republic	M/df	42
	Lake Harbour, Northwest Territories	R/sk	75
	Cornopass, Switzerland	R/rm	203
	Eifel District, Germany	M/vx	83
	Summit Rock, Oregon	M/vx	82
Howlite	Windsor & Iona, Nova Scotia	P/ev	97
	Boron, California	P/ev	3,4,6,139,163
Huebnerite	Pasto Bueno, Peru	P/hy-mg	6
	San Juan County, Colorado	P/hy-mg?	109,141
	Alpine County, California	P/hy	161
Hureaulite	Minas Gerais, Brazil	C/hp	50,135
	Black Hills, South Dakota	C/hp	135,170
	Mangualde, Beira, Portugal	C/hp	135
Hutchinsonite	Quiruvilca, La Libertad, Peru	P/hy-mg	118,210
	Binnental, Valais, Switzerland	P/hy-tc	6,69,188
Hyalophane	Busovaca, Bosnia	P/al	225
	Binnental, Valais, Switzerland	P/hy-tc	6,69,188
Hydromag-nesite	Staten Island, New York	P/hy-tc?	91,167
	Soghan mine, Kerman, Iran	P/hy-tc?	9
Hydrozincite	Mapimi, Durango, Mexico	C/ox	151
	Goodsprings, Nevada	C/ox	149
Ilvaite	South Mountain, Idaho	R/sk	165,204
	Seriphos, Greece	R/sk	61,204
	Elba, Italy	R/sk	204
Inesite	Trinity County, California	P/hy	51,154
	Kuruman District, Cape Province, South Africa	P/hy	79,200,220
Iron	Disko Island, Greenland	M + C/ox	148,159
Jadeite	Tawmaw, Burma	R/rm & C	100
	San Benito County, California	R/rm & C	100
Jamesonite	Noche Buena & Nieves, Zacatecas, Mexico	P/hy	151
	Wolfsberg, Harz, Germany	P/hy	148
Jeremejevite	Swakapmund, Namibia	M/gr	209
	Mt. Soktuj, Siberia, Russia	M/gr	181
Kernite	Kern County, California	P/ev	3,4,6,139,163
Kinoite	Christmas, Arizona	C/ox	221
Kornerupine	Harts Range, Northern Territory, Australia	R/rm	128
	Fiskenaesset, Greenland	R/rm	158,159
	Betroka, Tulear, Madagascar	R/rm	42
Kröhnkite	Chuquicamata, Chile	C/ox	33
Kulanite	Rapid Creek, Yukon Territory	P/hy-tc	176
Laumontite	Bombay, Maharashtra, India	P/hy-bs	40,196
	Paterson, New Jersey	P/hy-bs	156,157,196
	Minas Basin, Nova Scotia	P/hy-bs	196
	Bishop, California	P/hy	144,154
Lawsonite	Tiburon Peninsula, California	R/rm	154
Lazulite	North Groton, New Hampshire	C/hp	135,184
	Minas Gerais, Brazil	C/hp	135
	Rapid Creek, Yukon Territory	P/hy-tc	176
	Werfen, Salzburg, Austria	P/hy-tc	142,149
	White Mountains, California	P/hy	223
	Graves Mountain, Georgia	R/rm	35
Lead	Långban, Sweden	P/hy	134,148
Leadhillite	Granby, Missouri	C/ox	149
	Tsumeb, Namibia	C/ox	215
	Leadhills, Scotland	C/ox	64
Levyne	Spray, Oregon	P/hy-bs	196
	Faeroe Islands	P/hy-bs	15,196
Libethenite	Cornwall, England	C/ox	52
	Mindola, Kitwe, Zambia	C/ox	108
	Libethen, Slovakia	C/ox	149
Liddicoatite	Antsirabe, Madagascar	M/gr	6,219
Linarite	Red Gill, Cumbria, England	C/ox	36
	Graham County, Arizona	C/ox	96
	Bingham, New Mexico	C/ox	192
Ludlamite	Rapid Creek, Yukon Teritory	P/hy-tc	176
	Cobalt, Idaho	P/hy	165
	North Groton, New Hampshire	C/hp	135,184
	Municipio de Aquiles Serdán, Chihuahua, Mexico	P/hy	6,151
Magnetite	Balmat, New York	P/hy	30
	Chester, Vermont	R/rm	71
	French Creek, Pennsylvania	R/sk	63
	Bancroft, Ontario	R/cm	171
	Eifel District, Germany	M/vx	83
	Mineville, New York	M/df	91
	St. Gotthard Pass, Switzerland	P/al	5,203
Margarite	Chester, Massachusetts	P/hy-tc	123
Mercury	Almaden, Ciudad Real, Spain	P/hy	37
	Sonoma & Sanata Clara Counties, California	P/hy	154
	Idria, Slovenia	P/hy-mg	7
Microlite	Amelia Courthouse, Virginia	M/gr	133
	Minas Gerais, Brazil	M/gr	27
Milarite	Val Guif, Grisons, Switzerland	P/al	203
	Jaguaraçu, Minas Gerais, Brazil	M/gr	216
	Gunanajuato, Mexico	P/hy	6,151
Millerite	Halls Gap, Kentucky	P/hy-me	129
	Malartic, Quebec	R/vs + M/df?	189
	Siegerland, Germany	P/hy	19,148
	Orford Township, Sherbrooke County, Quebec	R/cm	193
	Gap mine, Lancaster County, Pennsylvania	M/df + P/hy	63,148

Species	Locality	Process/type	References
Molybdenite	Climax, Colorado	P/hy-mg	211
	Kingsgate, New South Wales, Australia	P/hy-mg	53
	Aldfield, Quebec	R/sk	199
	Renfrew area, Ontario	R/sk	199
Mottramite	Tsumeb, Namibia	C/ox	215
Narsarsukite	Narssârssuk, Greenland	M/ag	159
	Mont Saint-Hilaire, Quebec	M/ag & R/cm	85,125
Nickeline	Cobalt, Ontario	P/hy	14,78,148
	Port Radium, Northwest Territories	P/hy	197
	Freiberg and Schneeberg, Saxony, Germany	P/hy-mg	121,182
Okenite	Poona, Maharashtra, India	P/hy-bs	40
Orpiment	Hunan Province, China	P/hy	168
	Humboldt County, Nevada	P/hy	191
	Quiruvilca, La Libertad, Peru	P/hy	118
Osumilite	Sardinia, Italy	M/vx	81
	Eifel District, Germany	M/vx	81,83
	Lane County, Oregon	M/vx	81
Pachnolite	Ivigtut, Greenland	C/hp	6,159
Palygorskite	Metalline Falls, Washington	P/hy	22,164
Paravauxite	Llallagua, Bolivia	P/hy	6,8
Pargasite	Pargas, Finland	R/cm	212
Parisite	Muzo, Columbia	P/hy	6,206
	Quincy, Massachusetts	M/gr	149
	Mineral County, Montana	P/hy-mg?	113
Pentlandite	Sudbury, Ontario	M/df	48
	Malartic, Quebec	R/vs	189
Perovskite	Zlatoust, Ural Mtns., Russia	R/cm	5,181
	Magnet Cove, Arkansas	R/cm	148
	Gardiner Complex, Greenland	M/cb	94
	Jacupiranga, São Paulo, Brazil	M/cb	131
	Val Malenco, Lombardy, Italy	R/rd	13,70
	Eifel District, Germany	M/vx	83
Petalite	Minas Gerais, Brazil	M/gr	26
	Oxford County, Maine	M/gr	57,155
	Elba, Italy	M/gr	6,145
Petzite	Nagyag, Romania	P/hy-mg	6,148
	Goldhill, Colorado	P/hy	148,217
Phenakite	Minas Gerais, Brazil	M/gr	23
	Anjanabonoina, Madagascar	M/gr	219
	Chatham, New Hampshire	M/gr	179
	Mount Antero, Colorado	M/gr	86
	Takowaja, Urals, Russia	R/cm	181
	St. Gotthard area, Switzerland	P/al	203
Phillipsite	Melbourne, Victoria, Australia	P/hy-bs	196
	Monument, Oregon	P/hy-bs	196
Phosgenite	Monteponi, Sardinia, Italy	C/ox	5,149
	Cromford, Derbyshire, England	C/ox	20
	Laurium, Greece	C/ox (bi)	105
	Tsumeb, Namibia	C/ox	215
	Tiger, Arizona	C/ox	17
Phosphophyllite	Potosi, Bolivia	P/hy	5,6
	Hagendorf, Bavaria, Germany	C/hp	135
Phosphosiderite (Metastrengite)	Black Hills, South Dakota	C/hp	135,170
	Pleystein, Bavaria, Germany	C/hp	149
	Indian Mountain, Alabama	P/hy-me	12,67
Pollucite	Bernic Lake, Manitoba	M/gr	196
	Oxford County, Maine	M/gr	155,196
	Nuristan, Afghanistan	M/gr	10,196
	Island of Elba, Italy	M/gr	6,145,196
Polybasite	Arizpe, Sonora, Mexico	P/hy	151,202
	Freiberg, Saxony, Germany	P/hy-mg	6,121,182
	Guanajuato, Mexico	P/hy	6,151,202
Powellite	Keewenaw Peninsula, Michigan	P/hy-bs	213
	Nasik, Maharashtra, India	P/hy-bs	110
	Tungsten, Nevada	R/cm	149
	Randsburg, California	C/ox	154
Proustite	Chañarcillo, Atacama, Chile	P/hy-mg	6,34,202
	Freiberg, Saxony, Germany	P/hy-mg	6,121,182
	Pribram & Joachimstal, Bohemia, Czech Republic	P/hy-mg	6,190
	Cobalt, Ontario	P/hy	14,78,148
Pseudomalachite	Shaba Province, Zaire	C/ox	62
	Libethen, Slovakia	C/ox	149
Pumpellyite	Keewenaw Peninsula, Michigan	P/hy-bs	213
Purpurite	North Groton, New Hampshire	C/hp	41,135,184
Pyrargyrite	Freiberg, Saxony, Germany	P/hy-mg	6,121,182,202
	Andreasberg, Harz, Germany	P/hy-mg	6,121,202
	Fresnillo, Zacatecas, Mexico	P/hy	151,202
	Huancavelica, Peru	P/hy-mg	202,218
	Guanajuato, Mexico	P/hy	5,6,151,202
Pyrrhotite	Santa Eulalia, Chihuahua, Mexico	P/hy	151
	Riondel, British Columbia	P/hy-me?	73
	Nova Lima, Minas Gerais, Brazil	P/hy-mg	124
	Kapnic, Romania	P/hy-mg	6,148
	Trepca, Serbia	P/hy	120
	Sudbury, Ontario	M/df	48
	Malartic, Quebec	R/vs	189
Realgar	Hunan, China	P/hy	168
	Humboldt County, Nevada	P/hy	191
	King County, Washington	P/hy	5,45,164
	Kapnik, Romania	P/hy-mg	6,148
	Binnental, Valais, Switzerland	P/hy-tc	6,69,188
Rhodizite	Antandrokomby, Madagascar	P/gr	205
Roselite	Bou Azzer, Morocco	C/ox	6,150
Samarskite	Iveland, Satersdalen, Norway	M/gr	148,212
	Mitchell County, No. Carolina	M/gr	222
Sapphirine	Fiskenaesset, Greenland	R/rm	158,159
	Betroka, Madagascar	R/rm	42
Scholzite	Hagendorf, Bavaria, Germany	C/hp	135
	Reaphook Hill, South Australia	P/hy-me? + C/ox	95
Schorl	Bovey Tracey, Devon, England	P/hy-mg	52
	Oxford County, Maine	M/gr	56,155
	North Groton, New Hampshire	M/gr	41
	Chester, Vermont	R/rm	71
	Riverside County, California	M/gr	65
Scolecite	Nasik, Maharashtra, India	P/hy-bs	40,196
	Minas Basin, Nova Scotia	P/hy-bs	196
	Rio Grande do Sul, Brazil	P/hy-bs	196
	Cowlitz County, Washington	P/hy-bs	196
Scorodite	Tsumeb, Namibia	C/ox	215
	Concepción del Oro, Zacatecas, Mexico	C/ox	6,151
	Djebel Debar, Constantine, Algeria	C/ox	149
Serendibite	Johnsburg, New York	R/cm	178
Shattuckite	Bisbee, Arizona	C/ox	68

Species	Locality	Process/type	References
Shortite	Green River, Wyoming	P/ev	132
	Mont SaintHilaire, Quebec	R/cm	85,175
Siderite	Nova Lima, Minas Gerais, Brazil	P/hy-mg	124
	Mont Saint-Hilaire, Quebec	M/ag	85,125
	Durham, England	P/mvt	104
	Tavistock, Devon, England	P/hy + C/re	5,52
	Pikes Peak, Colorado	M/gr	55,140
	Allevard, Isère, France	P/al	6
	Rapid Creek, Yukon Territory	P/hy-tc	176
Siegenite	Viburnum Trend, Missouri	P/mvt	116
	Siegen, Westphalia, Germany	P/hy	19,148
Sodalite	Bancroft, Ontario	C/re	84
	Mont Saint-Hilaire, Quebec	M/ag	85,125
Sperrylite	Noril'sk, Siberia, Russia	M/df	46,137
	Sudbury, Ontario	M/df	48
	Transvaal, South Africa	M/df	5
Sphaero-cobaltite (Cobaltocalcite)	Shaba Province, Zaire	P/hy	119
	Concepción del Oro, Zacatecas, Mexico	P/hy	6,151
Stannite	Potosi & Oruro, Bolivia	P/hy	6,8,32
Stephanite	Arizpe, Sonora, Mexico	P/hy	151,202
	Freiberg, Saxony, Germany	P/hy-mg	121,182,202
	Pribram, Bohemia, Czech Republic	P/hy-mg	6,190
Strontianite	Hardin County, Illinois	P/mvt	6,122
	Francon quarry, Montreal, Quebec	M/cb	29
	Strontian, Scotland	P/hy	149
	Winfield, Pennsylvania	P/hy/me	44
	Jamesville, New York	C/bi	44
Sturmanite	Kuruman District, Cape Province, South Africa	P/hy	200
Sugilite	Kuruman District, Cape Province, South Africa	P/hy?	200
Sussexite	Franklin, New Jersey	R/sk	58,59,146
Sylvanite	Cripple Creek, Colorado	P/hy-mg	6,187,217
	Nagyag, Transylvania, Romania	P/hy-mg	6,148
Tantalite	Alto Ligonha, Mozambique	M/gr	6
	Taos County, New Mexico	M/gr	87
	San Diego County, California	M/gr	54
	Gilgit, Pakistan	M/gr	98
Tarbuttite	Broken Hill, Zambia	C/ox	143
	Reaphook Hill, South Australia	C/ox & P/hy	95
Tellurium	Vulcan, Colorado	P/hy	148,169
	Kalgoorlie, Western Australia	P/hy	148
	Moctezuma, Sonora, Mexico	P/hy	60
Tennantite	Tsumeb, Namibia	P/hy	215
	Binnental, Valais, Switzerland	P/hy-tc	6,69,188
	Concepción del Oro, Zacatecas, Mexico (incorrectly ascribed to Naica, Chihuahua)	P/hy	207
Tephroite	Franklin, New Jersey	R/rm	58,59,146
Thenardite	Camp Verde, Arizona	P/ev	195
	Searles Lake, California	P/ev	153,154
Thomsenolite	Ivigtut, Greenland	C/hp	6,159
Thomsonite	Goble, Oregon	P/hy-bs	196
	Renfrewshire, Scotland	P/hy-bs	196
	Asbestos, Quebec	R/rd	77
Thorianite	Fort Dauphin area, Madagascar	R/sk?	5,148
	Otter Lake, Quebec	R/cm	115
	Bancroft area, Ontario	R/cm, M/gr	102

Species	Locality	Process/type	References
Thorite	Bancroft, Ontario	R/cm	102,171
	Eifel District, Germany	M/vx	83
Torbernite	Shaba Province, Zaire	C/ox	6,62
	Mitchell County, North Carolina	C/hp	222
	Gunnislake, Cornwall, England	C/ox	52
Tridymite	Summit Rock, Oregon	M/vx	82
	Thomas Range, Utah	M/vx	82
	Eifel District, Germany	M/vx	83
Trona	Green River, Wyoming	P/ev	132
Tugtupite	Kvanefjeld, Greenland	M/ag + C/hp?	90,159
	Mont Saint-Hilaire, Quebec	M/ag + C/hp?	85
Tyuyamunite	Colorado Plateau area, Utah, Colorado, New Mexico, Arizona	C/ox	78,149
Ulexite	Boron, California	P/ev	3,4,6,139,163
Uraninite	Port Radium, Northwest Territories	P/hy	197
	Joachimstal, Bohemia, Czech Republic	P/hy-mg	6,78
Uranophane	Bancroft, Ontario	C/ox	102,171
	Grants, New Mexico	C/ox	78
	Kolwezi, Shaba, Zaire	C/ox	62
Vauxite	Llallagua, Potosi, Bolivia	P/hy	6,8,218
Veszelyite	Phillipsburg, Montana	C/ox	201
Villiaumite	Isle of Los, Guinea	M/ag	149
	Kvanefjeld, Greenland	M/ag	159
	Mont Saint-Hilaire, Quebec	M/ag	85,125
	Point of Rocks, New Mexico	M/vx	43
Vivianite	Potosi and Oruro, Bolivia	P/hy	6,8,32
	Cobalt, Idaho	P/hy	149,165
	Trepca, Serbia	P/hy	120
Wardite	Taquaral, Minas Gerais, Brazil	C/hp	6,24,135
	Rapid Creek, Yukon Territory	P/hy-tc	176
Wavellite	Llallagua, Potosi, Bolivia	P/hy	8
Whiteite	Rapid Creek, Yukon Territory	P/hy-tc	176
	Custer, South Dakota	C/hp	72
Willemite	Santa Eulalia, Chihuahua, Mexico	C/ox	151
	Tsumeb, Namibia	C/ox	215
	Mont Saint-Hilaire, Quebec	M/ag	85,125
Witherite	Hardin County, Illinois	P/mvt	6,122
	Alston, Cumbria, England	P/mvt?	5,6
	Hexam, Northumberland, England	P/mvt?	44
	Seneca County, New York	C/bi	44
Wurtzite	Llallagua, Potosi, Bolivia	P/hy-mg	6,8
	Butte, Montana	P/hy-mg	6,148
	Thomaston, Connecticut	P/hy-tc?	80
Zinnwaldite	Zinnwald, Bohemia, Czech Republic	P/hy-mg	6
	Pikes Peak, Colorado	M/gr	140
Zoisite	Merelani, Arusha, Tanzania	P/hy-tc+C/re	5,6,11,101
	Ducktown, Tennessee	R/vs	78
	Telemark, Norway	R/rm	212
	Alchuri, Baltistan, Pakistan	P/hy-tc?	138

Appendix References

1: Adolfsson, Stig G. 1973. "Håkansboda Copper and Cobalt Deposit, Sweden." *Mineralogical Record* 4 (1): 38-39.

2: Appel, Peter, W. Uitterdijk, and Aage Jensen. 1987. "Notes and New Techniques: A New Gem Material from Greenland: Iridescent Orthoamphibole." *Gems & Gemology* Spring: 36-42.

3: Aristarain, L. F., and C. S. Hurlbut, Jr. 1972. "Boron Minerals and Deposits, Part I." *Mineralogical Record* 3 (4): 165-172.

4: Aristarain, L. F., and C. S. Hurlbut, Jr. 1972. "Boron Minerals and Deposits, Part II." *Mineralogical Record* 3 (5): 213-220.

5: Bancroft, Peter. 1973. *The World's Finest Minerals and Crystals.* New York: Viking Press.

6: Bancroft, Peter. 1984. *Gem and Crystal Treasures.* Fallbrook, CA: Western Enterprises.

7: Bancroft, Peter, Joze Car, Mirjan Zorz, and Gregor Kobler. 1991. "Famous Mineral Localities: The Idria Mines, Slovenia, Yugoslavia." *Mineralogical Record* 22 (3): 201-208.

8: Bandy, Mark Chance. 1976. *Mineralogy of Llallagua, Bolivia.* Tucson Gem and Mineral Society Special Paper 1. Tucson.

9: Bariand, P., F. Cesbron, and H. Vachey. 1973. "Hydromagnesite from Soghan, Iran." *Mineralogical Record* 4 (1): 18-21.

10: Bariand, Pierre, and J. F. Poullen. 1978. "Famous Mineral Localities: The Pegmatites of Laghman, Nuristan, Afghanistan." *Mineralogical Record* 9 (5): 301-308.

11: Barot, N. R., and Edward W. Boehm. 1992. "Gem-Quality Green Zoisite." *Gems and Gemology* 28 (1): 4-15.

12: Barwood, Henry. 1974. "Iron Phosphate Mineral Locality at Indian Mountain, Alabama." *Mineralogical Record* 5 (5): 241-244.

13: Bedogné, Francesco, and Renato Pagano. 1972. "Mineral Collecting in Val Malenco." *Mineralogical Record* 3 (3): 120-123.

14: Berry, L. G., ed. 1971. "The Silver-Arsenide Deposits of the Cobalt-Gowganda Region, Ontario. *Canadian Mineralogist* 11 (1): 1-429.

15: Betz, Volker. 1981. "Famous Mineral Localities: Zeolites from Iceland and the Faeroes." *Mineralogical Record* 12 (1): 5-26.

16: Bideaux, Richard A. 1970. "Mineral Rings & Cylinders." *Mineralogical Record* 1 (3): 105-112.

17: Bideaux, Richard A. 1980. "Famous Mineral Localities: Tiger Arizona." *Mineralogical Record* 11 (3): 155-181.

18: Birch, W. D. 1988. "Zeolites from Phillip Island and Flinders, Victoria." *Mineralogical Record* 19 (6): 451-460.

19: Bode, Rainer, and Artur Wittern. 1989. *Mineralien und Fundstellen Bundesrepublik Deutschland.* Haltern: Doris Bode Verlag.

20: Burr, Peter S. 1992. "Notes on the History of Phosgenite and Matlockite from Matlock, England." *Mineralogical Record* 23 (5): 377-386.

21: Cairncross, Bruce. 1991. "The Messina Mining District, South Africa." *Mineralogical Record* 22 (3): 187-199.

22: Cannon, Bart. 1975. *Minerals of Washington.* Mercer Island, WA: Cordilleran Publishing.

23: Cassedanne, J. P. 1985. "Recent Discoveries of Phenakite in Brazil." *Mineralogical Record* 16 (2): 107-109.

24: Cassedanne, Jacques P., and Jeannine O. Cassedanne. 1973. "Minerals from the Lavra da Ilha Pegmatite, Brazil." *Mineralogical Record* 4 (5): 207-213.

25: Cassedanne, Jacques P., and Jeannine O. Cassedanne. 1977. "Axinite, Hydromagnesite, Amethyst and Other Minerals from near Vitória da Conquista (Brazil)." *Mineralogical Record* 8 (5): 382-387.

26: Cassedanne, J., and J. Cassedanne. 1981. "The Urubu Pegmatite and Vicinity." *Mineralogical Record* 12 (2): 73-77.

27: Cassedanne, J. P., and Jack Lowell. 1982. "Famous Mineral Localities: The Virgem da Lapa Pegmatites." *Mineralogical Record* 13 (1): 19-28.

28: Cesbron, F., and P. Bariand. 1975. "The Uranium-Vanadium Deposit of Mounana, Gabon." *Mineralogical Record* 6 (5): 237-249.

29: Chamberlain, Steven C. 1991. "Die Mineralien des Francon-Quarry, Montreal Island, Quebec, Canada." *Mineralien-Welt* June: 59-69.

30: Chamberlain, S. C., and G. W. Robinson. 1993. "Unusual Occurrence of Magnetite Crystals from the Balmat District, St. Lawrence County, New York." *Rocks and Minerals* 68 (3): 122-123 (abstract).

31: Cisneros, S. L., R. E. Witkowski, and D. L. Oswald. 1977. "Artinite from San Benito County, California." *Mineralogical Record* 8 (6): 457-460.

32: Cook, Robert B. 1975. "The Mineralogy of the Department of Oruro, Bolivia." *Mineralogical Record* 6 (3): 125-137.

33: Cook, Robert B. 1978. "Famous Mineral Localities: Chuquicamata, Chile." *Mineralogical Record* 9 (5): 321-333.

34: Cook, Robert B. 1979. "Famous Mineral Localities: Chañarcillo, Chile." *Mineralogical Record* 10 (4): 197-204.

35: Cook, Robert B. 1985. "Famous Mineral Localities: The Mineralogy of Graves Mountain Lincoln County, Georgia." *Mineralogical Record* 16 (6): 443-458.

36: Cooper, M. P., and C. J. Stanley. 1990. *Minerals of the English Lake District.* London: Natural History Museum.

37: Crawford, Jack W. 1988. "Famous Mineral Localities: The Almaden Mines, Cuidad Real, Spain." *Mineralogical Record* 19 (5): 297-302.

38: Crook, Wilson W., III. 1977. "The Clear Creek Pegmatite: A Rare Earth Pegmatite in Brunet County, Texas." *Mineralogical Record* 8 (2): 88-90.

39: Crowley, Jack A. 1977. "Minerals of the Blue Bell Mine, San Bernardino County, California." *Mineralogical Record* 8 (6): 494-496.

40: Currier, Rock H. 1976. "The Production of Zeolite Mineral Specimens from the Deccan Basalt in India." *Mineralogical Record* 7 (5): 248-264.

41: Dallaire, Donald A., and Robert W. Whitmore. 1990. "Mines and Minerals of North Groton, New Hampshire." *Rocks and Minerals* 65 (4): 350-360.

42: Dana, Edward Salisbury, and William E. Ford. 1957. *A Textbook of Mineralogy.* 4th ed. New York: John Wiley & Sons.

43: DeMark, R. S. 1984. "Minerals of Point of Rocks, New Mexico." *Mineralogical Record* 15 (3): 149-156.

44: Dietrich, R. V., and Steven C. Chamberlain. 1989. "Are Cultured Pearls Mineral?" *Rocks and Minerals* 64 (5): 386-392.

45: Dillhoff, Richard M., and Thomas A. Dillhoff. 1991. "Realgar from the Royal Reward Mine, King County, Washington." *Rocks and Minerals* 66 (4): 310-314.

46: Distler, V. V. 1992. "Platinum Mineralization of the Noril'sk Deposits." *Canadian Mineralogist* 30 (2): 480 (abstract).

47: Donnay, Gabrielle, C. O. Ingamells, and Brian Mason. 1966. "Buergerite, a New Species of Tourmaline." *American Mineralogist* 51 (1): 198-199.

48: Dressler, B. O., V. K. Supta, and T. L. Muir. 1991. "The Sudbury Structure." In *Geology of Ontario,* Edited by P.C. Thurston, H.R. Williams, R.H. Sutcliffe and G.M. Scott, Ontario Geological Survey Special Vol. 4, 593-625. Ministry of Northern Development and Mines, Toronto.

49: Dunn, Pete J., Peter B. Leavens, and Cynthia Barnes. 1980. "Magnesioaxinite from Luning, Nevada, and Some Nomenclature Designations for the Axinite Group." *Mineralogical Record* 11 (1): 13-15.

50: Dunn, Pete J., Peter B. Leavens, B. Darko Sturman, Richard V. Gaines, and Carlos do Prado Barbosa. 1979. "Hureaulite and Barbosalite from Lavra do Criminoso, Minas Gerais, Brazil." *Mineralogical Record* 10 (3): 147-151.

51: Dunning, G. E., and J. F. Cooper, Jr. 1987. "Inesite from the Hale Creek Mine, Trinity County, California." *Mineralogical Record* 18 (5): 341-347.

52: Embrey, P. G., and R. F. Symes. 1987. *Minerals of Cornwall and Devon.* London: British Museum (Natural History).

53: England, Brian M. 1985. "Famous Mineral Localities: The Kingsgate Mines, New South Wales, Australia." *Mineralogical Record* 16 (4): 265-289.

54: Foord, Eugene E. 1977. "The Himalaya Dike System: Mesa Grande District, San Diego County, California." *Mineralogical Record* 8 (6): 461-474.

55: Foord, Eugene E., and Robert F. Martin. 1979. "Amazonite from the Pikes Peak Batholith." *Mineralogical Record* 10 (6): 375-376.

56: Francis, Carl A. 1985. "Maine Tourmaline." *Mineralogical Record* 16 (5): 365-388.

57: Francis, Carl A. 1987. "Minerals of the Topsham, Maine, Pegmatite District." *Rocks and Minerals* 62 (6): 407-415.

58: Frondel, Clifford. 1972. *The Minerals of Franklin and Sterling Hill: A Check List.* New York: Wiley-Interscience.

59: Frondel, Clifford, and John L. Baum. 1974. "Structure and Mineralogy of the Franklin Zinc-Iron-Manganese Deposit, New Jersey." *Economic Geology* 69 (2): 157-180.

60: Gaines, Richard V. 1970. "The Moctezuma Tellurium Deposit." *Mineralogical Record* 1 (2): 40-43.

61: Gauthier, Gilbert, and Nicolaos Albandakis. 1991. "Minerals of the Seriphos Skarn, Greece." *Mineralogical Record* 22 (4): 303-308.

62: Gauthier, Gilbert, Armand François, M. Deliens, and P. Piret. 1989. "Famous Mineral Localities: The Uranium Deposits of the Shaba Region, Zaire." *Mineralogical Record* 20 (4): 265-288.

63: Geyer, Alan R., Robert C. Smith, II, and John H. Barnes. 1976. *Mineral Collecting in Pennsylvania.* 4th ed. Pennsylvania Geological Survey General Geology Report 33. Commonwealth of Pennsylvania Department of Environmental Resources Topographic and Geologic Survey, Harrisburg.

64: Gillanders, R. J. 1981. "Famous Mineral Localities: The Leadhills-Wanlockhead District, Scotland." *Mineralogical Record* 12 (4): 235-250.

65: Gochenour, Kenneth. 1988. "Black Tourmaline from Little Cahuilla Mountain, Riverside County, California." *Rocks and Minerals* 63 (6): 440-444.

66: Goldblum, Deborah R., and Mary Louise Hill. 1992. "Enhanced Fluid Flow Resulting from Competency Contrast within a Shear Zone: The Garnet Ore Zone at Gore Mountain, NY." *Journal of Geology* 100: 776-782.

67: Gordon, Jennings B., Jr., and Curtis L. Hollabaugh. 1989. "Phosphate Microminerals of the Indian Mountain Area." *Mineralogical Record* 20 (5): 355-362.

68: Graeme, Richard W. 1981. "Famous Mineral Localities: Bisbee, Arizona." *Mineralogical Record* 12 (5): 258-319.

69: Graeser, Stefan. 1977. "Famous Mineral Localities: Lengenbach, Switzerland." *Mineralogical Record* 8 (4): 275-281.

70: Gramaccioli, Carlo M. 1979. "Minerals of the Alpine Rodingites of Italy." *Mineralogical Record* 10 (2): 85-89.

71: Grant, Raymond W. 1968. *Mineral Collecting in Vermont.* Vermont Geological Survey Special Publication No. 2. Vermont Geological Survey, Montpellier.

72: Grice, Joel D., Pete J. Dunn, and Robert A. Ramik. 1989. "Whiteite-(CaMnMg), a New Mineral from the Tip Top Pegmatite, Custer, South Dakota." *Canadian Mineralogist* 27: 699-702.

73: Grice, J. D., and R. A. Gault. 1977. "The Bluebell Mine, Riondel, British Columbia, Canada." *Mineralogical Record* 8 (1): 33-36.

74: Grice, J. D., and R. A. Gault. 1981. "Edingtonite and Natrolite from Ice River, British Columbia." *Mineralogical Record* 12 (4): 221-226.

75: Grice, J. D., and R. A. Gault. 1983. "Lapis Lazuli from Lake Harbour, Baffin Island, Canada." *Rocks and Minerals* 58 (1): 12-19.

76: Grice, J. D., Robert A. Gault, and H. Gary Ansell. 1984. "Edingtonite: The First Two Canadian Occurrences." *Canadian Mineralogist* 22 (2): 253-258.

77: Grice, J. D., and R. Williams. 1979. "The Jeffrey Mine, Asbestos, Quebec." *Mineralogical Record* 10 (2): 69-80.

78: Guilbert, John M., and Charles F. Park, Jr. 1986. *The Geology of Ore Deposits.* New York: W. H. Freeman.

79: Gutzmer, Jens, and Bruce Cairncross. 1993. "Recent Discoveries from the Wessels Mine, South Africa." *Mineralogical Record* 24 (5): 365-368.

80: Henderson, William A., Jr. 1979. "Microminerals." *Mineralogical Record* 10 (4): 239-241.

81: Henderson, William A., Jr. 1981. "Microminerals." *Mineralogical Record* 12 (6): 381-385.

82: Henderson, William A., Jr. 1985. "Microminerals of the Western Volcanics." *Mineralogical Record* 16 (2): 137-135.

83: Hentschel, Gerhard. 1983. *Die Mineralien der Eifelvulkane.* Munich: Christian Weise Verlag.

84: Hewitt, D. F. 1961. *Nepheline Syenite Deposits of Southern Ontario.* Toronto: Ontario Department of Mines.

85: Horvath, Laszlo, and Robert A. Gault. 1990. "The Mineralogy of Mont Saint-Hilaire Quebec." *Mineralogical Record* 21 (4): 284-359.

86: Jacobson, Mark I. 1979. "Famous Mineral Localities: Mount Antero." *Mineralogical Record* 10 (6): 339-346.

87: Jahns, Richard H., and Rodney C. Ewing. 1977. "The Harding Mine, Taos County, New Mexico." *Mineralogical Record* 8 (2): 115-126.

88: Januzzi, Ronald E. 1966. *A Field Mineralogy of the Tilly Foster Iron Mine at Tilly Foster, Brewster, New York.* Brewster, NY: The Mineralogical Press.

89: Jaszczak, John A. 1991. "Graphite from Crestmore, California." *Mineralogical Record* 22 (6): 427-432.

90: Jensen, Aage, and Ole V. Petersen. 1982. "Tugtupite: A Gemstone from Greenland." *Gems & Gemology* 28: 90-94.

91: Jensen, David E. 1978. *Minerals of New York State.* Rochester, NY: Ward Press.

92: Jensen, Martin. 1985. "The Majuba Hill Mine: Pershing County." *Mineralogical Record* 16 (1): 57-72.

93: Jensen, Martin. 1993. "Update on the Mineralogy of the Majuba Hill Mine, Pershing County, Nevada." *Mineralogical Record* 24 (3): 171-180.

94: Johnsen, Ole, Ole Petersen, and Olaf Medenbach. 1985. "The Gardiner Complex, a New Locality in Greenland." *Mineralogical Record* 16 (6): 485-494.

95: Johnston, Christopher W., and Roderick J. Hill. 1978. "Zinc Phosphates at Reaphook Hill, South Australia." *Mineralogical Record* 9 (1): 20-25.

96: Jones, Robert W. 1980. "The Grand Reef Mine, Graham County, Arizona." *Mineralogical Record* 11 (4): 219-225.

97: Joyce, D. K., R. I. Gait, and B. D. Sturman. 1994. "The Morphology, Physical Properties and Occurrence of Howlite from Iona, Nova Scotia." *Rocks and Minerals* 69 (2): 119 (abstract).

98: Kazmi, Ali H., Joseph J. Peters, and Herbert P. Obodda. 1985. "Gem Pegmatites of the Shingus-Dusso Area: Gilgit, Pakistan." *Mineralogical Record* 16 (5): 393-412.

99: Kearns, Lance E. 1978. "The Amity Area, Orange County, New York." *Mineralogical Record* 9 (2): 85-90.

100: Keller, Peter C. 1990. *Gemstones and Their Origins.* New York: Van Nostrand Reinhold.

101: Keller, Peter C. 1992. *Gemstones of East Africa.* Phoenix: Geoscience Press.

102: Kennedy, Irwin. 1979. "Some Interesting Radioactive Minerals from the Bancroft Area, Ontario." *Mineralogical Record* 10 (3): 153-158.

103: Kidwell, Albert L. 1989. "Phosphate Minerals of Arkansas." *Rocks and Minerals* 64 (5): 189-195.

104: King, Robert J. 1982. "The Boltsburn Mine, Weardale, County Durham, England." *Mineralogical Record* 13 (1): 5-18.

105: Kohlberger, William. 1976. "Minerals of the Laurium Mines: Attica, Greece." *Mineralogical Record* 7 (3): 114-125.

106: Kokinos, Michael, and William S. Wise. 1993. "Famous Mineral Localities: The Gold Hill Mine, Tooele County, Utah." *Mineralogical Record* 24 (1): 11-22.

107: Kolesar, Peter. 1990. "Dyskrasit-Kristalle aus dem Bergbau-Revier Pribram in der Tschechoslowakei." *Lapis* 15 (9): 19-26.

108: Korowski, Stanley P., and Cor W. Notebaart. 1978. "Libethenite from the Rokana Mine, Zambia." *Mineralogical Record* 9 (6): 341-346.

109: Kosnar, Richard A., and Harold W. Miller. 1976. "Crystallized Minerals of the Colorado Mineral Belt." *Mineralogical Record* 7 (6): 278-307.

110: Kothavala, Rustam Z. 1982. "The Discovery of Powellite at Nasik, India." *Mineralogical Record* 13 (5): 303-309.

111: Kothavala, Rustam Z. 1991. "The Wagholi Cavansite Locality near Poona, India." *Mineralogical Record* 22 (6): 415-420.

112: Larson, Bill. 1977. "The Best of San Diego County." *Mineralogical Record* 8 (6): 507-518.

113: Lasmanis, R. 1977. "The Snowbird Mine, Montana's Parisite Locality." *Mineralogical Record* 8 (2): 83-86.

114: Lauf, Robert J. 1983. "Graphite from the Lead Hill Mine, Ticonderoga, New York." *Mineralogical Record* 14 (1): 25-30.

115: Leavitt, D. 1981. "Minerals of the Yates Uranium Mine, Pontiac County, Quebec." *Mineralogical Record* 12 (6): 359-363.

116: Le Font, Mark. 1984. "Siegenite from the Buick Mine, Bixby, Missouri." *Mineralogical Record* 15 (1): 3739.

117: Leicht, Wayne C. 1971. "Minerals of the Grandview Mine." *Mineralogical Record* 2 (5): 214-221.

118: Lewis, Richard W., Jr. 1956. "The Geology and Ore Deposits of the Quiruvilca District, Peru." *Economic Geology* 51: 41-63.

119: Lhoest, Joseph J., Gilbert Gauthier, and Vandall T. King. 1991. "Famous Mineral Localities: The Mashamba West Mine, Shaba, Zaire." *Mineralogical Record* 22 (1): 13-20.

120: Lieber, Werner. 1973. "Trepca and Its Minerals." *Mineralogical Record* 4 (2): 56-61.

121: Lieber, Werner, and Hermann Leyerzapf. 1986. "German Silver: An Historical Perspective on Silver Mining in Germany." *Mineralogical Record* 17 (1): 3-18.

122: Lillie, Ross C. 1988. "Minerals of the Harris Creek Fluorspar District, Hardin County, Illinois." *Rocks and Minerals* 63 (3): 210-226.

123: Lincks, G. Fred. 1978. "The Chester Emery Mines." *Mineralogical Record* 9 (4): 235-242.

124: Lucio, A., and Richard V. Gaines. 1973. "The Minerals of the Morro Velho Gold Mine, Brazil." *Mineralogical Record* 4 (5): 224-229.

125: Mandarino, J. A., and V. Anderson. 1989. Monteregian Treasures: *The Minerals of Mont Saint-Hilaire, Quebec.* Cambridge: Cambridge University Press.

126: Marty, Joe, Martin C. Jensen, and Andrew D. Roberts. 1993. "Minerals of the Centennial Eureka Mine, Tintic District, Eureka, Utah." *Rocks and Minerals* 68 (6): 406-416.

127: Mason, Brian. 1976. "Famous Mineral Localities: Broken Hill, Australia." *Mineralogical Record* 7 (1): 25-33.

128: McColl, Don, and Gladys Warren. 1984. "Kornerupine and Sapphirine Crystals from the Harts Range, Central Australia." *Mineralogical Record* 15 (2): 99-101.

129: Medici, John C. 1981. "The Halls Gap Millerite Locality." *Rocks and Minerals* 56 (3): 104-108.

130: Melanson, Frank, and George Robinson. 1982. "The Fluorite Mines of Madoc, Ontario." *Mineralogical Record* 13 (2): 87-92.

131: Menezes, Luiz Alberto Dias, and Joaniel Munhoz Martins. 1984. "The Jacupiranga Mine, São Paulo, Brazil." *Mineralogical Record* 15 (5): 261-270.

132: Milton, Charles. 1977. "Mineralogy of the Green River Formation." *Mineralogical Record* 8 (5): 368-379.

133: Mitchell, Richard S. 1977. "Some Noteworthy Minerals from Virginia." *Rocks and Minerals* 52 (5): 221-229.

134: Moore, Paul B. 1970. "Mineralogy & Chemistry of Långban-Type Deposits in Bergslagen, Sweden." *Mineralogical Record* 1 (4): 154-172.

135: Moore, Paul Brian. 1973. "Pegmatite Phosphates: Descriptive Mineralogy and Crystal Chemistry." *Mineralogical Record* 4 (3): 103-130.

136: Moore, Thomas P. 1987. "Notes from Germany." *Mineralogical Record* 18 (2): 159-163.

137: Moore, Thomas. 1992. "What's New in Minerals: Tucson Show." *Mineralogical Record* 23 (3): 273-281.

138: Moore, Thomas. 1993. "What's New in Minerals: Tucson Show." *Mineralogical Record* 24 (3): 219-230, 237.

139: Muessig, Siegfried. 1959. "Primary Borates in Playa Deposits: Minerals of High Hydration." *Economic Geology* 54: 495-501.

140: Muntyan, Barbara L., and John R. Muntyan. 1985. "Minerals of the Pikes Peak Granite." *Mineralogical Record* 16 (3): 217-230.

141: Murphy, Jack A. 1979. "The San Juan Mountains of Colorado." *Mineralogical Record* 10 (6): 349-361.

142: Niedermayr, Gerhard. 1986. "Mineral Localities in Austria." *Mineralogical Record* 17 (2): 105-110.

143: Notebaart, C. W., and S. P. Korowski. 1980. "Famous Mineral Localities: The Broken Hill Mine, Zambia." *Mineralogical Record* 11 (6): 339-348.

144: Novak, G. A., and S. G. Oswald. 1971. "Laumontite from the Pine Creek Mine, Bishop, California." *Mineralogical Record* 2 (5): 222.

145: Orlandi, Paolo, and Pier Bruno Scortecci. 1985. "Minerals of the Elba Pegmatites." *Mineralogical Record* 16 (5): 353-363.

146: Palache, Charles. 1935. *The Minerals of Franklin and Sterling Hill, Sussex County, New Jersey.* United States Geological Survey Professional Paper 180. United States Department of the Interior, Washington.

147: Palache, Charles. 1936. "Babingtonite and Epidote from Westfield, Massachusetts." *American Mineralogist* 21: 652-655.

148: Palache, Charles, Harry Berman, and Clifford Frondel. 1944. *The System of Mineralogy of James Dwight Dana and Edward Salisbury Dana, Yale University 1837-1892.* 7th ed. Vol. 1. New York: John Wiley and Sons.

149: Palache, Charles, Harry Berman, and Clifford Frondel. 1951. *The System of Mineralogy of James Dwight Dana and Edward Salisbury Dana, Yale University 1837-1892.* 7th ed. Vol. 2. New York: John Wiley and Sons.

150: Pallix, Gerard. 1978. "Famous Mineral Localities: Bou-Azzer, Morocco." *Mineralogical Record* 9 (2): 69-73.

151: Panczner, William D. 1987. *Minerals of Mexico.* New York: Van Nostrand Reinhold.

152: Pearl, Richard M. 1974. "Minerals of the Pikes Peak Granite." *Mineralogical Record* 5 (4): 183-189.

153: Pemberton, Earl H. 1975. "The Crystal Habits and Forms of the Minerals of Searles Lake, San Bernardino County, California." *Mineralogical Record* 6 (2): 74-83.

154: Pemberton, Earl H. 1983. *Minerals of California.* New York: Van Nostrand Reinhold.

155: Perham, Jane C. 1987. *Maine's Treasure Chest: Gems and Minerals of Oxford County.* West Paris, ME: Quicksilver Publications.

156: Peters, Joseph J. 1984. "Triassic Traprock Minerals of New Jersey." *Rocks and Minerals* 59 (4): 157-183.

157: Peters, Thomas A., Joseph J. Peters, and Julius Weber. 1978. "Famous Mineral Localities: Paterson, New Jersey." *Mineralogical Record* 9 (3): 157-179.

158: Petersen, Ole V., Ole Johnsen, and Aage Jensen. 1980. "Giant Crystals of Kornerupine." *Mineralogical Record* 11 (2): 93-96.

159: Petersen, Ole V., and Karsten Secher. 1993. "The Minerals of Greenland." *Mineralogical Record* 24 (2): 4-67.

160: Pohl, Demetrius, Renald Guillemette, and James Shigley. 1982. "Ferroaxinite from New Melones Lake, Calaveras County, California, a Remarkable New Locality." *Mineralogical Record* 13 (5): 293-302.

161: Prenn, Neil, and Peggy Merrick. 1991. "The Monitor-Mogul Mining District, Alpine County, California." *Mineralogical Record* 22 (1): 29-40.

162: Pring, Allan. 1988. "Minerals of the Moonta and Wallaroo Mining Districts, South Australia." *Mineralogical Record* 19 (6): 407-416.

163: Puffer, J. H. 1975. "The Kramer Borate Mineral Assemblage, Boron, California." *Mineralogical Record* 6 (2): 84-91.

164: Ream, Lanny R. 1985. *Gems and Minerals of Washington.* Renton, WA: Jackson Mountain Press.

165: Ream, Lanny R. 1989. *Idaho Minerals.* Coeur d'Alene, ID: L. R. Ream Publishing.

166: Ream, Lanny R. 1991. "Two Autunite Localities in Northeastern Washington, The Daybreak and Triple H & J Mines." *Rocks and Minerals* 66 (4): 294-297.

167: Reasenberg, Julian. 1968. "New Artinite Find on Staten Island, New York." *Rocks and Minerals* 43 (9): 643-647.

168: Ren, Kai-Wen. 1980. *Minerals in China.* Shanghai, China: Shanghai Scientific and Technical Publishers.

169: Rickard. T. A.; annotated by Arthur E. Smith and Richard A. Kosnar. 1983. "Across the San Juan Mountains." *Mineralogical Record* 14 (4): 243-249.

170: Roberts, Willard Lincoln, and George Rapp, Jr. 1965. *Mineralogy of the Black Hills.* South Dakota School of Mines and Technology, Bulletin No. 18. Rapid City, SD.

171: Robinson, George, and Steven C. Chamberlain. 1982. "An Introduction to the Mineralogy of Ontario's Grenville Province." *Mineralogical Record* 13 (2): 71-86.

172: Robinson, George W., and Joel D. Grice. 1993. "The Barium Analog of Brewsterite from Harrisville, New York." *Canadian Mineralogist* 31 (3): 687-690.

173: Robinson, George W., and Vandall T. King. 1990. "What's New in Minerals?" *Mineralogical Record* 21 (5): 481-492, 501.

174: Robinson, George W., and Vandall T. King. 1993. "What's New in Minerals?" *Mineralogical Record* 24 (5): 381394.

175: Robinson, George W., Vandall T. King, Eric Asselborn, Forrest Cureton, Rudy Tschernich, and Robert Sielecki. 1992. "What's New in Minerals?" *Mineralogical Record* 23 (5): 423-437.

176: Robinson, George W., Jerry Van Velthuizen, H. Gary Ansell, and B. Darko Sturman. 1992. "Mineralogy of the Rapid Creek and Big Fish River Area, Yukon Territory." *Mineralogical Record* 23 (4): 4-47.

177: Rossman, George R., and Richard L. Squires. 1974. "The Occurrence of Alstonite at Cave in Rock, Illinois." *Mineralogical Record* 5 (6): 266-269.

178: Rowley, Elmer B. 1987. "Serendibite, Sinhalite, Sapphirine, and Grandidierite from the Adirondack Mountains, at Johnsburg, New York." *Rocks and Minerals* 62 (4): 243-246.

179: Samuelson, Peter B., Kenneth H. Hollmann, and Carlton L. Holt. 1990. "Minerals of the Conway and Mount Osceola Granites of New Hampshire." *Rocks and Minerals* 65 (4): 286-296.

180: Sansoni, Gerhard. 1993. "Über Akanthit, insbesondere aus dem Freiberger Revier." *Mineralien-Welt* 4 (2): 36-40.

181: Scalisi, Philip, and David Cook. 1983. *Classic Mineral Localities of the World: Asia and Australia.* New York: Van Nostrand Reinhold.

182: Schröder, Boris, and Ulrich Lipp. 1990. "Der Uranerzbergbau der SDAG-Wismut im Roum Schneeberg-Aue-Schlema und seine Mineralien (II)." *Mineralien-Welt* Heft 3 (November-December): 20-44.

183: Seemann, Robert. 1986. "Famous Mineral Localities: Knappenwand, Untersulzbachtal, Austria." *Mineralogical Record* 17 (3): 167-181.

184: Segeler, Curt G., Anthony R. Kampf, William Ulrich, and Robert W. Whitmore. 1981. "Phosphate Minerals of the Palermo No.1 Pegmatite." *Rocks and Minerals* 56 (5): 197-214.

185: Skobel', L. S., I. I. Nekhanenko, and N. P. Popova. 1993. "Axinitfunde in der Lagerstätte Puiva, Polarural." *Mineralien-Welt* 4 (5): 33-37.

186: Smith, Arthur E. 1991. "Texas Mineral Locality Index." *Rocks and Minerals* 66 (3): 196-224.

187: Smith, Arthur E., Jr., Ed Raines, and Leland Feitz. 1985. "Great Pockets: The Cresson Vug, Cripple Creek." *Mineralogical Record* 16 (3): 231-238.

188: Stalder, H. A., P. Embrey, S. Graeser, and W. Nowacki. 1978. *Die Mineralien des Binntales.* Bern: Naturhistorisches Museum der Stadt Bern.

189: Staveley, R. C. 1976. "Minerals of Marbridge Mines Limited." *Mineralogical Record* 7 (4): 174-178.

190: Stobbe, J. 1981. "Famous Mineral Localities: Príbram, Czechoslovakia." *Mineralogical Record* 12 (3): 157-165.

191: Stolburg, Craig S., and Gail E. Dunning. 1985. "The Getchell Mine, Humboldt County, Nevada." *Mineralogical Record* 16 (1): 15-23.

192: Taggard, Joseph E., Abraham Rosenzweig, and Eugene E. Foord. 1989. "Famous Mineral Localities: The Hansonburg District, Bingham, New Mexico." *Mineralogical Record* 20 (1): 31-46.

193: Tarassoff, P. 1993. "History and Mineralogy of the Orford Nickel Mine, Orford Township, Quebec." In *Proceedings of the 20th Annual Mineralogical Symposium,* 27 (abstract). Rochester, NY: Rochester Academy of Science.

194: Thomas, William J., and Ronal B. Gibbs. 1983. "Famous Mineral Localities: The New Cornelia Mine, Ajo, Arizona." *Mineralogical Record* 14 (5): 283-298.

195: Thompson, Robert J. 1983. "Camp Verde Evaporites." *Mineralogical Record* 14 (2): 85-96.

196: Tschernich, Rudy W. 1992. Zeolites of the World. Phoenix: Geoscience Press.

197: Tyson, Rod. 1989. "The Port Radium District, Northwest Territories, Canada." *Mineralogical Record* 20 (3): 201-208.

198: Van Velthuizen, Jerry. 1993. "The Parker Mine, Notre Dame du Laus, Quebec." *Mineralogical Record* 24 (5): 369-373.

199: Vokes, F. M. 1963. *Molybdenum Deposits of Canada.* Geological Survey of Canada Economic Geology Report No. 20. Geological Survey of Canada, Ottawa.

200: Von Bezing, K. L., Roger D. Dixon, Demetrius Pohl, and Greg Cavallo. 1991. "The Kalahari Manganese Field: An Update." *Mineralogical Record* 22 (4): 279-297.

201: Waisman, Dave. 1992. "Minerals of the Black Pine Mine, Granite County, Montana." *Mineralogical Record* 23 (6): 477-483.

202: Wallace, Terry C., Mark Barton, and Wendell E. Wilson. 1994. "Silver and Silver-bearing Minerals." *Rocks and Minerals* 69 (1): 16-38.

203: Weibel, M. 1966. *A Guide to the Minerals of Switzerland.* New York: Interscience, Wiley.

204: Weiner, K. L., and Rupert Hochleitner. 1987. "Steckbrief: Ilvait." *Lapis* 12 (4): 7-9.

205: Weiner, K. L., and Rupert Hochleitner. 1990. "Steckbrief: Rhodizit." *Lapis* 15 (2): 7-9.

206: White, John S. 1971. "What's New in Minerals." *Mineralogical Record* 2 (5): 231.

207: White, John S. 1972. "Tennantite-Tetrahedrite from Naica, Chihuahua, Mexico." *Mineralogical Record* 3 (3): 115-119.

208: White, John S. 1974. "Mineral Notes: New Minerals." *Mineralogical Record* 5 (2): 74.

209: White, John S. 1975. "What's New in Minerals?" *Mineralogical Record* 6 (1): 38-40.

210: White, John S., and J. A. Nelen. 1985. "Hutchinsonite from Quiruvilca, Peru." *Mineralogical Record* 16 (6): 459-460.

211: White, W. H., A. A. Bookstrom, R. J. Kamilli, M. W. Ganster, R. P. Smith, D. E. Ranta, and R. C. Steininger. 1981. "Character and Origin of Climax-Type Molybdenum Deposits." *Economic Geology* 75th Anniversary Vol.: 270-316.

212: Wilke, Hans-Jürgen. 1976. *Mineral-Fundstellen.* Vol. 4, Skandinavien. Munich: Christian Weise Verlag.

213: Wilson, Marc L., and Stanley J. Dyl, II. 1992. "The Michigan Copper Country." *Mineralogical Record* 23 (2): 4-72.

214: Wilson, W. E. 1976. "Famous Mineral Localities: Saint John's Island, Egypt." *Mineralogical Record* 7 (6): 310-314.

215: Wilson, Wendell E., ed. 1977. "Tsumeb! The World's Greatest Mineral Locality." *Mineralogical Record* 8 (3): 1-128.

216: Wilson, W. E. 1981. "What's New in Minerals?" *Mineralogical Record* 12 (3): 177-186.

217: Wilson, W. E. 1982. "The Gold-containing Minerals: A Review." *Mineralogical Record* 13 (6): 389-400.

218: Wilson, W. E. 1982. "What's New in Minerals?" *Mineralogical Record* 13 (1): 39-42.

219: Wilson, Wendell E. 1989. "The Anjanabonoina Pegmatite, Madagascar." *Mineralogical Record* 20 (3): 191-200.

220: Wilson, W. E., and Pete J. Dunn. 1978. "Famous Mineral Localities: The Kalahari Manganese Field." *Mineralogical Record* 9 (3): 137-153.

221: Wilson, W. E., and J. S. White. 1976. "What's New in Minerals?" *Mineralogical Record* 7 (2): 55-59.

222: Wilson, W. F., and B. J. McKenzie. 1978. *Mineral Collecting Sites in North Carolina.* Information Circular 24. Raleigh, NC: North Carolina Geological Survey.

223: Wise, William S. 1977. "Mineralogy of the Champion Mine, White Mountains, California." *Mineralogical Record* 8 (6): 478-486.

224: Worner, H. K., and R. W. Mitchell, eds. 1982. *Minerals of Broken Hill.* Melbourne: Australia Mining & Smelting.

225: Zebec, Vladimir, and Marin Soufek. 1986. "Hyalophan von Busovaca, Jugoslawien." *Lapis* 11 (1): 28-31.

Acknowledgments

In the year it has taken me to write this book, I have received much encouragement, advice, and assistance from many individuals, to whom I am extremely grateful. First and foremost, I owe a great deal of thanks to my wife, Susan, for the original illustrations on which the figure illustrations were based, and for her encouragement, critical reading, and infinite patience while I spent many an evening and weekend at the word processor. Second, I would like to thank Jeffrey Scovil, with whom it was a pleasure to work and whose exceptional photographic talent has added much to the book. I am also indebted to Michael Hamilton of the Geological Survey of Canada, who provided the photograph of labradorite [PLATE 10]. The suggestions of Steven Chamberlain, T. Scott Ercit, Joel Grice, William Henderson, Richard Herd, Louis Moyd, Peter N. Nevraumont, Terri Ottaway, Jeffrey Scovil, Art Soregaroli, Richard Taylor, and Brad Van Diver greatly improved the manuscript. Patricia M. Leasure and Linda Cunningham at Simon and Schuster were early and enthusiastic supporters of the book. I would also like to thank José Conde for his exquisite book design, and Stephanie Hiebert for her expert copyediting. Finally, I would like to thank Dawn Arnold and Cathy Ripley for their technical support throughout the project, and my assistant, Michel Picard, for the extra work he assumed in my absence.

Index

Acanthite, 311
Achroite, 92. *See also* Tourmaline
Acid, 275, 316, 317, 323, 355
Acid rain, 317
Actinolite, 368, 377
Adamite, 330
Aegirine, 262, **263**, 385
Agalmatolite, 199, 201. *See also* Pyrophyllite; Talc
Agate, 18, 26, 28, 154, **154**, 155, **156**, 157, 230, **288**, 290, 291. *See also* Chalcedony
Agpaite, 262-266
Agricola, Georgius, 281
Akori, 185. *See also* Coral
Alabaster, 20, 199, 201. *See also* Gypsum
Albertus Magnus, 21
Albite, 248, 255, **256**, **257**, 263, **265**, 391
Alexandrite, 30, 70, 72, 73, 74, **74**, 75, 79, 372. *See also* Chrysoberyl
Allochromatic color, 27, 202
Alluvial deposit, 30, 202. *See also* Placer deposits
Almandine, 25, 168, **168**, **254**, **364**, 365, 385. *See also* Garnet
Almandine spinel, 78. *See also* Spinel
Alpine cleft, 301, 303, 381
Aluminum, 222, 237, 258, 261, 262, 278, 279, 291, 305, 317, 343, 347, 362, 363, 365, 366, 368, 373, 377, 387, 397, 401
Amazonite, 18, 20, **130**, 131, **256**. *See also* Feldspar
Amber, 20, 22, 180-183
- classification, 182
- evaluation, legends, and occurrences, 183
- historic notes, 182

Amblygonite, 193, **194**, 195, 200, 261, 343
Amethyst, **23**, 29, 31, 79, 144, **144**, 146, **146**, **147**, **149**, **261**, **288**, **289**, 291. *See also* quartz
Ametrine, 146, **146**
Amphiboles, 239, 245, 262, 360, 372, 377
Amphilobolite, 372
Amulet, 20. *See also* Talisman
Analcime, **264**, 265, 291
Anapaite, **396**, 397
Anatase, 301
Andadite, 383
Andalusite, 87, 94, 227, 361, 362, 363
Andesite, 237
Andradite, 164, 377. *See also* Garnet
Anglesite, **328**, 329
Anhydrite, 315, 397
Anions, 271, 306, 307, 311-312, 316, 323, 329, 330, 342, 354, 355
Antimony, 295, 297, 329, 338, 341, **342**, 353
Apatite, 69, 87, 94, 245
Apophyllite, **226**, 291
Aquamarine, 26, **26**, 30, 31, **58**, **62**, **64-65**, **66**, **68**, 252, 255. *See also* Beryl
Aqueous solutions, 228, 230, 231, 233, 271, 275, 276, 279, 280, 281, 288, 291, 294, 297, 306, 316, 330, 336, 354, 360, 399, 400
Aragonite, 174, 230, 231, 276, 279, 398
Arfvedsonite, 262
Arkansas, Quachita Mountains (Hot Springs), 299, 300
Armstrong Diamond, 13, **41**
Arsenic, 294, 295, 298, 305, 329, 330
Asbestos, 145, 146
Asphalt, 285, 287
Assimulation, 240 (Figure D), 241, 266
Asterism, 28, 202. *See also* Star stones
Aureole, 371, 377, 386
Aurichalcite, 330, **331**
Austria, Swiss Alps—Utersulzbachtal, 305
Autunite, 315
Aventurine, 141, 146. *See also* Quartz
Azurite, 116, **197-198**, 201, **228**, 231, 325, 326, 329

Baker, George F., 13
Balas ruby, 78. *See also* Spinel
Barite, 200, **276**, 279, **286**, 287, 353
Barium, 239, 262, 305
Barrow, George, 365
Basalt, 222, 226, 237, 238, 239, 244, 245, 288, **290**, 291, 392
Bases, 316
Basin, 273, 274, 283
Bauxite, 317
Bedrock, 317
Bement, Charles S., 11
Benitoite, 55, **191**, 200, **304**, 305
Beryl, 24, 56-69, 94, 231, **250**, 251, **252**, 343, **370**, 371, 372, **390**, 391
- aquamarine cat's eye, **66**
- aquamarine crystal, **62**, **68**
- aquamarine jewelry, **64-65**
- emerald crystal, **60**
- emerald pendant, **61**
- evaluation, 69
- gemstones confused with beryl, 69
- heliodor crystal, **61**, **63**
- historic notes, 62
- legends, 65
- morganite, **69**
- morganite crystal, **63**
- occurrences, 66
- Patricia Emerald, **57**
- varieties, 61.
- *See also* Emerald

Beryllium, 240, 241, 248, 251, 262, 343, 347, 371, 392
Betrandite, 343
Bickmore, Albert S., 14
Biogenic origin, 395, 398
Biotite, 261, 377
Birefringence, 28, 202. *See also* Refractive index
Bixbite (red beryl), 61. *See also* Beryl
Bixbyite, 251
Black Prince's Ruby, 48, 76
Black smokers, 216, 392
Bloodstone (heliotrope), 155. *See also* Chalcedony
Blue John (Derbyshire spar), 194. *See also* Fluorite
Book of Marvels, 48
Book of Wings, The, 85
Boot, Anselmus Boetius de, 21, 51, 65, 102, 111, 122, 168
Borates, 273, 274
Borax, 273, 306, 314, 315
Bornite, **294**, 295
Boron, 240, 251, 255, 267, 274, 343, 372
Botryoid, 279
Bournonite, 295
Bowen, Norman, 238, 240
Bowenite, 141, 201
Boyle, Robert, 20
Brazil, Minas Gerais, 244, 248, 252, 256, 258, 345
Brazilian Princess Topaz, 20, 84, **85**
Breccia, 293, **325**
Brilliance, 28, 202
Brine, 266, 283, 285, 287, 306, 392
Bromine, 266
Brookite, 301
Brushite, 397

Cabochon, 31, 202
Cairngorm, 143, 145, 148. *See also* Quartz
Calcite, 22, **23**, 194, **196**, 200, **224**, 225, 227, **228**, 230, 231, **246**, 263, **274**, 275, 276, 279, 282, 283, 285, **286**, 287, **310**, 311, 317, 320, 347, **348**, 349, 363, 365, 368, **369**, 373, **375**, 378, **379**, 391
Calcium, 222, 230, 231, 238, 239, 275, 287, 291, 301, 305, 315, 327, 349, 363, 365, 377, 387, 395, 397
Caplan, Allan, 80
Carat, 30, 43, 170, 174, 202
Carbon, 244, 247, 275, 277, 279
Carbonate, 241, 244, 245, 246, 247, 263, 265, 275, 279, 283, 287, 329, 330, 365, 400
Carbonatite, 227, 241, 245-248, 262
Cardano, Geronimo, 21, 39, 85, 102, 131
Carnelian, 18, 20, 152, **153**, 155, **155**, 157. *See also* Chalcedony
Cartier, 128, 172
Cassiterite, 261, **292**, 298, 373
Catapleiite, 266
Cations, 271, 306, 311, 319, 323, 329, 330, 343, 354, 355
Cat's eye, 26, 28, 70, **71**, 73, 75, 99, 145, 146, 192, 193. *See also* Chrysoberyl; Chatoyancy
Cat's eye stones, **66**, 75
Cerium, 260
Cerussite, **329**
Ceylonite, 78. *See also* Spinel
Chalcanthite, **339**
Chalcedony, 152-160
- evaluation, 160
- historic notes, 157
- legends, 158
- occurrences, 159
- varieties, 155.
- *See also* Quartz

Chalcocite, 295
Chalcopyrite, **228**, 231, **282**, 283, 285, **294**, 295, 298, 329, 373, 392
Chariot of the Muses cameo shell, **178**
Charoite, 384, 385
Chatoyancy, 28, 31, 46, 59, 73, 75, 92, 146, 202. *See also* Star stones, Cat's eye, Cat's eye stones
Chemical alteration of preexisting minerals, 230, 231-233, 311, 312
Chemosynthesis, 395
Childrenite, 347
Chloride, 271, 283, 286, 287, 306
Chlorine, 266, 293
Chlorite, 301, 347, 361, 365
Chlorspinel, 78. *See also* Spinel
Chromian Diopside, **244**
Chromite, 241
Chromium, 231, 244, 252, 255, 305, 372, 383

Chromophores, 255
Chrysoberyl, 22, **23**, 55, 70-75, **71**, **72**, 87, **252**, 255, 372
evaluation, 75
gemstones confused with chrysoberyl, 75
historic notes, 73
legends, 74
occurrences, 75
Chrysocolla, 113, 293, 325, 326, 330
Chrysolite, 107. *See also* Peridot
Chrysoprase, 155, 157. *See also* Chalcedony
Chrysotile asbestos, 268
Citrine, 22, **23**, **26**, 27, 32, 145, **149**, 151. *See also* Quartz
Clay, 229, 279, 283, 291, 317, 363, 365
Cleavage, 29, 202
Cleopatra, Queen, 372
Clinochlore, **305**, 381, **384**
Clinozoisite, 383
Cobalt, 294, 298, 338
Cobaltite, 338
Coesite, 401
Colemanite, 273, **274**
Collinsite, 301
Colombia, Muzo and Chivor Districts, 390, 391, 392
Color change, 46, 69, 72, 75. *See also* Pleochroism
Columbite, 248
Compaction, 230
Composite stones, 33, **33**
Compounds, 224, 225, 273, 287
Convection, 220-222
Cookeite, 343
Copper, 228, 230, 255, 281, 283, **290**, 294, 295, 298, 306, **324**, 325, 326, 329, 330, 390, **398**
Coral, 184-185
Cordierite, 55, 347, **364**, 365
Corundum, 24, 43-55, 366, **367**, 368
DeLong Star Ruby, **53**
evaluation, 55
gemstones confused with corundum, 55
historic notes, 48
legends, 51
Midnight Star Sapphire, **46**
Montana "Yogo" sapphires, **49**
occurrences, 52
padparadscha sapphire, 8, 47, **48**, 52
rubies, **50**
ruby crystal, **51**
sapphire crystals, **54**
Star of India Sapphire, 44, **45**
varieties, 47
Crandallite, 278
Cristobalite, **288**, 401
Crocoite, **334**, 338
Crust, 219, 220, 221, 222, 223, 237, 244, 249, 255, 280, 281
Crystal, 217, 224, 225, 226, 227, 228, 236, 237, 238, 239, 246, 247, 249, 250, 251, 252, 254, 255, 256-261, 263, 265, 266, 267, 271, 273, 277, 281, 285, 286-287, 290, 291, 293
definition, 24, 202
crystal habit, 24, 202
crystal symmetry, 24, **25**, 200-201
crystal system, 24, **25**, 202-203
crystal twins, 24, 72
Crystallization, 226, 227, 229, 237, 238, 239 (Figure C), 241, 243, 247, 249, 251, 260, 262, 263, 265, 266, 267, 271, 273, 281, 288, 290, 291, 343
from molten rock, 230, 231, 233, 236-267
recrystallization by heat and pressure, 230, 231, 233, 281. *See also* Recrystallization
Cubanite, **390**
Cubic zirconia, 33, 43
Cuprite, **324**, 325, **326**, **398**
Cuprosklodowskite, **337**
Curious Lore of Precious Stones, The, 20
Cutting styles, **31**, **32**
Cylinder seals, 18, 20

Damigeron, 18, 51
Danburite, 87
Datolite, 291
De Beers Consolidated Mines, Ltd., 38
De Gemmis et Coloribus, 21
De Mineralibus, 21
DeLong Star Ruby, 15, **52**
DeLong, Edith Haggin, 13, 15
Demantoid, 69, 164, **164**, 165. *See also* Garnet
Dendrites, **276**, 279
Diamond, 20, 26, 28, 29, 30-32, 34-43, 225, 227, 231, **242**, 243, **244**, 391
Aurora collection, **35**
Armstrong Diamond, **41**
diamond crystals, **37**, **42**
evaluation, 43
synthetic diamond and cubic zirconia, 43
Golden Maharaja Diamond, **38**
historic notes, 38
legends, 39
kimberlites, **37**, 42
Lounsbery necklace, **40**
occurrences, 42
royal patronage, 38
sources of production, 38
Tiffany Diamond, **43**
Dichroism, 202. *See also* Pleochroism
Differentiation, 240, 241, 246, 247, 248, 262, 266, 267, 281, 293, 295
Diffusion, 240 (Figure D), 241, 267, 371
Dike, 240 (Figure D), 249, 251, 266, 385
Diopside, 193, 200, **244**, **346**, 347, 365, **366**, 368, **375**, 377, 381, **382**, **384**
Dioscorides, 116
Dispersion, 28, **28**, 202
Dissociation, 271-272
Dolomite, **284**, 285, 287, **348**, 349, 350, 365
Dolostone, 283, 285, 349, 350, 351, 366, 387
Doublets, 33, 113, 125. *See also* Composite stones
Dravite, 90, 91, 372. *See also* Tourmaline
Dumortierite, 116
Durability, 29
Eagle Diamond, 15
Eclogite, 244
Edenite, 261, 377
Elbaite, 63, 90, 92, **91-93**, **253**, 255, 261
crystal form, 94, 98.
See also Tourmaline
Elements, 224, 225, 231, 262, 265, 266, 293, 295, 305, 323, 329, 371
Elpidite, 263, 265
Emerald, 26, 27, 31, **57**, **60**, **61**, **67** 231, 252, 255
Soudé emerald doublet, **33**.
See also Beryl
Emerson, R. W., 126
Energy
activation, in chemical reactions, 312, 354, 355
kinetic and potential, 312
England, Cornwall and Devon, 292, 294, 295, 330, 333, 336
Eosphorite, **345**, 347
Epidote, 301, **303**, 305, 377, 383
Equilibrium, 311-313, 323, 353, 355, 362
Erythrite, **338**
Euclase, 69, 192, **192**, 200, **248**
Eucryptite, 343
Eudialyte, 262, 263
Evans, Joan, 20
Evaporite, 230, 273, 283
Exposition Universelle, Paris, 9-10

Faceting, 32, **32**
round brilliant, **30**
popular cuts, **31**
Fault, 222, 223 (Figure A), 273, 293, 387, 393
Fayalite, 105. *See also* Peridot
Feldspar, 126-131, 229, **238**, 238, 239, 249, 255, 256, 261, 262, 263, 317, 345, 365, 377
gemstones confused with feldspar, 131
historic notes, 128
legends, occurrences, evaluation, 131
varieties, 130
Fire, 28, 202
Fluor Silicic Edenite, **375**
Fluorapatite, 261, **292**, 295, 377, 378, **379**
Fluorine, 239, 247, 251, 265, 292, 293, 295, 307, 375, 377
Fluorite, **23**, 194, 200, **284**, 285, **295**, 301, 353
Forsterite, 105, 365. *See also* Peridot
Franklinite, 368, **369**

Gahnite, 77. *See also* Spinel
Gahnospinel. *See* Spinel
Galena, **284**, 285, 295, 298, **328**, 329, 351, **368**, 392
Garnet, **25**, 162-171, 243, 244, 250, 251, 254, 255, 359, 377, 383
evaluation, 170
doublet, 33
gemstones confused with garnet, 171
historic notes, 167
legends, 168
occurrences, 170
varieties, 165
Garnierite, 317
Gem
cutting, 31-32
definition, 22, 202
evaluation and properties, 31
formation, 30
market, 30
optical properties, 26-28
trade, ancient, 20
Gemmarum et Lapidum Historia, 21
Gemstone, 22, 202
formation, 30
Geode, 230, 287
Geothermal gradient, 219-220
Germany-Czech Republic, Erzgebirge district, 298
Gneiss, 320, 365
Goethite, **232**, 233, 279, 320, 322, **340**, 341

Gold, **224**, 225, 231, 244, 281, 295, **296**, 297, 305, 329
Golden Maharaja Diamond, **38**, 39
Goshenite, 61. *See also* Beryl
Gossan, 322
Granite, 222, 223, 229, 230, 237, 248, 249-261, 267, 281, 292, 295, 298, 320, 343, 345, 365, 371, 372, 391
Gratacap, Louis Pope, 14
Grossular, 164, 377, 381, **382**, 383, **384**. *See also* Garnet
Groundwater, 317, 320, 322, 325, 326, 329, 330, 333, 336, 353
Guggenheim, Harry F., 13
Gummite, 343, **344**, 345
Gypsum, 199, 201, **272**, 273, 283, 306, 315, 317, 391, 397

Halite, 231, **270**, 271, 273, 283, 313
Hamlin, Elijah, 95
Hardness, 29, 202. *See also* Mohs hardness scale
Harlequin Prince Opal, 13, **122**
Harlow, George E., 14, **17**
Hawk's eye, 145. *See also* Quartz
Heat and pressure, 230, 231, 233, 281, 359, 360-362, 363, 365, 366, 368, 371, 373, 383, 385, 386, 387, 391, 401
Hedenbergite, 377
Heliodor, **26**, **58**, **61**, **63**. *See also* Beryl
Heliodore, 255
Hematite, 26, 155, 196, 201, 279, 301, **302**, 303, 319, 320, 322
Hemimorphite, 326, **327**, 329
Hessonite, 165, **168**. *See also* Garnet
Heterosite, 343
Heulandite, 291
Hiddenite, 69, **168**, 192, 200, 255. *See also* Spodumene
Historia Naturalis, 20-21
Hollandite, 279
Holmquisite, 372
Howlite, 113
Hutton, James, 217
Hydration-dehydration reactions, 315, 354
Hydrogen, 295, 298
Hydrothermal solutions, 230, 258, 280, 281, 287, 290, 291, 293, 295, 297, 298, 299, 301, 305, 306, 307, 381, 383, 385, 387, 392, 393
Hydroxides, 320, 330
Hydroxlapatite, 397, 398
Hydroxylherderite, **345**, 347

Iceland spar, 194. *See also* Calcite
Idiochromatic color, 26, 202
Ilmenite, 238, 241, 343
Imitation (gem), 33, 202
Imperial topaz, **81**, **83**, 87. *See also* Topaz
Indicolite, 92. *See also* Tourmaline
Ions, 271, 273, 279, 283, 285, 287, 291, 293, 298, 305, 307, 311, 313, 316, 317, 328, 329, 330, 349, 354, 371, 397
Iridescence, 26, 126, 128, 130, 131, 155
Iron, 222, 223, 239, 252, 261, 276, 279, 288, 291, 295, 297, 301, 305, 311, 319, 320, 329, 350, 363, 373, 377, 387, 391, 392, 395
Isotopes, 219-220, 243, 244, 295, 298, 391, 397

Jade, 18, 20, 22, 132-141
  chinese carving, **132**, **136**, **139**, **140**
  evaluation, 140
  gemstones confused with jade, 141
  historic notes, 136
  Imperial jade, 140, **140**
  legends, 138
  nephrite, 133, 385
  occurrences, 139
  serpentine, **141**
  triplets, 33
Jadeite, 18, **19**, **20**, **26**, 3, **134**, **159**. *See also* Jade
Jasper, 24, 152-160. *See also* Chalcedony
Jesup, Morris K., 9-10
Jet, 22, 186-187
Jordanite, **304**

Kakovin, Jakov I., 66
Kämmererite, **305**
Kaolinite, 363
Kermesite, 341, **342**
Kimberlite, 37, 42, 170, **242**, 243, 244
Koh-i-noor Diamond, 39
Korite (amolite), **183**
Kornerupine, 192, 200
Kunz, George F., 9-11, **9**, 13, 20, 43, 44, 62, 88, 95, 100, 138, 192
Kunzite, **26**, 29, 69, **190**, 191-192, **192**, 200, 255. *See also* Spodumene
Kyanite, 55, 69, **226**, 227, **361**, 362, 363, 365

Labradorite, 26, 126, **127**, 237, **238**. *See also* Feldspar
Lalique, René, 122
Lamproite, 243, 244
Lapidaries, 20-21
Lapis lazuli, 18, 20, 22, 114-117, **114**, **115**, **117**
  evaluation, 117
  gemstones confused with lapis lazuli, 116
  occurrences, 117
Laurionite, 397
Lava, 217, 226, 230, 236, 237, 238, 283, 288, 290, 291 (Figure G). *See also* Magma
Lazulite, 116, 301
Lazurite, 115, **366**, 368. *See also* Lapis lazuli
Lead, 219, 283, 285, 287, 295, 306, 329, 330, 350, 397
Legrandite, 330, **332**
Leonardus, Camillus, 21, 51
Lepidolite, 261
Leucite, 237
Liddicoat, Richard T., 93
Liddicoatite, 93. *See also* Tourmaline
Limestone, 227, 230, 231, 247, 275, **276**, 283, 284, 287, 317, 325, 330, 350, 351, 359, 363, 365, 368, 373, 375, 377, 380, 385, 387, 391, 392, 395, 397
Liroconite, 330, **333**
Lithiophylite, 261, 345
Lithium, 239, 241, 251, 252, 255, 261, 343, 371, 372
Lounsbery diamond necklace, **40**
Ludlamite, 343
Luster, 28, 202
Luxite, 377
Lyell, Sir Charles, 217

Magical Jewels of the Middle Ages and the Renaissance, 20

Magma, 223, 225, 230, 231, 233, 236, 237, 238, 239, 240, 241, 243, 244, 245, 246, 247, 248, 249, 250, 251, 255, 258, 262, 265, 266, 267, 275, 280, 281, 290, 292, 293, 295, 297, 298, 305, 359, 360, 361, 371, 372, 373, 377, 380, 385, 387, 391, 392, 393, 399. *See also* Lava
Magnesite, 350
Magnesium, 222, 239, 244, 252, 287, 291, 305, 349, 363, 365, 368, 372, 387
Magnetite, **241**, 245, 319, 320, 373, 377
Malachite, 26, 113, **197-199**, 199, 201, **228**, 325, **326**, 329
Malaia, 162, 165. *See also* Garnet
Manganese, 251, 252, 255, 261, 266, **276**, 277, 279, 368, 373, 400
Manganite, 279
Manson, D. Vincent, 14, **16**
Mantle, 218, 219, 220, 221, 222, 223, 237, 244, 245, 362, 392
Marble, 18, 194, **304**, 320, 350, 359, 363, 373, 377, 380
Marbode, 21, 51, 65, 85, 102, 107, 122, 131, 158
Marcasite, 196
Marine evaporite, 273
Mason, Brian H., 14, 15
Materia Medica, 116
Matrix, 237, 378
Meionite, **376**, 377
Mesolite, **226**, 291
Meta-autunite, 315
Metamorphic rock, **226**, 227, 230 (Figure B), 231, 233, 299, 312, 347, 362, 365, 380
Metamorphism, 365, 383
  contact, 361, 371, 372 (Figure I), 386, 387
  regional, 361
  retrograde, 347
Metasyenite, 385
Metatorbernite, **314**, 315
Meteoric solution, 230, 280, 281, 288, 290, 295, 298, 306, 399
Meteorite, **218**, 219, 230, 311
Miarolitic cavity, 249, 251, 256, 263, 266
Mica, **226**, 239, 245, 249, 255, 258, 261, 262, 343, 347, 360, **364**, 365, 372, 377, 391
Michigan, Keweenaw Peninsula, 307
Microcline, 130, 255, 256, 343. *See also* Feldspar
Midnight Star Sapphire, **16**
Milky quartz, 145, 146. *See also* Quartz
Millerite, **318**, 319
Mimetite, **336**, 337, 338
Mineral group, 25, 77, 90, 128, 164
Minerals
  defined, 224, 203, 225
  indicator, 361, 362
  primary, 320, 323, 398
  secondary, 320, 329, 330, 341, 342, 343, 397
Mississippi Valley Type (MVT) deposits, 230, 283, 284, 285, 287, 307, 392
Mohs Hardness Scale, 29
Moldavite, 107
Molybdenite, 258, **259**, 261
Monazite, **260**
Monetite, 397
Moonstone, 28, **129**. *See also* Feldspar
Mordenite, **288**, 291
Morgan, J.P., 9-11, **12**, 62
Morgan, J.P., Jr., 13

Morganite, **23**, **26**, **58**, **59**, 61, **63**, 66, 69, **69**, 255. *See also* Beryl
Morion, 145. *See also* Quartz
Moss agate, 155, **156**. *See also* Chalcedony
Mudstone, 301
Murph the Surf, 15
Muscovite, **258**, 261

Nacre, 174. *See also* Pearl
Namibia, Tsumebl, 329, 330
Natrolite, 263, 291
Nepheline, 262, 263, 385
Nephrite, 24, 29, **133**, **138**, **139**, 385. *See also* Jade
Neptunite, **304**, 305
Neutralization, 316, 317, 320, 330, 392
New Jersey, Franklin, 368
New York, Sterling mine, Antwerp, 319-320
Nickel, 239, 294, 298, 317, 338
Nickel-Iron meteorite, **218**, 219
Nickel-Skutterudite, **294**
Niobium, 240, 245, 246, 248, 251, 262

Obsidian, 20, 28, 29, **159**, 187, 200
Odontolite, 113
Olivenite, **330**
Olivine, 105, **218**, 219, 238, 239 (Figure C), 244, 245, 317, 347, 377, 381, 385. *See also* Peridot
On Stones, 20
Ontario (Southeastern Ontario—Quebec area), Bancroft, 373
Onyx, 154, 156. *See also* Chalcedony
Onyx marble, 194. *See also* Calcite
Opal, 26, 28, 29, 118-125, 231, 274, **275**, 401
doublets, 33, 125
evaluation, 125
gemstones confused with opal, 125
Harlequin Prince, **122**
historic notes, 122
legends, 122
varieties, 121
Orient, 174. *See also* pearl
Orta, Garcia de, 21, 39, 51
Orthoclase, 174, 255, 303. *See also* Feldspar
Oxidation, 228, 230, 231, 319, 320, 323, 325, 328, 330, 337, 338, 341, 342, 343, 349, 354, 355, 391
reduction reactions, 319-322, 395
Oxide zone, 320, 322, 325, 326, 329, 330, 335, 341, 354
Oxides, 233, 240, 261, **276**, 277, 279, 288, 292, 295, 311

Padparadscha sapphire, 8, 47, **48**, 52. *See also* Corundum
Paragenetic sequence, 341-342
Pas de Danse chalcedony carving, **160**
Patricia Emerald, 56, **57**
Pearl, **19**, 22, 28, 29, 172-179, **173**, **177-179**
evaluation, 178
historic notes, 175
legends, 176
occurrences, 176
Pectolite, 291, 383
Pegmatite, 30, 59, 75, 85, 95, 103, 131, 150, 170, 203, 230, 248, 249-261, 262-266, 267, 293, 343, 345, 371, 372, 391
Penfieldite, 397

Peridot, **23**, 30, 69, 94, **104**, 105-107, **106**, **244**
historic notes, 106
legends, occurrences, and evaluation, 107
Peridotite, **218**, 222, 244, 368, 387
Peristerite, 130. *See also* Feldspar
Peters, Joe, 14
Phlogopite, **375**, 377
Phosphate, 260, 261, 278, 279, 397, 398
Phosphorus, 245, 278, 279, 292, 301, 373, 397
Phosphosiderite, 397
Phyllite, 365
Piezoelectricity, 92, 145
Pillow, 290
Placer deposits, 30, 42, 52, 66, 170, 203, 230, 244
Plagioglase, 238, 239, 263
Plasma, 155, 157. *See also* Chalcedony
Plate tectonics, 220-223, 223 (Figure A), 225, 229, 233, 237, 243, 281, 283, 298, 299, 301, 304-305, 306, 307, 359, 362, 371, 381, 387, 392, 393, 395, 399
Platinum, 238, 241
Play of color, 27, 32, 120-122, 125, 203
Playas, 273, 274, 306, 313
Pleochroism, 27, 46, 72, 92, 191, 203
Pliny the Elder, 20-21, 38, 48, 80, 84, 93, 111, 122, 131, 185, 187
Plummbogummite, 398
Polo, Marco, 21, 48, 73, 185
Porphyry copper, 230, 298
Potassium, 219, 222, 238, 239, 255, 258, 261, 263, 291, 363, 373, 387
Pough, Frederic H., 14, **15**
Prase, 155, 157. *See also* Chalcedony
Praziolite, 145. *See also* Quartz
Precipitation, from aqueous solutions, 230, 231-233, 273, 274, 275, 276, 279, 284, 285, 287, 288, 290, 291, 292, 294, 295, 306, 307, 313, 320, 322, 323, 326, 354, 392, 393, 400
Prehnite, **159**, 291, 383
Prinz, Martin, 14, **16**
Products of chemical reactions, 312
Proustite, 312
Pseudochromatic color, 26, 128, 203
Pseudomorph, 315, 320, 329, 341, 351, 353
Pyrite, 115, 196, 201, 233, 279, **297**, 320, **321**, 322, 329, 330, **340**, 341, **351**, 392, 395
Pyrochlore, **246**
Pyrolusite, 279
Pyromophite, **336**, 337, 338
Pyrope, 79, 163, 164, **169**, 244. *See also* Garnet
Pyrophyllite, 199, 201, 363
Pyroxene, 219, 238, 239, 245, 262, 217, 377, 381
Pyrrhotite, 320

Quartz, 18, 28, 29, 142-151, 229, 239, 249, 255, 263, 275, 287, **292**, **294**, 295, **297**, 298, 299, **300**, 301, 305, **330**, **338**, 341, 347, **350**, **353**, 361, 363, 365, 391, 401
citrine, 261
evaluation, 151
historic notes, 147
legends, 148
occurrences, 150
properties, 144-145
rose, 149, **258**, 261, **345**
Quartz, *continued*
smokey, **23**, 27, **149**, 256, **257**
varieties, 145.
See also Chalcedony ; Jasper
Quartzite, 363
Quebec
Asbestos, 383
Francon Quarry, Montreal, 245-247
Mont Saint-Hilaire, 262, 263, 265, 266, 380

Ragiel, 85
Reactants in chemical reactions, 312, 313, 323
Recrystallization, 230, 231, 233, 299, 358-359, 360-361, 362, 363, 365, 366, 368, 371, 375, 377, 386, 387, 395
Reduction, 319, 322, 343, 354, 355
Reflectivity, 28
Refractive index (R.I.), **27**, 28, 203
Rhodes, Cecil, 38
Rhodochrosite, **23**, 194, **195**, 200, **277**, 279, **295**, 399, **400**
Rhodolite, 69, 162, 163, 167, **167**. *See also* Garnet
Rhodonite, 199, 201, **368**
Rhyolite, 231, **250**, 251, 255
Riebeckite, **353**
Rock crystal, **10**, 18, 20, 31, **143**, **144**, **148**. *See also* Quartz
Rock, definition, 24
Rockbridgeite, **279**, 343
Rodingites, 381, 383, 385 (Figure J)
Romanechite, 279
Rosasite, 330
Rose quartz, 149, **258**, 261, **345**. *See also* Quartz
Rubellite, 92. *See also* Tourmaline
Rubicelle, 78. *See also* Spinel
Ruby, 22, 26, 27, 28, 31, **50**, **51**, 52, **52**, **79**, **244** *See also* Corundum
Rutilated quartz, **21**, 145, 146, 151. *See also* Quartz
Rutiles, 192, 200, 245, 261

Sagenite, 146. *See also* Quartz
Sal Ammoniac, **394**, 397
Sandstone, 301, 363
Sapphire, **19**, 22, 26, 27, 28, 31, **46**, **47**, **48**, **49**, 52, **53**, **54**, 69, 79, 87, 244. *See also* Corundum
Sapphire spinel, 78. *See also* Spinel
Sard, 155, 157. *See also* Chalcedony
Sardonyx, 155. *See also* Chalcedony
Satin spar, 199, 201. *See also* Gypsum
Saturation, 273, 280, 291, 295, 306, 313
Scapolite, 193, 200, 377
Scheelite, 295, 298, 373
Schettler emerald, 56, **67**
Schettler, Elizabeth Cockcroft, 13, 15
Schist, **226**, **364**, 365, 372, 391
Seaman, Dave, 14
Secondary enrichment zones, 322
Sedimentary rock, 227, 229, 230 (Figure B), 233, 266, 279, 305, 312, 361, 391
Sediments, 222, 223, 225, 229, 230, 237, 266, 280, 283, 287, 306, 322, 368, 380, 392, 393
Segregation, 240 (Figure D), 241, 251
Selenite, 199, 201
Septaria, 287
Serandite, **264**, 265, 266
Serpentine, 141, **141**, 201, 317, 347, 368

Serpentinite, 305, 372, 385, 387
Shakespeare, 122
Shale, 301, 320, **321**, 322, 363, 365, 391
Shalerite, 392
Siberite, 92. *See also* Tourmaline
Sibiconite, 342
Siderite, 301, 322
Silica, 227, 262, 274, 275, 291, 301, 365, 368, 401
Silicate, 239, 240, 241, 245, 265, 305, 317, 326, 329, 365, 368, 373, 375
Silicon, 219, 222, 237, 239, 258, 261, 291, 305, 362, 365, 373, 387, 301
Sill, 240 (Figure D), 247, 249
Sillimanite, 227, 365
Silver, 281, 294, 295, 298, **310**, 311, 397, 399, **400**
Simulant, 32-33, 203
Sinhalite, 107, 192, 201
Skarn, 373, 377, 380, 383, 386, 387
Skutterudite, 338
Smithsonite, **326**, 329
Smoky quartz, **23**, 27, **149**, 256, **257**. *See also* Quartz
Soapstone, 199. *See also* Talc
Sodalite, 116, **159**
Sodalite, 262
Sodium, 222, 238, 239, 245, 255, 262, 263, 265, 266, 271, 291, 363, 373, 387
Solubility, 271-272, 273, 275, 279, 281, 283, 285, 306, 316, 323, 329
Some considerations Touching the Usefulnesse of Experimental Natural Philosophy, 20
Soudé emerald doublet, 33, **33**
South Africa, Kimberlay, 243
Specific gravity, 28, 203
Speculum Lapidum, 21
Spessartine, 25, **162**, **165**, **171**, 231, **250**, 251, 255. *See also* Garnet
Sphalerite, **284**, 285, 295, 298, **304**, 329, 330, 351, 373
Sphene, 193. *See also* Titanite
Spinel, 55, 66, 69, 76-79, **76**, **78**, **79**, 238, 244, 368
  Black Prince's Ruby, 48, 76
  evaluation, 79
  gemstones confused with spinels, 79
  historic notes, 77
  legends, 78
  occurrences, 78
  Timur ruby, 77
Spodumene, **26**, 94, 191, 200, 343. *See also* Kunzite
Stability, 29. *See also* Durability
Stalactite, **232**, 233, **274**, 276, 284, 288, 320, 399
Star of India (sapphire), 8, 15, 44, **45**
Star Stones, 26, 28, 44, **45**, 46, 51, 52, 55, 75, 77, 164, 170, 192, 193
Staurolite, 226, 365
Steatite, 18, 199, 201
Steinbüch, 21, 158
Stibiconite, 342, **352**, 353
Stibnite, **297**, 341, **342**, 353, 399
Stilbite, **288**, 291
Stillimanite, 362
Stilpnomelane, **318**, 319, 320
Stishovite, 227, 401
Strahlers, 301
Strengite, **279**, 343, 397
Strontium, 245, 247, 262
Struvite, 398
Subduction, 222, 223, 225, 237
Sublimation, 397
Succinite, 181. *See also* Amber
Sulfate, 233, 273, 283, 285, 287, 315, 317, 320, 322, 329, 354, 395, 397
Sulfide, 233, 241, 258, 285, 287, 295, 297, 298, 305, 311, 317, 320, 322, 325, 329, 330, 336, 354, 355, 391, 392, 393 (Figure K), 395, 397
Sulfur, 225, 231, 236, 279, 285, 293, 307, 311, 312, 319, 390, 392, **394**, 395, 397
Sunstone, 130. *See also* Feldspar
Sylvite, 273
Synchysite, 263, **265**
Synthetic (gem), 32-33, 203

Talc, 199, 201, 347, 365, 368, 385
Talisman, 20, 74, 122, 136, 142, 148, 187
Tantalum, 240, 251
Tanzanite, 55, **189**, 191, 200. *See also* Zoisite
Tavernier, Jean Baptiste, 21, 38, 84
Tennantite, 329, 330
Tenorite, **325**
Tetrahedra, 239
Tetrahedrite, **294**, 295, 329
Theophrastus, 20, 39, 65, 116
Thompson, Gertrude Hickman, 13
Thompson, William Boyce, 13
Thorium, 219, 240, 245, 260, 261, 262
Thulite, 199, 200. *See also* Zoisite
Tiffany, Charles L., 9-13, 88
Tiffany & Co., 9-11, 88, 128, 162, 175, 191
Tiffany Diamond, **43**
Tiger's eye, 145, 146, **150**, 353. *See also* Quartz
Tin, 240, 261, 292, 293, 295, 298
Tincalconite , **314**, 315, 351
Titanite, 301, **303**, 377
Titanium, 245, 261, 262, 305, 373
Titantie, 193, **193**, 200
Topaz, **23**, 26, **26**, 29, 30-32, 55, 69, 80-87, **81-86**, 231, 250, 251, **254**, 255, 295
  Brazilian Princess, **85**
  evaluation, 87
  gemstones confused with topaz, 87
  historic notes, 84
  Imperial topaz, 81, **83**, 87
  legends, 85
  occurrences, 85
Topazolite, 165. *See also* Garnet
Torbernite, 314, 315
Toughness, 29, 135, 154
Tourmaline, **19**, **26**, 28, 30, 55, 69, 87, 88-99, **90**, 252, 255, 261, 343, 372, 377
  Catherine the Great Ruby, 95
  dravite, 90
  elbaite, 90, **91**, 92, **97-99**
  evaluation, 99
  gemstones confused with tourmaline, 94
  Historic Mt. Mica crystals, **96**
  historic notes, 93-95
  Liddicoatite, 93
  occurrences, 95
  Uvite, 90
Tridymite, 301
Triphylite, **260**, 261, 343, 345, 347
Triplet, 33, 75, 125, 140. *See also* Composite stones
Tristate District (Kansas, Oklahoma, Missouri), 285
Trystine, 146, **146**
Tsavorite, 63, 162, 165, **166**. *See also* Garnet
Tungsten, 240, 292, 293, 295, 298
Turquoise, 20, 26, 29, 69, 108-113, **109-111**, **159**, 330, **333**
  gemstones confused with turquoise, 20, 26, 29, 113
  historic notes, 111
  legends, occurrences and evaluation, 113
Tuthankhamen, 20, 128
Twin crystal, 24, 203
Twinning, 24

Ulexite, 75
Ulvite, 374
Uniformitarianism, 219
Uraninite, 343, **344**, 345, 377, **378**
Uranium, 219, 240, 260, 261, 293, 295, 298, 338, 343, 373
Uvarovite, **374**, 377
Uvite, 90, 91, 377. *See also* Tourmaline
Uvoarovite, 377

Valentinite, 341, 342
Vanadinite, **337**, 338
Vanadium, 252, 255, 372
Variety, 25, 203
Variscite, **112**, 113, 201, **278**, 279, 397
Verdelite, 92. *See also* Tourmaline
Vermiculite, 347
Vesuvianite, 377, 381, **383**
Vivianite, 279, **300**, 301, 312, 343, 391, 397
Volmar, 21, 158
Vugs, 350

Wardite, 278, 301
Wavellite, **278**, 279
Weddellite, 398
Weloganite, **246**, 247
Whitlock, Herbert P., **14**, 17
Willemite, 368
Wolframite, 295, 298
Wollastonite, 365, 377
Wulfenite, **335**, 338

Yukon, Rapid Creek—Big Fish River area, 300, 301

Zeolite, 230, **288**, 290, 291
Zinc, 283, 285, 287, 295, 306, 329, 330, 350, 368, 397
Zincite, 368, **369**
Zircon, 55, 100-104, **101**, **103**, **244**, 377, **378**
  evaluation, 103
  historic notes, 102
  legends, 102
  occurrences, 103
  properties, 102
Zirconium, 240, 247, 262, 263, 265
Zoisite, **26**, 181, 200. *See also* Tanzanite
Zussmanite, 368